Advanced Ceramic Materials - Emerging Technologies

Edited by Amparo Borrell Tomás and Rut Benavente Martínez

Published in London, United Kingdom

Advanced Ceramic Materials - Emerging Technologies
http://dx.doi.org/10.5772/intechopen.1004508
Edited by Amparo Borrell Tomás and Rut Benavente Martínez

Contributors
Abderrahim Lahlahi, Afshin Amiri-Mogh, Alexandru Gyorgy Okos, Ali Alizadeh, Ali Karatutlu, Alina Pruna, Amparo Borrell, Ariana Gjoni, Ashley Bonilla, Bülend Ortaç, Carlos Eduardo Francci, Chunchun Li, Cristina Florentina Ciobota, David Busquets-Mataix, Diego Fraga, Edson Leroy Meyer, Fabiola Colmenero Fonseca, Faisal Rafiq Adikan, Farshad Soleimani, Fayeda Srarfi, Guobin Zhu, Hamed Naderi-Samani, Hamid-Reza Bahari, Hasan Abbaszadeh, Issa Samb, Jaime González Cuadra, Jinxue Ding, Juan Carda, Juan Francisco Palomino Bernal, Khedidja Makhloufi, Khouloud Moualhi, Maria D. Salvador-Moya, María Dolores Salvador Moya, Mehri Mashhadi, Mevludin Shabani, Mihail Botan, Mouldi Zouaoui, Mylena Régis, Oludolapo Akanni Olanrewaju, Olusegun Adigun Afolabi, Pablo Serna-Gallén, Pouria Dehghani, Radu-Robert Piticescu, Reza Shoja Razavi, Robinson Cadena, Rut Benavente, Samara Silva, Samuel Porcar, Santiago Toca, Seyed Salman Seyed Afghahi, Shah Aarif Ul Islam, Siyu Xiong, Vaishali Poddar, Wei Li, Xiaowei Zhu

First published in London, United Kingdom, 2025 by IntechOpen
IntechOpen is the global imprint of INTECHOPEN LIMITED, registered in England and Wales, registration number: 11086078, 167-169 Great Portland Street, London, W1W 5PF, United Kingdom

For EU product safety concerns: IN TECH d.o.o., Prolaz Marije Krucifikse Kozulić 3, 51000 Rijeka, Croatia, info@intechopen.com or visit our website at intechopen.com.

British Library Cataloguing-in-Publication Data
A catalogue record for this book is available from the British Library

Advanced Ceramic Materials - Emerging Technologies
Edited by Amparo Borrell Tomás and Rut Benavente Martínez
p. cm.

This title is part of the Materials Science Book Series, Volume 1
Topic: Metals and Nonmetals
Series Editor: Chonghe Li
Topic Editor: Joan Josep Roa Rovira

Print ISBN 978-1-83634-017-1
Online ISBN 978-1-83634-016-4
eBook (PDF) ISBN 978-1-83634-018-8
ISSN 3049-8856

IntechOpen Book Series
Materials Science
Volume 1

Aims and Scope of the Series

Materials science has always occupied an extremely high position in the human development process. As we explore the oceans of stars, various industries have put forward more stringent requirements for the performance of materials, forcing us to pay more and more attention to the development of new materials. At the same time, the formation of a data-driven scientific paradigm is dramatically shortening the development cycle of new materials. The huge data generated by synergistically combining theories, high-throughput experiments, high-throughput computation, and artificial intelligence is greatly contributing to our ability to utilize materials science to solve real-world problems. The three topics of this book series - Metals and Nonmetals; Composite Materials; and Surface Science - will address important areas of advancement in materials science. There will be a range of interesting works published under these topics.

Meet the Series Editor

Prof. Chonghe Li received his Ph.D. from the Chinese Academy of Sciences in 1995. From 1995 to 2000, he worked as a researcher at the Shanghai Institute of Metallurgy, Chinese Academy of Sciences, where he also served as director of the research laboratory. In 2000, he was appointed Professor at the Institute of High-Performance Computing in Singapore, where he worked on computation and simulation in materials science until 2004. Since then, he has been a professor at the School of Materials Science and Engineering, Shanghai University, China, as well as the director of the Shanghai Specialty Casting Engineering and Technology Research Center. Prof. Li's research focuses on titanium alloys, titanium-aluminum single crystals, intermetallic compounds, theoretical calculations, alloy design, and special refractory materials. His broad scientific expertise is well recognized by the scientific community around the world. He is a member of the editorial board of the journal Metals. As an author, he has published more than 200 peer-reviewed papers, 2 books, and over 40 patents.

Meet the Volume Editors

Dr. Amparo Borrell has a Physics degree and a Materials Science Ph.D. from the Spanish National Research Council (CSIC). From 2018-2022, she was a Ramón y Cajal researcher. Currently, she is a senior Materials Science and Engineering lecturer at the Polytechnic University of Valencia in Spain. Her main research lines are ceramics processing by non-conventional sintering techniques. She is the author and co-author of about 100 papers in international scientific journals and has participated in 28 research projects. She has participated in 130 congresses and several technology transfer activities that have generated 3 patents. She has been a visiting researcher at the University of São Paulo (Brazil) and the University of Modena e Reggio Emilia (Italy).

Dr. Rut Benavente is a Senior Lecturer at the Polytechnic University of Valencia. She has a degree in Chemistry from the University of Valladolid and a Ph.D. in Materials Science from the Polytechnic University of Valencia in Spain. Her main scientific activity is focused on the field of Science and Engineering of Advanced Ceramic Materials with specific functionalities, covering different aspects, ranging from the processing of materials to the study of properties, mainly mechanical and microstructural. She is the author of 35 scientific publications and has collaborated on more than 60 contributions to national and international scientific congresses.

Contents

Preface

In a world increasingly driven by technological innovation, advanced ceramic materials have become key players in numerous breakthroughs in strategic sectors such as energy, health, electronics and environmental sustainability. This book, *Advanced Ceramics Materials – Emerging Technologies*, seeks to explore and highlight the transformative impact of these emerging technologies, offering a comprehensive overview of the most recent developments, their innovative applications and prospects in this dynamic field.

The evolution of ceramic materials, from traditional to advanced ceramics, has witnessed exponential growth in terms of properties and functionalities. Materials such as piezoelectric ceramics, porous ceramics, and nanostructured ceramic composites have revolutionized applications in microelectronics, clean energy generation, and biomedicine. This book not only explores the fundamental properties of these materials but also analyses their interactions with other disciplines, demonstrating how the convergence of materials science, nanotechnology and engineering has overcome technical barriers and opened new frontiers.

In recent years, research and development in advanced ceramic materials have led to innovations that push the boundaries of the imaginable. From functional ceramics designed to withstand extreme conditions to nanostructured composites with unique properties, the possibilities are as exciting as they are challenging. This book addresses recent advances from a technical perspective and explores how these technologies are shaping a more sustainable, efficient and technologically advanced future.

Ceramic materials are destined to play a central role in this exciting journey into the future, and this research is a testament to their continued importance in modern science and technology. Their exceptional properties and versatility make them indispensable in addressing some of the most pressing challenges of our time, from advancing clean energy solutions to enabling next-generation biomedical applications. This book celebrates the remarkable potential of ceramic materials and the innovative technologies that continue to push their boundaries.

This preface offers a brief glimpse into the diverse and groundbreaking research currently shaping the field of ceramic materials. As we stand on the brink of discoveries that will transform construction, technological innovation, and daily life, ceramic materials emerge as pivotal to this evolution. The ongoing advancements highlighted in this book emphasize their profound significance in modern science and technology, reaffirming their central role in shaping a more innovative and sustainable future.

Amparo Borrell Tomás and Rut Benavente Martínez
Institute of Materials Technology,
Valencia Polytechnic University,
Valencia, Spain

Chapter 1

Advancements in Hard Turning: Leveraging Ceramic Cutting Tools for Sustainable and High-Quality Metalworking

Mevludin Shabani and Ariana Gjoni

Abstract

This section of the book delves into the critical role of ceramic cutting tools in modern metalworking, specifically in hard turning. It sets specific objectives to explore the fundamental principles, properties, and advancements of these tools. Distinguishing hard turning from conventional methods, it evaluates the benefits and challenges, focusing on key factors such as tool wear and surface integrity. A detailed analysis of ceramic materials used in tools, including coatings, emphasizes their specific properties for hard-turning applications. The evolution of ceramic tool technology is traced through innovative designs and surface treatments, improving operational efficiency. Strategies for optimizing cutting parameters and extending tool life have been elaborated, highlighting environmental benefits and sustainability. Quality assurance methods ensure precision in hard-turning operations, as demonstrated across various industries. Looking ahead, the chapter concludes by outlining emerging trends and potential research areas, affirming the transformative impact of ceramic tools on the future of metalworking.

Keywords: ceramic cutting tools, hard turning, tool wear, surface integrity, technological advancements

1. Introduction

1.1 Overview of hard turning as a machining process

Hard turning is a precise machining process intended for efficiently and accurately machining hardened materials, usually those with a hardness exceeding 45 HRC on the Rockwell scale. Unlike grinding (which used to be the only method for final surface processing until the introduction of the hard-turning technique), it offers numerous benefits, including reduced setup time, cost-effectiveness in terms of energy, and the elimination of coolants, making it both economical and environmentally friendly [1].

The process has become more prominent in various industries due to advances in tool materials and cutting-edge geometries. This allows manufacturers to achieve superior surface finishes and precise tolerances directly from the machining process. In addition, hard turning decreases the need for numerous needless operations, which improves productivity and reduces manufacturing costs. However, besides the useful benefits, there are challenges such as tool wear, residual stresses, and thermal effects on the workpiece to pay attention to in the process planning.

The evolution of hard turning continues to drive innovations in machining strategies, ensuring its continued relevance and effectiveness in modern manufacturing.

1.2 The importance of ceramic cutting tools in modern metalworking

Ceramic cutting tools have become increasingly important in modern metalworking due to their unique properties and the specific challenges they address in machining applications. Cutting tool development is important for enhancing the mechanical properties, especially in high-speed machining with a long duration [2].

Ceramics offer superior hardness, wear resistance, and thermal stability compared to traditional tool materials like carbides and high-speed steels. These characteristics make ceramics especially suitable for machining hard materials such as hardened steels and super alloys. Conventional tools may wear out quickly or struggle to maintain dimensional accuracy and surface finish when working with up-mentioned materials (**Figure 1**).

Ceramic cutting tools exhibit low friction coefficients. Low affinity with metal and low coefficients of friction can significantly reduce cutting forces and cutting temperatures, ensuring machining accuracy [2]. Moreover, ceramics are characterized by their ability to maintain physical-mechanical properties, high hardness, and wear resistance at elevated temperatures. In ceramic cutting, cutting speed significantly influences surface quality, with wear indicators assessing the cutting edge's condition [1].

Figure 1.
Appearance of cutting board holder CSRNR 25 × 25 M12H3 from the company HERTEL (owned by the faculty of mechanical engineering in Skopje) [1].

The ability of ceramics to withstand high operating temperatures makes them a leading engineering material. Their stiffness and strength are comparable to that of the metals and are generally chemically unreactive [3].

2. Fundamentals of hard turning

2.1 Definition and characteristics of hard turning

Hard turning is a precision machining process specifically designed for the machining of hardened materials; unlike traditional turning processes that primarily deal with softer materials, hard turning employs advanced cutting tools, often ceramics, to achieve high material removal rates and superior surface finishes. The process of hard machining, of which the dominant is the machining operation performed on the hardened materials, is defined as the process of single-point cutting of part pieces that have hardness values over 45 HRC (**Figures 2–4**).

Characterized by high cutting speeds and low feed rates, hard turning operates under dry or minimal lubrication conditions to reduce environmental impact and operational costs. The objective of machining technology is to manufacture at the lowest possible cost, but simultaneously with the highest dimensional accuracy. Concurrently, the machining process should be eco-friendly, clean, and sustainable [4]. In metal-cutting operations, the cutting temperature field during cutting determines key process issues such as many parameters, including accuracy of the

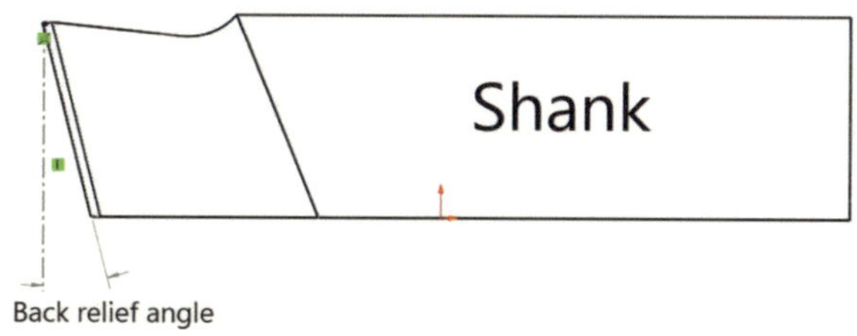

Figure 2.
Geometry of the cutting tool (back relief angle).

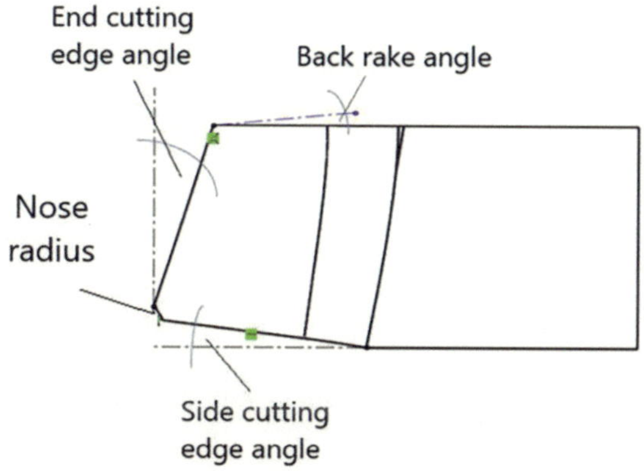

Figure 3.
Geometry of the cutting tool (side rake edge angle, back relief angle, side cutting edge angle, nose radius).

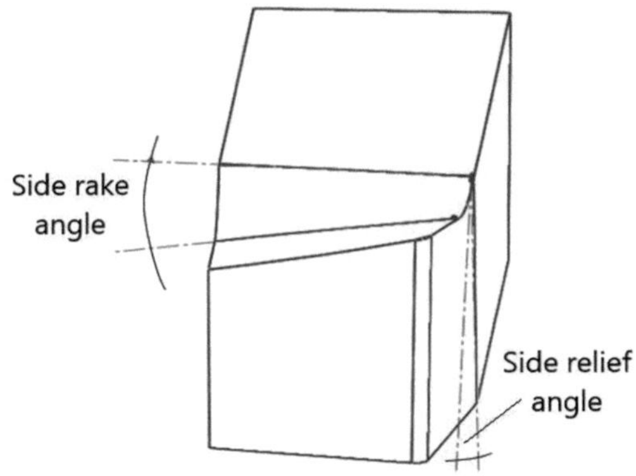

Figure 4.
Geometry of the cutting tool (side rake angle, side relief angle).

machined surface, tool wear, tool life, mechanics of chip formation, surface quality, cutting forces, and cutting parameters, as well as process efficiency [5].

2.2 Comparison with traditional turning processes: Hard turning

Hard turning has emerged as a competitive alternative to traditional turning processes such as grinding, offering advantages in terms of cost-effectiveness, efficiency, and environmental impact. This chapter examines the key differences between hard turning and traditional turning methods, highlighting the technological advancements and benefits that have made hard turning increasingly popular in manufacturing industries.

As mentioned previously [1], hard turning provides numerous advantages, such as reduced setup time, cost-effectiveness in terms of energy, and the elimination of coolants, making it both economical and environmentally friendly.

In terms of machining performance, hard turning achieves comparable or superior surface finishes to grinding, especially with advancements in tooling technology and machine capabilities. The precision and dimensional accuracy achievable through hard turning have made it a preferred method for applications requiring tight tolerances and high surface integrity.

While traditional turning processes continue to serve specific applications effectively, hard turning has emerged as a versatile and efficient alternative offering significant advantages in terms of cost, time, and environmental impact.

The adaptability and cost-effectiveness of hard turning make it a compelling choice, particularly considering the ongoing advancements in tooling, materials, and machining strategies. As manufacturing technologies evolve, the role of hard turning is expected to expand further, supported by continuous innovations in the field.

2.3 Advantages and challenges of hard turning

In the following, we will outline the primary advantages and disadvantages that need to be considered when designing the machining process, including the selection of cutting tools:

2.3.1 Advantages

- Achieves high precision and excellent surface finish in a single setup
- Reduces operational costs
- Minimizes energy consumption
- Requires less coolant and produces fewer emissions than grinding

2.3.2 Challenges

- Intensive tool wear
- Residual stresses

In conclusion, while hard turning offers significant advantages in terms of efficiency, cost-effectiveness, and environmental impact, it requires careful consideration of tool wear management and process optimization to maximize its benefits in precision machining of hardened materials.

3. Ceramic cutting tools: Properties and types

3.1 Overview of ceramic materials used in cutting tools

Ceramic materials play a crucial role in modern cutting tool applications because of their exceptional hardness, wear resistance, and ability to withstand high temperatures. Alumina (Al_2O_3), silicon nitride (Si_3N_4), and silicon carbide (SiC) are some of the primary ceramics used in cutting tools, each offering unique advantages.

Ceramic materials have been used to meet many engineering requirements [6], such as in energy production and the aerospace industry [7, 8], because of their chemical stability and relatively high hardness [6–8], lower density compared to metals [3], high mechanical resistance [1, 3] and resistance to high temperatures [6–9]. Compared to other materials, ceramics are also less prone to damage caused by corrosive environments [7].

Silicon nitride ceramics exhibit excellent fracture toughness and thermal shock resistance, essential for high-performance machining operations. Silicon carbide ceramics represent a class of ceramic materials characterized by good mechanical properties, thermo-mechanical stability, and fracture toughness (**Figure 5**).

Despite their higher initial cost and susceptibility to impact damage, ceramic cutting tools offer substantial benefits in terms of extended tool life, enhanced productivity, and reduced environmental impact in machining processes.

In conclusion, ceramic cutting tools represent a vital advancement in machining technology, providing superior performance in demanding applications where high precision and durability are essential.

Figure 5.
Appearance of cutting plates SNGN 120708-120,712-120,716 from mixed ceramics MC 2 (Al_2O_3 + TiC) from the company HERTEL [1].

3.2 Properties that make ceramics suitable for hard turning

Ceramic materials exhibit several properties that make them highly suitable for hard-turning applications, particularly in machining operations involving hardened steels and other challenging materials. Key properties include high hardness, excellent wear resistance, thermal stability, and chemical inertness.

The wear of ceramic materials is an issue of great industrial importance [6] because wear leads to the frequent replacement of components [10].

The proper selection of wear-resistant materials increases the useful life of equipment and considerably reduces maintenance costs [6, 10, 11]. Alumina is the most commonly used material in applications requiring wear resistance [6, 12–15], making this material particularly important in a wide range of applications [16]. Its various attractive properties include high chemical inertia [13], considerable fracture toughness, relatively low cost [6, 12], great hardness [6–8, 13, 16], high erosion resistance [6–8, 15, 16], good behavior at high temperatures [6, 7, 16, 17], high electric resistance, and great availability [17].

Si_3N_4-TiN composites have extensive manufacturing applications because of their first-rate mechanical and thermal properties. Conversely, they are very demanding on machines that use conventional practices because of their superior strength and hardness [18].

Ceramic materials like silicon nitride and silicon carbide are essential for minimizing tool wear and extending the lifespan of tools, which improves machining efficiency and cost-effectiveness. Due to their thermal stability, they can withstand high-speed cutting temperatures without degrading or softening.

This is important for maintaining precision in machining, ensuring dimensional accuracy, and achieving the desired surface finish. Additionally, ceramics' chemical inertness reduces the risk of reactions with workpiece materials and cutting fluids. This property not only extends tool life but also maintains machining accuracy over long periods. Overall, these characteristics make ceramics highly beneficial for demanding machining applications, where durability, precision, and cost-efficiency are essential considerations.

3.3 Types of ceramic cutting tools (inserts, coatings, etc.)

Ceramic cutting tools encompass various types of inserts, coatings, and materials designed to optimize performance in machining applications. Here's a comprehensive overview of each type:

1. *Ceramic inserts*: Ceramic inserts are cutting tool inserts made from ceramic materials such as alumina (Al_2O_3), silicon nitride (Si_3N_4), or silicon carbide (SiC). These inserts are typically used in turning, milling, and boring operations. They offer high hardness, excellent wear resistance, and thermal stability, making them suitable for machining hardened steels and heat-resistant alloys.

2. *Coated ceramics*: Coated ceramic cutting tools involve applying thin layers of materials like titanium nitride (TiN), titanium carbonitride (TiCN), or aluminum oxide (Al_2O_3) onto ceramic inserts. These coatings improve tool life, enhance wear resistance, and reduce friction during cutting processes. Coated ceramics are effective in high-speed machining applications where prolonged tool life and increased productivity are critical.

3. *Cermet inserts*: Cermet inserts are a hybrid of ceramic and metallic materials, combining ceramic particles (typically titanium carbide TiC) with a metallic binder (often cobalt or nickel). Cermet inserts offer a balance between hardness and toughness, providing improved fracture resistance compared to conventional ceramics. They are suitable for high-speed machining of steel and cast iron.

4. *Whisker-reinforced ceramics*: Whisker-reinforced ceramic cutting tools incorporate ceramic matrices reinforced with whisker-like fibers of materials such as silicon carbide (SiC) or alumina (Al_2O_3). These fibers enhance the toughness and mechanical strength of the ceramic, improving resistance to chipping and fracture during machining operations. Whisker-reinforced ceramics are used in demanding applications requiring high cutting speeds and precision.

5. *Composite ceramics*: Composite ceramic cutting tools combine different ceramic materials or incorporate ceramic particles within a metallic or ceramic binder. This approach aims to optimize properties such as hardness, toughness, and thermal stability, tailoring the tool's performance to specific machining conditions. Composite ceramics offer versatility in machining various materials while maintaining durability and reliability.

Each type of ceramic cutting tool is engineered to address specific challenges in machining operations, ranging from high-speed cutting to precision finishing, and contributes to improved efficiency and cost-effectiveness in manufacturing processes.

4. Performance metrics in hard turning

4.1 Key performance indicators (KPIs) for evaluating hard-turning operations

Key Performance Indicators (KPIs) for evaluating hard-turning operations encompass several critical metrics that gauge the efficiency, quality, and effectiveness of the machining process. Here are the essential KPIs:

1. *Surface roughness (Ra)*: Surface roughness indicates the quality of machined surfaces. Lower Ra values signify smoother surfaces, crucial for functional components. Monitoring Ra helps ensure parts meet design specifications.

$$R_a = \frac{1}{l_r} \cdot \int_{l_r}^{0} |z(x)| \cdot dx \tag{1}$$

It's important to note that there are numerous hardness parameters, but typically, the parameter Ra is used as a representative measure for the surface hardness of machined parts.

2. *Dimensional accuracy*: KPIs related to dimensional accuracy assess how closely machined parts conform to specified dimensions. This metric is vital in industries requiring tight tolerances, such as aerospace and automotive.

3. *Tool wear rate*: Tool wear rate measures the degradation of cutting tools during machining. It influences production costs and downtime for tool changes. Lower tool wear rates indicate longer tool life and improved cost-effectiveness.

4. *Material removal rate (MRR)*: MRR quantifies the volume of material removed per unit time. Higher MRR values indicate greater productivity, but must be balanced with tool life and surface quality considerations.

5. *Energy efficiency*: Evaluating energy consumption metrics helps optimize operational costs and environmental impact. Efficient hard-turning processes reduce energy use while maintaining performance standards.

6. *Chip control effectiveness*: Effective chip control involves managing the formation and removal of chips during cutting. Proper chip control reduces tool wear, improves surface finish, and prevents chip entanglement, optimizing machining operations.

7. *Process stability*: Process stability metrics evaluate the consistency and reliability of hard-turning operations, considering factors such as vibrations, tool deflection, and thermal effects that affect stability and need to be managed to ensure consistent part quality and machining performance.

8. *Environmental impact*: Assessing the environmental impact of hard-turning operations involves evaluating factors such as emissions, coolant usage, and waste generation. Sustainable manufacturing practices strive to minimize environmental harm while upholding operational efficiency.

These KPIs provide a comprehensive framework for evaluating and optimizing hard-turning operations, ensuring enhanced productivity, reduced costs, and superior quality in machined components.

4.2 Impact of tool wear, surface integrity, and tool life

The impact of tool wear, surface integrity, and tool life is critical in the context of machining processes like hard turning, influencing both productivity and the quality of machined components.

Tool wear is a natural process that develops on cutting tools during machining operations because of mechanical, chemical, and thermal loads. Several wear mechanisms appear in the cutting zone due to these external effects, i.e., flank wear (VB) and crater wear, which were highly investigated in the past studies [19].

Tool wear directly affects machining performance and cost-efficiency. As cutting tools wear down during operations, their ability to maintain dimensional accuracy and surface finish decreases. This leads to higher scrap rates, more frequent tool changeovers, and potential delays in the production process.

Additionally, high tool wear rates result in increased machining costs due to the need for frequent tool replacements and maintenance.

Surface integrity can be defined as the combination of mechanical, metallurgical, topographical, thermal, and chemical features of the surface of a component obtained from a particular manufacturing process and can be related to the performance during the intended application of the same component [20].

Surface integrity refers to the quality of the machined surface, which includes factors such as roughness, residual stresses, and microstructural changes. When surface integrity is poor, it can pose a risk to the performance and longevity of components, especially in critical applications where fatigue resistance and dimensional stability are important. It is essential to optimize cutting parameters in order to achieve the desired surface integrity. This includes minimizing heat generation and controlling chip formation to reduce surface defects.

Tool life has a significant impact on machining efficiency and cost-effectiveness. Extending tool life reduces downtime for changes, improves process stability, and decreases production costs. Key factors influencing tool life include the material properties of the cutting tool, cutting conditions such as speed and feed rate, and the effectiveness of cooling and lubrication systems in managing heat and wear. With the quest for manufacturing sustainability with respect to energy, process, material, and environment friendliness as well as the clamor for circular economy, which emphasizes zero tolerance for waste, there is a need for a critical review of the life cycle of machine tools employed for machining operations and product development.

Understanding and effectively managing these factors is crucial for optimizing machining processes. Strategies aimed at reducing tool wear, improving surface integrity, and prolonging tool life include utilizing advanced tool materials, applying coatings such as titanium nitride (TiN) or diamond-like carbon (DLC), and optimizing machining parameters. Coatings like TiN or DLC enhance tool hardness and wear resistance, reduce friction, and contribute to superior surface finishes. These approaches collectively enhance machining efficiency and contribute to cost-effectiveness in manufacturing operations.

5. Advances in ceramic cutting tool technology

5.1 Advances in ceramic cutting tool technology: Evolution of ceramic materials for cutting tools

The technology of ceramic cutting tools has advanced significantly due to the need for better machining capabilities in industries that require high precision, durability, and efficiency. Ceramics like alumina (Al_2O_3), silicon nitride (Si_3N_4), and silicon carbide (SiC) have been continuously developed to improve their mechanical properties and broaden their use in cutting tools.

Initially, ceramics encountered challenges due to their natural brittleness and susceptibility to fracturing under mechanical stresses. However, strategic advancements in ceramic processing techniques, such as hot pressing, sintering, and the inclusion of reinforcing agents like whiskers or fibers, have notably enhanced their toughness and reliability, for example, the strategy that involves the utilization of whiskers as reinforcements to fabricate high-strength CMCs through a composite preparation route. Undoubtedly, this strategy relies on whisker synthesis, whisker preform forming, and the interphase and matrix preparation processes [21].

Furthermore, advancements in ceramic technology include the development of composite ceramics, which integrate ceramic materials with metallic binders or other ceramics. These composites offer a balanced combination of hardness, toughness, and thermal stability, expanding their suitability for diverse machining operations such as hard turning and high-speed milling.

Moreover, ceramic cutting tools have evolved significantly with the adoption of advanced coating technologies. Coatings like titanium nitride (TiN) or aluminum oxide (Al_2O_3) augment tool hardness, wear resistance, and thermal insulation, thereby prolonging tool life and enhancing machining performance.

Overall, these innovations in ceramic cutting tools underscore their increasing significance in modern manufacturing. They contribute to heightened productivity, reduced operational costs, and improved environmental sustainability through extended tool longevity and enhanced machining efficiency.

5.2 Innovative designs and geometry

Innovative advancements in ceramic cutting tool technology have transformed machining capabilities by addressing challenges such as tool wear, chip control, and surface finish in demanding applications. These developments aim to optimize tool performance, extend tool longevity, and enhance overall machining efficiency.

A notable breakthrough involves the creation of intricate tool designs tailored for specific machining tasks. Modern designs incorporate optimized *rake and clearance angles* to minimize cutting forces and heat generation, crucial for preserving dimensional accuracy and surface integrity. These geometries also improve chip control, reducing entanglement and enhancing chip evacuation from the cutting zone, thereby boosting machining stability and surface finish (**Figure 6**).

Additionally, ceramic tool designs have evolved with the integration of advanced cooling and lubrication mechanisms. Internal coolant channels and specialized coatings enhance heat dissipation and reduce friction, playing a pivotal role in extending tool life and maintaining consistent performance, particularly in high-speed machining operations.

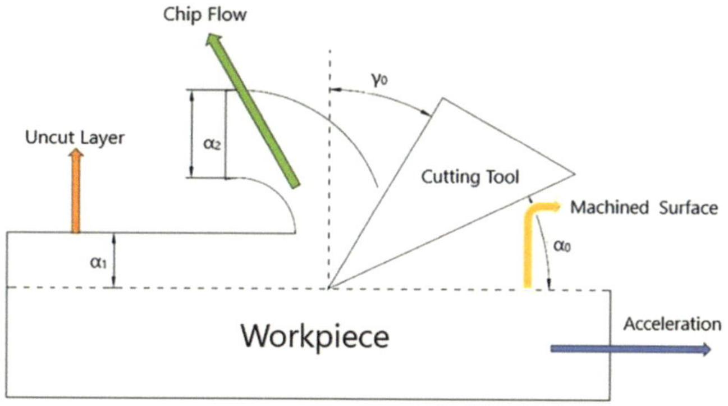

Figure 6.
Basic cutting process geometry parameters. α_1 - uncut chip thickness; α_2 - chip thickness; α_0 - clearance angle (orthogonal); γ_0 - rake angel (orthogonal).

Furthermore, innovations in ceramic tool geometries include the adoption of cutting-edge preparations such as wavy or serrated edges. These configurations facilitate smoother chip flow and diminish cutting forces, resulting in improved tool durability and surface finish quality.

Moreover, the utilization of multi-phase ceramics and nanocomposite materials has enabled the development of cutting tools with superior mechanical properties. These advanced materials offer heightened hardness, toughness, and thermal stability, enabling more aggressive machining parameters and enhancing overall productivity.

In conclusion, innovative designs and geometries in ceramic cutting tool technology continue to push the boundaries of machining capabilities, offering solutions that improve efficiency, reduce costs, and meet the growing demands of modern manufacturing environments.

5.3 Coatings and surface treatments to enhance tool performance

Coatings and surface treatments are essential for boosting the performance and longevity of cutting tools, particularly in demanding machining tasks. By applying advanced coatings such as titanium nitride (TiN), titanium aluminum nitride (TiAlN), or diamond-like carbon (DLC), tools gain enhanced wear resistance and reduced friction. These treatments significantly extend tool life, allowing for longer periods of operation without the need for replacement. Additionally, they contribute to a superior surface finish quality on machined parts, which is crucial in precision manufacturing. Ultimately, these technologies lead to increased efficiency, cost savings, and improved product quality in various industrial applications.

Titanium nitride (TiN) is often used as a coating for cutting tools due to its high hardness and wear resistance. These coatings protect tool surfaces from abrasive wear, reduce friction during cutting, and ultimately extend tool life and improve machining efficiency.

Similarly, titanium carbonitride (TiCN) coatings offer enhanced performance in high-speed machining applications by combining the wear resistance of titanium nitride with improved toughness. TiCN coatings are especially effective in cutting operations that require higher cutting speeds and feeds.

Diamond-like carbon (DLC) coatings represent advancements in tool surface treatments. DLC coatings are exceptionally hard, have low friction, and exhibit excellent resistance to wear and corrosion. These properties make DLC-coated tools suitable for machining abrasive materials and for extending tool life in dry-cutting conditions.

Advanced surface treatments involve integrating nanoparticles into conventional coatings like TiN or Al_2O_3 to create nanocomposite coatings. These coatings offer superior hardness, thermal stability, and wear resistance compared to traditional coatings, improving tool performance in extreme machining conditions.

Moreover, using surface treatments such as chemical vapor deposition (CVD) and physical vapor deposition (PVD) technologies allows for precise control over coating thickness and composition, optimizing tool performance for specific machining needs.

In conclusion, coatings and surface treatments continue to evolve, providing cutting tools with enhanced properties that contribute to increased productivity, reduced tooling costs, and improved machining quality in various industrial applications.

6. Machining strategies for hard turning with ceramics

6.1 Cutting parameter optimization

Optimizing cutting parameters for hard turning with ceramics is crucial for efficient machining processes and achieving the desired surface finish and tool life. Several key parameters influence the performance of ceramic cutting operations, including cutting speed, feed rate, depth of cut, tool geometry, and cutting fluid application (**Figure 7**).

1. *Cutting speed*: Determining the appropriate cutting speed is crucial in hard turning with ceramics. Higher cutting speeds generally result in increased material removal rates but must be balanced to avoid excessive tool wear and thermal damage.

2. *Feed rate*: The feed rate directly impacts the tool's engagement with the workpiece. A higher feed rate can enhance productivity but may lead to increased tool wear and surface roughness if not optimized.

3. *Depth of cut*: Controlling the depth of cut influences cutting forces and tool life. Optimal depths of cut for ceramics are typically smaller compared to conventional machining materials due to their brittleness and susceptibility to fracture.

4. *Tool geometry*: Tool geometry, including rake angle, clearance angle, and cutting-edge preparation, significantly affects cutting forces and surface integrity. Proper tool geometry selection can minimize cutting forces and improve surface finish.

5. *Cutting fluid application*: Despite ceramics being typically machined dry, some studies suggest the controlled application of cutting fluids can mitigate tool wear and improve surface quality under specific conditions.

In conclusion, optimizing cutting parameters for hard turning with ceramics requires a balanced consideration of these factors to achieve desired machining outcomes effectively. Experimental studies and computational simulations play crucial roles in determining the optimal parameter combinations tailored to specific ceramic materials and machining conditions.

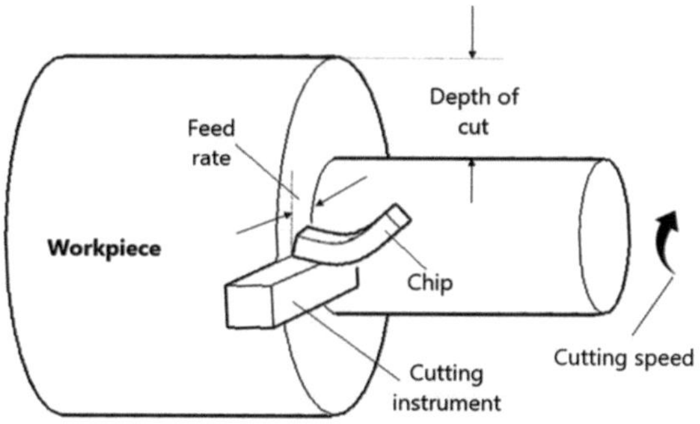

Figure 7.
Cutting parameters for turning operations.

6.2 Tool path strategies for maximizing tool life and surface quality

Tool path strategies in hard turning with ceramics are crucial for optimizing tool life and achieving superior surface quality. Several strategies focus on minimizing tool wear and enhancing machining efficiency:

1. *Constant cutting force strategy*: This approach maintains a consistent cutting force by adjusting the tool path, reducing tool wear, and ensuring predictable machining conditions.

2. *Trochoidal tool path*: Utilizing circular or trochoidal tool paths can distribute wear evenly across the cutting edge, mitigating tool wear and improving surface finish.

3. *Adaptive control strategies*: These strategies adjust the tool path in real-time based on cutting conditions, reducing vibrations and optimizing cutting forces to extend tool life.

4. *Multi-step machining*: Breaking down the machining process into multiple steps with varying parameters can reduce stress concentrations and thermal gradients, enhancing both tool life and surface integrity.

Implementing these tool path strategies requires a thorough understanding of ceramic properties, cutting dynamics, and machining conditions. It is essential to conduct experimental validation and computational simulations to adjust these strategies for specific applications and materials. This ensures consistent and efficient ceramic machining processes.

7. Sustainable aspects of hard turning with ceramics

7.1 Environmental benefits of ceramic tools over traditional methods

Ceramic tools have numerous environmental advantages over traditional methods. This is primarily due to their exceptional durability, reduced material consumption, and lower energy requirements. Unlike traditional cutting tools made from carbide or high-speed steel, ceramic tools are extremely hard and resistant to wear, resulting in a longer tool life. This means fewer tool changes and less material required over the tool's lifetime. Additionally, using ceramic tools for machining often requires lower cutting forces and energy inputs compared to conventional methods. This reduced energy consumption leads to lower greenhouse gas emissions and lower energy costs during manufacturing operations. Furthermore, ceramic tools contribute to improved material utilization and waste reduction. Their ability to maintain sharp cutting edges for extended periods minimizes material waste generated from rework and scrap, promoting more sustainable manufacturing practices. In conclusion, ceramic tools offer significant environmental benefits and have the potential to enhance manufacturing sustainability by optimizing resource utilization, reducing energy consumption, and minimizing waste generation compared to traditional tooling materials.

7.2 Energy efficiency and material savings, and waste reduction and recycling consideration

In the realm of sustainable manufacturing, hard turning with ceramics offers significant advantages in terms of energy efficiency, material savings, waste reduction, and recycling considerations.

Energy efficiency and material savings: Ceramic tools are characterized by high hardness and wear resistance, resulting in longer tool life compared to traditional materials such as carbides or high-speed steels. This extended tool lifespan reduces the frequency of tool changes, leading to lower energy consumption during replacement processes and reduced downtime. Furthermore, the use of ceramic tools in machining processes often entails lower cutting forces and requires less energy input due to their superior cutting properties. As a result, energy consumption per component produced is reduced, contributing to overall energy efficiency in manufacturing operations.

Waste reduction and recycling consideration: Ceramic tools contribute to waste reduction through several mechanisms. Their durability and ability to maintain sharp cutting edges for prolonged periods minimize material waste generated from tool wear and rework. This characteristic not only reduces material consumption but also decreases the amount of machining scrap typically produced with less durable tooling materials. Additionally, ceramics can be recycled or repurposed after their operational life as cutting tools. Processes such as reclaiming ceramic materials from worn-out tools for reuse or recycling into other applications can further minimize environmental impact and enhance resource efficiency.

8. Quality assurance and control in hard turning

8.1 Quality assurance and control in hard turning

Quality assurance and control in hard turning involve rigorous monitoring and control of process variables, as well as meticulous inspection techniques to ensure surface integrity and dimensional accuracy. Here are some methods and techniques employed in this context:

8.1.1 Methods for monitoring and controlling process variables

1. *Cutting parameters monitoring*: Continuous monitoring of cutting parameters such as cutting speed, feed rate, and depth of cut is essential to maintain stable machining conditions and optimize tool life.

2. *Tool wear and condition monitoring*: Techniques like acoustic emission sensing, vibration analysis, and tool condition monitoring systems help detect tool wear and predict tool failure, enabling timely tool changes to maintain machining accuracy.

3. *Temperature monitoring*: Infrared thermography and embedded thermocouples are used to monitor cutting temperatures, which impact tool wear, surface finish, and material integrity during hard-turning operations.

8.1.2 Inspection techniques for surface integrity and dimensional accuracy

1. *Surface roughness measurement*: Surface profilometers and coordinate measuring machines (CMMs) are used to assess surface roughness, ensuring compliance with specified requirements and optimizing post-machining processes (**Figure 8**).

2. *Roundness and cylindricity inspection*: Dedicated roundness testers and CMMs equipped with appropriate software are employed to inspect the dimensional accuracy of cylindrical components produced through hard turning.

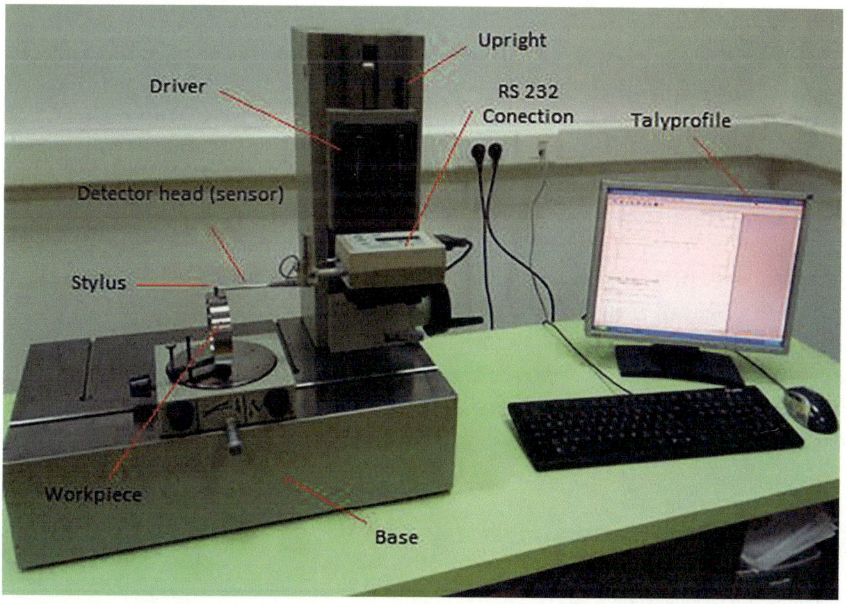

Figure 8.
Computerized measuring device Surtroniic 3+ from Taylor Hobson and TalyProfile software [1].

3. *Non-destructive testing (NDT)*: Techniques such as ultrasonic testing and eddy current testing are utilized to detect subsurface defects and ensure component integrity without compromising dimensional.

Implementing these methods and techniques requires a systematic approach to quality management in hard-turning processes. Continuous improvement initiatives based on real-time data and feedback from inspection results help optimize machining operations and enhance overall product quality.

9. Applications of hard turning with ceramic tools

9.1 Industries and sectors benefiting from ceramic tool advancements

Industries across various sectors have increasingly benefited from advancements in ceramic tool technology, leveraging their unique properties to enhance productivity, reduce costs, and improve product quality. Here are two genuine case studies demonstrating the application of ceramic tools in different industries:

1. *Aerospace industry*: The aerospace sector demands high-performance materials and stringent machining tolerances. Ceramic tools have been instrumental in machining aerospace components made from difficult-to-machine materials such as titanium alloys and composites. For instance, Rolls-Royce, a prominent aerospace manufacturer, has utilized ceramic cutting tools to improve efficiency in machining turbine blades and other critical components. Ceramic tools offer superior wear resistance and thermal stability, enabling higher cutting speeds and reducing machining times while maintaining dimensional accuracy.

2. *Medical device manufacturing*: In the medical device industry, where precision and biocompatibility are paramount, ceramic tools have become indispensable. Case in point: Zimmer Biomet, a leading medical device manufacturer, employs ceramic cutting tools for machining orthopedic implants made from materials like titanium and cobalt-chromium alloys. These tools ensure exceptional surface finish and dimensional accuracy critical for implant components, enhancing biocompatibility and patient safety. The durability and wear resistance of ceramic tools also contribute to reducing tool changeovers and associated downtime, improving overall production efficiency.

These case studies illustrate how ceramic tool advancements are tailored to meet specific challenges in industries requiring high precision, durability, and efficiency, thereby driving technological innovation and economic competitiveness.

10. Challenges and future directions

10.1 Challenges and future directions

Advancements in hard turning with ceramic cutting tools require improved tooling materials to enhance tool life and reduce costs. Future directions include exploring nanocomposite ceramics for superior toughness and wear resistance, as well as developing advanced coatings to further enhance tool performance. Additionally, integrating real-time monitoring and adaptive control systems will optimize machining parameters dynamically for sustainable and high-quality metalworking processes. These efforts aim to consolidate ceramic tools as integral components of efficient and environmentally responsible manufacturing practices.

10.2 Current limitations and challenges in using ceramic tools

Using ceramic cutting tools in hard turning has advantages but also presents challenges. Ceramics are brittle, prone to chipping, and require specialized equipment for shaping and grinding. Achieving stable cutting conditions at high speeds is difficult due to the risk of thermal damage and tool wear. Researchers are exploring advanced ceramic compositions and coatings to enhance toughness and reduce brittleness. These efforts are crucial for expanding the application range of ceramic tools in metalworking.

These references provide comprehensive insights into the challenges and ongoing research efforts related to the use of ceramic cutting tools in hard turning processes.

10.3 Future trends and innovations in hard-turning technology

Future trends in hard-turning technology are focused on improving machining efficiency, sustainability, and precision. Key advancements include the use of advanced ceramic composites and nanomaterials to enhance toughness and wear resistance. Research is also concentrating on developing advanced coatings and surface treatments to increase tool durability in high-speed, high-temperature conditions. Integrating AI and machine learning enables real-time monitoring and adaptive control, optimizing cutting parameters, and improving machining accuracy. Additive manufacturing is being explored to create customized tool geometries and complex ceramic components, which enhance flexibility and efficiency in production. These trends are addressing challenges such as tool wear and thermal damage, advancing hard-turning capabilities for the aerospace, automotive, and medical device industries.

10.4 Potential areas for further research

Further research in hard turning with ceramic tools presents several promising areas that can advance machining capabilities, improve efficiency, and expand application domains. Here are some potential areas for future research:

1. *Advanced ceramic materials*: Investigating new compositions of ceramics, including nanocomposites and hybrid materials, to enhance toughness, wear resistance, and thermal stability. This could broaden the range of materials that can be effectively machined with ceramic tools.

2. *Surface coatings and treatments*: Developing advanced coatings and surface treatments to improve tool life, reduce friction and wear, and enhance cutting performance in challenging machining environments.

3. *Tool geometry optimization*: Exploring novel tool geometries and edge preparations tailored for specific machining conditions to minimize cutting forces, tool wear, and improve surface finish.

4. *Process monitoring and control*: Integrating advanced sensors, real-time monitoring techniques, and machine learning algorithms for predictive maintenance, adaptive control of cutting parameters, and optimization of machining processes.

5. *Additive manufacturing (AM) applications*: Utilizing additive manufacturing techniques to produce ceramic tools with complex geometries and customized designs, optimizing tool performance, and reducing lead times.

6. *Environmental and sustainability aspects*: Investigating the environmental impact of ceramic tool manufacturing and machining processes, with a focus on sustainability through material recycling, waste reduction, and energy-efficient machining strategies.

7. *Multi-scale modeling and simulation*: Developing comprehensive multi-scale models and simulations to predict tool behavior, optimize process parameters, and understand the fundamental mechanisms governing ceramic tool wear and fracture.

8. *Applications in emerging industries*: Exploring new applications of hard turning with ceramic tools in emerging industries such as renewable energy (e.g., wind turbines), biomedical devices, and electronics manufacturing.

9. *Integration with industry 4.0 technologies*: Leveraging Industry 4.0 concepts such as digital twins, IoT (Internet of Things), and cyber-physical systems to create smart machining environments and enhance overall manufacturing efficiency and quality.

The research in these areas has the potential to overcome current limitations in hard turning with ceramic tools, drive technological innovation, and contribute to sustainable and high-performance machining practices across various industrial sectors. By focusing on these areas, researchers can address key challenges and advance the capabilities of ceramic tools, making them more versatile, efficient, and reliable in demanding machining applications.

10.5 Cutting-edge research

In this next section, we will present a comprehensive overview of various research projects investigating advanced studies on the wear mechanisms of ceramic tools utilized in metal machining. These sources showcase the most recent advancements and perspectives on how ceramic tools function in different machining conditions, providing insights into their longevity, wear characteristics, and overall efficiency.

The paper “The critical raw materials in cutting tools for machining applications” reviews advancements in cutting tools under extreme conditions, emphasizing ceramic alternatives to tungsten carbide. It covers hybrid machining processes, CRM substitutes, tool recycling, and Industry 4.0 applications to enhance tool longevity and sustainability, aiming to reduce reliance on critical raw materials like tungsten and cobalt [22].

The review “High-performance coatings for cutting tools” highlights the evolution of hard coatings for cutting tools, noting that 85% of cemented carbide tools are coated. It emphasizes the role of nanostructured PVD coatings in enhancing efficiency and productivity. The paper underscores ongoing innovation in coating materials and tool design to meet growing industry demands [23]. The study "Tool wear mechanisms in the machining of steels and stainless steels" reviews tool wear mechanisms in machining, focusing on ceramics' role in metalworking. It identifies attrition as a significant wear mechanism for ductile materials like steel, with additional mechanisms such as abrasion and diffusion. The research utilizes SEM and EDS analysis to enhance understanding of tool wear in various cutting conditions [24]. The study "A review on cutting of industrial ceramic materials" reviews advancements in non-conventional and hybrid machining techniques for ceramics, highlighting their excellent high-temperature properties and hardness. It addresses challenges related to ceramics' brittleness and discusses recent innovations aimed at improving machinability and surface integrity, enhancing their precision, and their application in manufacturing industries [3]. The research study "Mathematical Modeling of the Bearing Ratio Curve Rmr (50% Rz), through Investigation of the Effect of Process Parameters in Hard Turning of Steel C55 (DIN) with Mixed Ceramics MC2 (Al_2O_3+ TiC)." develops a mathematical model using design of experiment methods to analyze hard turning of Steel C55 with mixed ceramics MC2 (Al_2O_3 + TiC). By predicting the bearing ratio curve parameter Rmr and using ANOVA for validation, the study provides insights into how process parameters impact surface finish, leveraging ceramics to enhance machining consistency and surface quality [1].

11. Conclusion

The introduction to hard turning establishes its critical role in modern metalworking, highlighting ceramic cutting tools as pivotal for achieving efficient machining processes. This chapter sets clear objectives for exploring the fundamentals, properties, and advancements of ceramic tools in hard-turning applications. Fundamentals delve into the definition, characteristics, and comparative analysis of hard turning versus traditional methods, addressing both advantages and challenges. The section on ceramic cutting tools provides an extensive overview of materials used, their specific properties conducive to hard turning, and various types such as inserts and coatings. Performance metrics crucial for evaluating hard-turning operations, including key indicators like tool wear, surface integrity, and tool life, are examined. Advances in ceramic tool technology trace their evolution, emphasizing innovative designs, geometries, and surface treatments aimed at enhancing performance. Strategies for optimizing cutting parameters and maximizing tool life are described, supported by practical examples and case studies demonstrating effective implementation. Environmental sustainability considerations highlight the benefits of ceramic tools over traditional methods, including energy efficiency, material savings, and waste reduction strategies. Quality assurance methods for monitoring process variables and ensuring dimensional accuracy are discussed, underscoring the precision demanded in hard-turning operations. Industries benefiting from ceramic tool advancements are identified through specific case studies, illustrating successful implementations across various applications. Current challenges and future directions in ceramic tool technology explore limitations, future trends, innovations, and potential research areas to enhance hard-turning processes. The conclusion synthesizes key insights, emphasizing the transformative role of ceramic tools in shaping the future of metalworking, with a focus on ongoing advancements and their potential impact on the field.

12. Final thoughts

Ceramic cutting tools are critical for the future of metalworking, providing unparalleled benefits in modern manufacturing. Their extreme hardness, thermal stability, and resistance to wear are essential for machining tough materials such as aerospace alloys and advanced composites. They deliver superior finishes, longer tool life, and help reduce environmental impact through energy efficiency and waste reduction. Looking ahead, the potential for hard turning with ceramic tools is promising. Ongoing advancements in ceramic materials, coatings, and tool designs will improve performance and application versatility. Additive manufacturing will allow for the production of complex ceramic tools tailored to specific tasks. The integration of AI and machine learning in process control will enhance efficiency and accuracy. Despite challenges such as brittleness and high costs, ongoing research aims to develop tougher ceramics and cost-effective production methods. Industry 4.0 technologies offer opportunities for smarter machining environments. Overall, ceramic tools are poised to revolutionize metalworking and drive sustainable, high-performance operations globally. Collaborative efforts in research and technology promise a bright future, ensuring that ceramic tools continue to innovate and shape the manufacturing landscape.

Acknowledgements

The author acknowledges the use of AI for language polishing of the manuscript.

Author details

Mevludin Shabani and Ariana Gjoni*
UBT Higher Education Institution, Pristina, Kosovo

*Address all correspondence to: ariana.gjoni@ubt-uni.net

References

[1] Shabani M, Tomov M. Mathematical modeling of the bearing ratio curve Rmr (50% Rz), through investigation of the effect of process parameters in hard turning of steel C55 (DIN) with mixed ceramics MC2 (Al_2O_3+ TiC). Advances in Tribology. 2024;**2024**(1):5541719

[2] Grigoriev SN, Fedorov SV, Hamdy K. Materials, properties, manufacturing methods and cutting performance of innovative ceramic cutting tools– A review. Manufacturing Review. 2019;**6**:19

[3] Rakshit R, Das AK. A review on cutting of industrial ceramic materials. Precision Engineering. 2019;**59**:90-109

[4] Krolczyk GM, Maruda RW, Krolczyk JB, Wojciechowski S, Mia M, Nieslony P, et al. Ecological trends in machining as a key factor in sustainable production–A review. Journal of Cleaner Production. 2019;**218**:601-615

[5] Zhang S, Liu Z. A new approach to cutting temperature prediction considering the diffusion layer in coated tools. International Journal of Machine Tools and Manufacture. 2009;**49**(7-8):619-624

[6] Medvedovski E. Wear-resistant engineering ceramics. Wear. 2001;**249**:821-828

[7] Zhou J, Bahadur S. Erosion characteristics of alumina ceramics at high temperatures. Wear. 1995;**181-183**:178-188

[8] Zhang Y, Cheng YB, Lathabai S. Erosion of alumina ceramics by air- and water-suspended garnet particles. Wear. 2000;**240**:40-51

[9] Zhang FC, Luo HH, Wang TS, Roberts SG, Todd RI. Influence factors on wear resistance of two alumina matrix composites. Wear. 2008;**265**:27-33

[10] Sahin Y, Durak O. Abrasive wear behaviour of austempered ductile iron. Materials and Design. 2007;**28**:1844-1850

[11] Hejwowski T. Erosive and abrasive wear resistance of overlay coatings. Vacuum. 2009;**83**:166-170

[12] Medvedovski E. Alumina-mullite ceramics for structural applications. Ceramics International. 2006;**32**:369-375

[13] Baudín C, Tricoteaux A, Joire H. Improved resistance of alumina to mild wear by aluminium titanate additions. Journal of the European Ceramic Society. 2014;**34**:69-80

[14] Esposito L, Tucci A. Microstructural dependence of friction and wear behaviours in low purity alumina ceramics. Wear. 1997;**205**:88-96

[15] Latella BA, O'Connor BH. Effect of porosity on the erosive wear of liquid-phase-sintered alumina ceramics. Journal of the American Ceramic Society. 1999;**82**(8):2145-2149

[16] Ćurković L, Kumić I, Grilec K. Solid particle erosion behaviour of high purity alumina ceramics. Ceramics International. 2011;**37**:29-35

[17] Kumar AS, Durai AR, Sornakumar T. Wear behaviour of alumina-based ceramic cutting tools on machining steels. Tribology International. 2006;**39**:191-197

[18] Selvarajan L, Rajavel R, Venkataramanan K, Srinivasan VP. Experimental investigation on surface

morphology and recasting layer of Si_3N_4-TiN composites machined by die-sinking and rotary EDM. Ceramics International. 2023;**49**(5):8487-8501

[19] Kuntoğlu M, Sağlam H. Investigation of progressive tool wear for determining of optimized machining parameters in turning. Measurement. 2019;**140**:427-436

[20] Thakur A, Gangopadhyay S. State-of-the-art in surface integrity in machining of nickel-based super alloys. International Journal of Machine Tools and Manufacture. 2016;**100**:25-54

[21] Lv X, Ye F, Cheng L, Zhang L. Novel processing strategy and challenges on whisker-reinforced ceramic matrix composites. Composites Part A: Applied Science and Manufacturing. 2022;**158**:106974

[22] Rizzo A, Goel S, Luisa Grilli M, Iglesias R, Jaworska L, Lapkovskis V, et al. The critical raw materials in cutting tools for machining applications: A review. Materials. 2020;**13**(6):1377

[23] Bobzin K. High-performance coatings for cutting tools. CIRP Journal of Manufacturing Science and Technology. 2017;**18**:1-9

[24] Diniz AE, Machado ÁR, Corrêa JG. Tool wear mechanisms in the machining of steels and stainless steels. The International Journal of Advanced Manufacturing Technology. 2016;**87**:3157-3168

Chapter 2

Photon-Powered Ceramics: New Frontiers in Material Science and Applications

Pablo Serna-Gallén, Robinson Cadena, Samuel Porcar, Jaime González Cuadra, Abderrahim Lahlahi, Santiago Toca, Diego Fraga and Juan Carda

Abstract

This chapter provides a comprehensive review of light-mediated processes in advanced ceramics, emphasizing their role in developing new materials with enhanced properties. The discussion covers photocatalysis for environmental remediation and energy production, ceramic materials in photovoltaic cells for sustainable energy, and the role of ceramics in photonic devices, such as waveguides, lasers, and optical fibers. The analysis also includes ceramic applications in laser technology, focusing on their optical properties, and biocompatible and bioactive ceramic materials in biotechnology for drug delivery, sensors, and theragnosis. Additionally, the chapter examines how light-mediated processes contribute to sustainability and the circular economy by enhancing material properties at the nanoscale and promoting energy-efficient production and recycling methods. The aim is to highlight the transformative potential of light interactions in ceramics, driving advancements in energy efficiency, environmental protection, and medical technology, and inspiring future research and applications in these areas.

Keywords: multifunctional ceramics, photonics, nanotechnology, energy ceramics, biomaterials, emerging technologies

1. Introduction

Light-mediated processes in advanced ceramics are recognized as a main pillar of renewable technologies, which can strongly contribute to fulfilling the roadmap and the Strategic Energy Technology Plan (SET Plan) targets [1]. These processes are also aligned with the United Nations and European Sustainable Development Objectives (SDGs) [2].

The claim for advanced materials with needed economic feasibility, cost-efficiency, and higher tolerances has tiled the way for new multifunctional elements for fabrication. This is achieved by using enhanced production materials and technologies. Advanced materials, intended to meet specific requirements through designed properties, are an area of enormous importance for many researchers from the material science discipline. These compounds are typically traditional materials whose properties have been enhanced and include nanomaterials, semiconductors, and biomaterials.

Ceramic materials based on light-mediated processes have attracted great interest from researchers after demonstrating the possibility of creating multifunctional ceramic devices. Numerous research developments in the literature focus on acquiring electrical [3], optical [4], and luminescent [5] properties to release their perspective in various technological applications.

This chapter aims to present a thorough review of light-mediated processes in advanced ceramics, emphasizing their role in developing new materials with enhanced properties. Initially, recent advances in photocatalytic materials, such as titanium dioxide (TiO_2), are shown. The incorporation of photoactive additives in different materials that optimize the use of light for disinfection treatments is very attractive when it comes to guaranteeing the quality of life of the people in such types of buildings. These additives should react with natural or artificial light, without the need for ultraviolet (UV) radiation, so that their bactericidal and fungicidal properties act, as it happens in other antibacterial treatments. Their antibacterial activities are directly related to a photoredox-catalyzed process that produces free reactive oxygen species (ROS), succeeding in the induction of oxidative stress and ultimately resulting in bacterial death.

Subsequently, the chapter is focused on the advances in the development of alternative wide band gap absorbers with low-temperature processing. These materials are inspired by the very recent developments of quasi-one-dimensional (Q1-D) $Sb_2(S,Se)_3$, with applications for advanced photovoltaics (PV) thoughts such as tandem and semitransparent devices, thanks to the exceptional combination of optical, electrical, and structural properties.

Afterward, a perspective on current advances and emerging challenges in photonics is also evaluated, where lanthanide ions, such as Eu^{3+}, Tb^{3+}, Er^{3+}, and Yb^{3+}, play a crucial role in the development of multifunctional materials with luminescence properties. Then, the laser innovation process aims to address issues about improving the functional performance of various devices, including ceramic surfaces. Besides, light processes for biological applications are described focusing on recent advances in photothermal and photodynamic therapies. Finally, the chapter also highlights the importance of sustainability in the production of advanced ceramic materials.

2. Pioneering photocatalysis in ceramics

The fusion of art and science has rarely been as transformative as in the field of ceramics enhanced through photocatalysis. This section delves into the pioneering advancements that have integrated photocatalytic properties into ceramic materials, ushering in a new era of functionality and innovation. Historically, ceramics have been celebrated for their esthetic and utilitarian roles, gracing ancient pottery and modern-day architectural marvels alike. However, the introduction of photocatalysis has revolutionized their applications, extending their capabilities far beyond traditional boundaries.

Photocatalysis, a process in which light energy catalyzes chemical reactions, has found a powerful ally in ceramics. These materials, known for their durability and versatility, have become a canvas for cutting-edge scientific endeavors [6].

The connection of these fields promises not only to enhance the properties of ceramics but also to address pressing environmental and technological challenges. From self-cleaning surfaces to air purification and antibacterial functionalities, photocatalytic ceramics are paving the way for smarter and more sustainable solutions in several industries [7].

This part of the chapter will explore the key milestones in the development of photocatalytic ceramics, tracing their journey from experimental concepts to practical implementations. It will highlight the contributions of visionary researchers who have pushed the boundaries of what ceramics can achieve. By examining the fundamental principles of photocatalysis and the unique properties of ceramic materials, it will uncover how these two domains synergize to create materials that are not only esthetically beautiful but also highly functional.

2.1 Advances in antibacterial ceramics

Bactericidal ceramics that leverage photocatalysis have emerged as promising materials in the fight against microbial contamination. These ceramics are typically embedded with photocatalytic materials like titanium dioxide (TiO_2), which, upon exposure to light, generate reactive oxygen species (ROS) capable of degrading organic pollutants and killing bacteria [8, 9].

2.1.1 Photocatalytic mechanism

The primary mechanism for ROS generation involves the absorption of light by the photocatalyst, which excites electrons and creates electron-hole pairs. These pairs react with water and oxygen to form ROS, such as hydroxyl radicals and superoxide anions, which possess strong oxidative properties capable of breaking down microbial cell walls and degrading organic matter [10].

Many materials, such as tin oxide (SnO_2) and titanium dioxide, can generate ROS when exposed to light, especially UV light. This process involves the absorption of photons, which excites electrons from the valence band to the conduction band, leaving behind holes in the valence band. Additionally, ROS can be generated through chemical reactions, such as the Fenton reaction, where hydrogen peroxide reacts with ferrous ions to produce hydroxyl radicals [11]. Applying electricity to certain materials can also induce electrochemical reactions that produce ROS.

2.1.2 Mechanism of bacterial killing

The bactericidal action of ROS operates through several interconnected mechanisms that ultimately lead to bacterial cell death. One primary mechanism is membrane damage, where ROS induce lipid peroxidation, disrupting the bacterial cell membranes [8]. This disruption compromises membrane integrity, causing leakage of cellular contents and resulting in cell death.

Additionally, ROS oxidize essential proteins, disrupting their function. This oxidative damage affects vital bacterial processes, such as enzyme activity, nutrient transport, and cell signaling, further incapacitating the bacteria [12, 13]. Finally, ROS can inflict oxidative damage on DNA, leading to mutations and hindering DNA replication and transcription. This damage prevents the bacteria from reproducing and repairing itself, culminating in bacterial death. Through these combined actions, ROS effectively neutralize bacterial threats by attacking multiple cellular targets [14].

2.1.3 Advantages of ROS for antimicrobial activity

The use of ROS for antimicrobial activity offers significant advantages. First, ROS exhibit a broad-spectrum effectiveness, targeting a wide range of bacteria, including both Gram-positive and Gram-negative species [15]. This wide applicability makes ROS a versatile tool in combating diverse bacterial infections.

Second, bacteria are less likely to develop resistance to ROS compared to traditional antibiotics. This characteristic is particularly valuable in the current medical landscape, where antibiotic-resistant strains pose a significant challenge. The ability of ROS to attack multiple cellular targets simultaneously reduces the likelihood of resistance development, ensuring the continued efficacy of ROS-based antimicrobial methods [16]. These advantages underscore the potential of ROS as a powerful and reliable alternative in the fight against bacterial infections.

2.1.4 Applications

The antimicrobial properties of ROS have been harnessed in various practical applications to enhance public health and safety. In the medical field, ROS-generating materials are applied as coatings for medical devices, effectively preventing bacterial colonization and infections [17]. This innovation is crucial in reducing hospital-acquired infections and improving patient outcomes.

In water treatment systems, photocatalytic materials are employed to disinfect water by killing harmful bacteria, ensuring the provision of safe and clean drinking water. These materials leverage the power of light to generate ROS, which then eliminate microbial contaminants efficiently [9].

Moreover, surfaces coated with ROS-generating materials offer continuous disinfection, providing an additional layer of protection in hospitals and public spaces. This continuous antimicrobial action helps maintain a hygienic environment, reducing the spread of infectious diseases. For optimal antibacterial efficacy, it is essential to select appropriate materials and methods that maximize ROS generation and ensure effective interaction with bacterial cells. By tailoring these parameters, the use of ROS can be fine-tuned to meet specific needs in various applications, enhancing their overall effectiveness and reliability.

2.2 Perspectives

Antibacterial ceramics offer significant advantages. First, they exhibit broad-spectrum effectiveness, targeting a wide range of bacteria, including both Gram-positive and Gram-negative species. Second, bacteria are less likely to develop resistance to these materials compared to traditional antibiotics, as they attack multiple cellular targets simultaneously, ensuring continued efficacy. These advantages highlight the potential of antibacterial ceramics as a powerful and reliable alternative in combating bacterial infections in various settings, such as hospitals, kitchens, and bathrooms.

3. Next-generation photovoltaic ceramic solutions

Solar photovoltaic technology has been gaining prominence in global energy production in recent years. Despite the dominance of silicon and halide perovskites,

$CH_3NH_3PbX_3$ (X = I, Br, or Cl), in the field of photovoltaic solar cells, due to their high efficiency (26.1% in both cases) and low production costs, ceramic materials could eventually play a crucial role in the race to surpass or complement these technologies [18]. Historically, the most popular ceramic alternatives have been cadmium indium gallium selenide (CIGS), cadmium telluride (CdTe), and copper zinc tin sulfide (CZTS). These either struggle to compete with the technologies due to the high cost of raw materials (as is the case with CIGS and CdTe) or lack sufficient efficiency to be economically feasible (CZTS) [19].

To address these challenges, recent developments have introduced new ceramic materials that are cost-effective to produce and exhibit significantly enhanced performance. These advancements render these ceramics feasible alternatives or integral components in the development of tandem photovoltaic solar cells, thereby improving the efficiency of current systems [20]. In this way, oxide, sulfide, or selenide absorbers rich in materials abundant in the Earth's crust, such as $CuSbS_2$, Cu_2SnS_3, $Cu_2ZnSnSe_4$, $CuSbSe_2$, Cu_2O, SnS, FeS_2, Sb_2S_3, $Bi_2S_3Sb_2(S,Se)_3$, and Sb_2Se_3, have been investigated as substitutes for thin-film solar cells that are efficient, affordable, eco-friendly, and durable solar cells [20, 21]. Out of all the possibilities, binary compounds present great potential for future development because of their higher simplicity and scalability.

3.1 Antimony chalcogenides and selenides

Compounds Sb_2S_3, Sb_2Se_3, and $Sb_2(S,Se)_3$ naturally occur as the mineral stibnite, featuring an orthorhombic structure and belonging to the *Pnma* space group [18, 20, 22–24]. The Sb_2X_3 (X = S or Se) structure is characterized by X-Sb-X chains running along the c-axis (ribbons). Consequently, these compounds possess a crystal structure composed of one-dimensional (1D) ribbons formed from Sb_4X_6 units, with the ribbons held together by weak X-X interactions. Sb_2S_3 and Sb_2Se_3 present 2.2 eV and 1.7–1.8 eV band gaps, respectively, with an absorption coefficient > 10^4 cm^{-1} in the visible solar spectra [23]. However, the main drawback of antimony sulfoselenide solar cell materials is their atypical intrinsic defects caused by the low symmetry of their near one-dimensional crystal structure. This defectiveness results in a substantial voltage deficit that constrains their maximum power conversion efficiency [18]. Low-dimensional crystal-structural (LDCS) materials exhibit electronic properties, such as conductivity and charge mobility, which vary with direction. Consequently, a good alignment between the thin-film layers and the substrate is crucial for optimizing charge transport efficiency [20].

3.2 Bismuth and tin sulfides

Bi_2S_3 is naturally found as the mineral bismuthinite, with an orthorhombic structure and a space group *Pmcn* [18]. Bi_2S_3 presents a 1.3 eV band gap and an absorption coefficient of near 10^5 cm^{-1} in at 600 nm. The Bi_2S_3 has a lower energy in the conduction band compared to those in the antimony ones, which is less efficient for electron injection into the buffer layer [18, 25]. As shown in **Figure 1**, despite it shows lower efficiency compared with the other materials for the common thin-film solar cells, it also seems that good results are being achieved using this material for dye-synthetized solar cells (DSSCs).

Similar to the antimony compounds, SnS has an orthorhombic crystal structure in the *Pnma* space group, where the unit cell comprises double layers stacked along

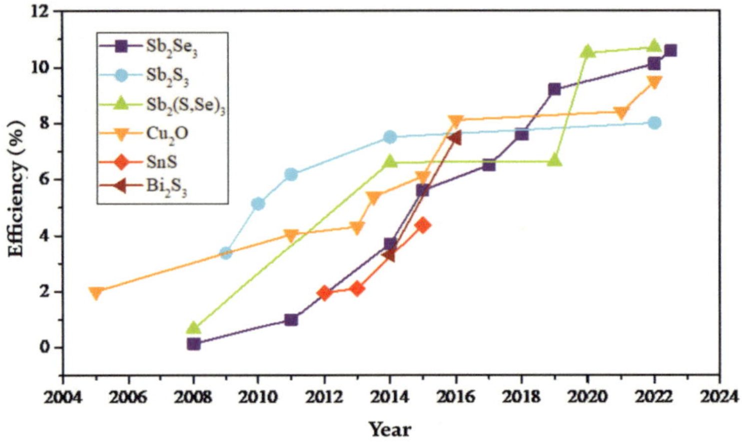

Figure 1.
Record efficiencies per year for Sb_2Se_3, Sb_2S_3, $Sb_2(S,Se)_3$, Cu_2O, SnS, and Bi_2S_3.

the *a*-axis with weak van der Waals-like coupling, and each atom in the layers forms covalent bonds with three surrounding atoms [26, 27]. SnS exhibits a direct band gap of 1.3 eV and an indirect band gap of 1.0 eV, along with an absorption coefficient that surpasses 10^4 cm^{-1} in the visible light spectrum. Defect calculations for this material indicate that its p-type behavior is due to the facile creation of tin vacancies (V_{Sn}), which act as hole acceptors. As previously mentioned, the crystal orientation of LDCS is vital for achieving high efficiency [27]. Also, different binary phases can coexist depending on the preparation conditions, affecting the absorber material's properties and efficiency. Moreover, alterations in stoichiometry can result in a variation that produces a change in the conductivity type, resulting in either p-type or n-type behavior [19].

3.3 Cuprous oxide

Cuprous oxide (Cu_2O) has a cubic crystal lattice where each oxygen atom is surrounded by four copper atoms, and each copper atom is linked to two oxygen atoms within a $Pn\bar{3}m$ space group [28]. Cu_2O is a p-type semiconductor with a wide direct band gap of 2.0–2.4 eV. It features excellent mobility, affordability, a long minority carrier diffusion length, and a significant absorption coefficient in the visible spectrum (near 10^5 cm^{-1}) making it suitable for solar cell devices, especially for tandem solar cells (like Sb_2S_3). Additionally, Cu and CuO impurities seem to play a very important role in achieving good efficiency [29].

3.4 Perspectives

Although there are other series of highly promising compounds in this field, these ceramic binary compounds will generate significant scientific interest in the coming years. This attention is not only due to their simplicity and high stability, which reduce manufacturing costs but also because of the projected improvements in efficiency that are currently being achieved and are expected to continue in the future.

4. Photonics: Shaping the future of multifunctional ceramics

The field of photonics, which involves the generation, manipulation, and detection of light, has experienced significant advancements in recent years, profoundly impacting the development and applications of various materials. Among these materials, multifunctional ceramics have benefited greatly from these photonic advancements. This section explores the recent progress in photonics as applied to multifunctional ceramics, highlighting key examples to illustrate these developments and their broader implications.

4.1 Optical properties of lanthanide-doped ceramics

One of the major advances in photonics for multifunctional ceramics is the enhancement of luminescent properties. Lanthanide ions, such as Eu^{3+}, Tb^{3+}, Er^{3+}, and Yb^{3+}, are known for their sharp emission lines, long luminescent lifetimes, and high quantum yields, which make them ideal for various applications, including light-emitting diodes (LEDs), organic LEDs (OLEDs), and luminescent thermometry, among others. Lanthanide ions exhibit distinct electronic transitions within the 4*f* orbitals, resulting in characteristic emission spectra. For example, Eu^{3+} ions are well known for their red emission and site-selective properties, which are widely used in display technologies and lighting [30]. Tb^{3+} ions, on the other hand, emit green light and are utilized in phosphors for cathode-ray tubes (CRTs) and X-ray imaging systems (e.g., oxysulfide Gd_2O_2S:Tb) due to their high luminescence efficiency and fast response time, providing clear and detailed images [31].

4.1.1 LEDs or OLEDs? Key differences and applications

As previously introduced, light-emitting diodes (LEDs) and organic light-emitting diodes (OLEDs) are two prominent applications of lanthanide-doped ceramics. LEDs are semiconductor devices that emit light when an electric current passes through them [32]. They are known for their energy efficiency, long lifespan, and robustness. Materials, such as the yttrium aluminum garnet doped with trivalent cerium ions (YAG:Ce^{3+}), are commonly used in white LEDs, where the ceramic phosphor converts blue LED light into white light through a combination of blue, green, and red emissions [33]. Specifically, Ce^{3+} ions are excellent for this application due to their ability to absorb blue light and re-emit it as a broad spectrum that appears white to the human eye. Additionally, phosphors doped with Eu^{2+} can emit intense blue or green light, further enhancing the color quality and efficiency of LEDs by providing specific wavelength emissions [33].

In contrast, OLEDs are based on organic compounds that emit light in response to an electric current. They offer advantages like flexible displays, wider viewing angles, and higher contrast ratios compared to LEDs. Lanthanide complexes can be incorporated into OLEDs to achieve specific emission colors and improve device efficiency. For instance, OLEDs utilizing Eu^{3+} complexes can produce high-quality red emission, which significantly enhances the color gamut of displays by providing a vivid red that is difficult to achieve with other materials. However, OLEDs typically have a shorter lifespan and are more susceptible to moisture and oxygen degradation compared to LEDs. This susceptibility is a significant drawback in applications where durability and longevity are critical [32].

Despite these differences, both technologies leverage the unique properties of lanthanides to enhance their performance. The choice between LEDs and OLEDs ultimately depends on the specific requirements of the application, balancing factors, such as durability, efficiency, display quality, and color rendering.

4.1.2 Luminescent nanothermometry

Luminescent nanothermometry is an emerging application of lanthanide-doped ceramics, leveraging their temperature-dependent luminescent properties. This technique involves measuring temperature changes based on the variations in luminescence intensity, emission wavelength, or lifetime of lanthanide ions. It is particularly valuable in biomedical applications, where precise and non-invasive temperature monitoring is crucial [34].

Significant advancements in luminescent nanothermometry have been made, particularly in the development of near-infrared (NIR) emitting nanothermometers and upconversion nanoparticles (UCNPs). Common UCNPs are composed of a host crystal such as metal fluorides, or ceramics based on oxides, phosphates, and vanadates doped with lanthanide ions, either individually or in combinations [35].

For example, Er^{3+}/Yb^{3+} co-doped ceramics can serve as effective luminescent nanothermometers. These materials exhibit upconversion luminescence, where the emission intensity ratio between two bands varies with temperature. Such properties allow for accurate temperature measurements at the nanoscale, essential for monitoring intracellular processes and improving hyperthermia cancer treatments, as exemplified by L.F. Dos Santos *et al.* [36].

Furthermore, integrating luminescent nanothermometers with drug delivery systems has led to multifunctional hybrid materials that enable simultaneous temperature sensing and controlled drug release, highlighting their potential in precision therapeutic applications [37].

4.2 Nonlinear optical properties

The nonlinear optical properties of multifunctional ceramics have also been a focus of recent photonic research. Nonlinear optics involves the interaction of intense light with materials, leading to phenomena, such as frequency doubling, self-focusing, and optical solitons. Ceramics with enhanced nonlinear optical properties are crucial for applications in laser technology, optical switching, and signal processing [38].

For example, lithium niobate ($LiNbO_3$) ceramics are renowned for their strong nonlinear optical coefficients, making them ideal for frequency conversion applications, such as second-harmonic generation (SHG). The material's wide transparency range, from the visible to the infrared spectrum, coupled with its high electro-optic and acousto-optic coefficients, further enhances its utility in various photonic applications. Advances in the fabrication of domain-engineered $LiNbO_3$ ceramics have led to improved phase-matching conditions, significantly increasing the efficiency of nonlinear optical processes [39].

Domain engineering techniques, such as electric field poling, allow for the creation of periodically poled lithium niobate (PPLN) structures, which optimize the quasi-phase matching (QPM) conditions. This innovation enables efficient frequency doubling of laser light and facilitates the generation of new wavelengths not easily accessible with other materials [40].

4.3 Perspectives

Continued research and innovation in photonics are expected to drive further technological progress, enabling new applications like high-efficiency photonic sensors and advanced light-manipulating devices, while also enhancing the performance of existing technologies. This convergence not only showcases the synergy between material science and optical engineering but also paves the way for next-generation devices and systems. For instance, we can anticipate the development of highly sensitive diagnostic tools in biomedicine, more efficient energy harvesting systems, and sophisticated environmental sensing technologies.

5. Laser innovations in ceramic engineering

Laser processing has garnered significant attention for its ability to create functional surface patterns. The number of review articles is increasing due to the significant advancements in this field [41]. This non-contact method employs high-energy laser ablation to remove materials without the need for molds or masks. It offers numerous advantages, including precise results, simple operation, excellent control, adaptability to various materials, reproducibility, and sustainability. Because of its exceptionally high energy density, which exceeds the damage limits of most compounds, laser processing is capable of treating nearly all materials, including metals, semiconductors, glass, and polymers. By modifying variables such as wavelength, pulse width, pulse rates, and scanning speed, the surface structure can be readily controlled. Furthermore, advances in ultrafast lasers, like picosecond and femtosecond variants, enable the creation of intricate surface patterns, such as micropillars, cavities, and porous structures, as well as multiscale structures ranging from large to nanoscale dimensions.

5.1 Modern laser processing techniques for functional surface pattern

Several laser processing technologies exist, each suited for specific applications. The most frequently mentioned techniques in the literature include:

- *Direct laser ablation (DLA)*: This method efficiently removes materials using laser energy to create micro/nanostructures without the need for molds. It has been enhanced by innovations, such as Laser-induced plasma micromachining (LIPMM) and double-pulse laser ablation (DPLA) [42].

- *Direct laser interference patterning (DLIP)*: This technique leverages the interference of laser beams to create periodic, three-dimensional (3D) structures that can be controlled. This method is suitable for both large-scale and nanoscale processing [43].

- *Pulsed laser deposition (PLD)*: Utilizing a high-intensity laser to create a plasma plume for material deposition, this method allows for control over surface morphology and composition, although achieving uniformity can be challenging [44].

- *Selective laser melting (SLM):* This technique is a 3D printing process that uses a laser to melt and sinter metal powder layer by layer to create metal or ceramic parts. This technique builds complex geometries with precise control of material properties [45].

5.2 Laser-processed functional surface structures

A variety of structures can be achieved through the processing methods outlined earlier. These are extensively cataloged in the referenced review article [46] and summarized below:

- *Laser-induced periodic surface structures (LIPSS)*: Laser-induced surface periodic structures (LIPSS): These are nanoscale patterns that form on the surfaces of materials when irradiated with ultrafast laser pulses. These periodic structures are the result of interference between laser light and surface electromagnetic waves. LIPSS can improve surface properties, such as wettability, friction, optical or magnetic properties.

- *Microchannels or microgrooves*: Created through direct laser ablation, these structures are crucial for fluid transport in confined spaces. They are widely used in devices, such as microcoolers, microreactors, microfuel cells, and microfluidic chips due to their large surface area, compact size, and efficient flow distribution.

- *Microhole structures*: Typically produced by direct laser ablation, these structures are widely used in many fields. The dimensions and surface morphologies of the microholes are closely related to laser parameters, such as fluence and pulse width.

- *Micropillars*: These structures are tiny pillar-like formations produced on the surface of a material using advanced laser fabrication techniques. These structures have applications in a variety of fields, such as optics, electronics, and biology, due to their ability to manipulate light, electrical signals, and biological interactions on a microscopic scale.

- *Porous structures*: Surface porous structures created by laser ablation are patterns on a material resulting from removing material with laser pulses, forming a surface with high specific area and numerous micropores. These structures enhance heat transfer and modify wettability.

- *Hierarchical structures*: These structures consist of complex multilevel patterns in a material that combine micro- and nano-sized features. These structures improve properties, such as adhesion, wettability, heat transfer, and optical properties (**Figure 2**).

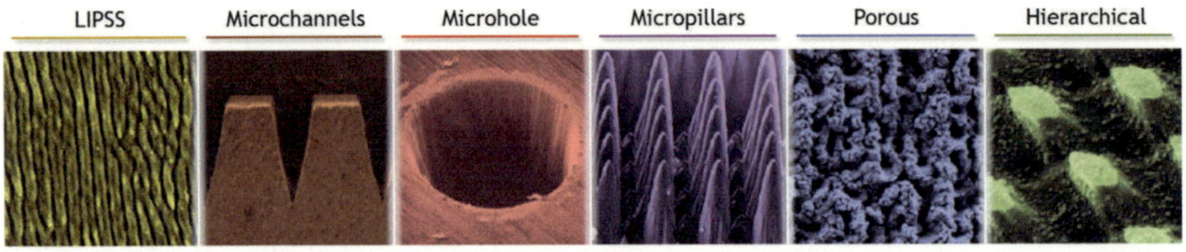

Figure 2.
Scanning electron microscopy (SEM) images of the most common laser-created functional surface structures. Adapted with permission from reference [47].

5.3 Advanced surface structures and their uses

With the advancement of the laser techniques discussed earlier, a wide range of macro-, micro-, and nano-scale surface structures have been designed and fabricated to enhance the functional performance of various devices. These structures can be applied in numerous functional areas, including optical, superhydrophobic, electronic, tribological, and biomedical applications. This section delves into the primary optical applications of laser-processed surfaces, focusing on two main categories: light trapping and antireflection structures, and surface coloring structures. Additional optical applications and their processing characteristics are reviewed comprehensively by Wang *et al.* [47].

In direct connection with Section 3, it must be emphasized that light-trapping structures like microcones or micropillars enhance solar photovoltaic efficiency by increasing light absorption. For example, Nayak *et al.* [48] used femtosecond lasers to create effective light traps on silicon, boosting the efficiency up to 14.2% with uniform micropillar arrays, and Zhao *et al.* [49] achieved a 50% efficiency increase by creating LIPSS on SIGaAs wafers. In addition, laser processing also creates antireflective surfaces for solar, aerospace, and military applications, improving optical device efficiency. Challenges include optimizing these structures, especially on curved surfaces, and maintaining performance across diverse conditions.

Alternatively, surface coloration manipulates light reflection, scattering, interference, or diffraction on microscopic or nanoscopic scales, giving materials like metal or silicon specific colors. Common applications include displays, anticounterfeiting, and encryption [50]. Laser-induced periodic surface structures (LIPSS) are crucial for creating structural color by acting as diffraction gratings under white light. Studies, as those reported by Vorobyev and Guo [51] on aluminum, Li *et al.* [52], and Liu *et al.* [53] on stainless steel and copper, demonstrate LIPSS' potential for coloration. However, challenges remain in optimizing parameters like incident light angle and scanning intervals to control color output, especially in liquids, where uniformity and vibrant colors are observed, but precise control is still evolving. In the other work, Rico *et al.* [54] colored ceramic surfaces by laser-induced decomposition of coordination complexes forming surface plasmon resonance (SPR) structures. These structures are similar to LIPSS-type structures, except that the SPR structures have a random structural ordering.

5.4 Perspectives

Laser processing technologies offer a highly controllable and adaptable method for creating functional surface structures with applications in optics, thermal management, electronics, tribology, biomedical fields, and superhydrophobic surfaces. Despite their advantages, such as high precision, the ability to work at micro- and nanoscales, flexibility, and environmental friendliness, they still face significant challenges. These challenges include the need for more robust theories on the formation mechanisms of structures, morphology control, improving durability, and achieving the integration of multiple functions into a single structure. Future research should focus on better understanding of the laser-material interaction processes, developing precise theoretical models, creating methods to ensure stability and durability of the structures, and advancing the industrialization of these technologies for mass production.

6. Ceramics applied to biotechnology and nanomedicine

Ceramic compounds are experiencing significant growth in their application within biotechnology and nanomedicine. This expansion is due to their unique properties that include biocompatibility with the human body, high mechanical strength, optical properties, chemical stability, and the ability to be designed at nanometric scales. In this section, the properties of ceramics for biological applications will first be described, followed by their applications in biotechnology and nanomedicine, and finally, the prospects facing bioceramics.

6.1 Ceramic properties for biological applications

The biocompatibility of bioceramic materials has been extensively studied, highlighting their seamless interaction with living tissues without causing adverse reactions or rejection [55, 56]. This compatibility is crucial in medical implants, particularly in orthopedic and dental applications, where seamless incorporation with bone structures is essential.

The optical properties of ceramics play a crucial role in biotechnology and nanomedicine. Their high transparency in the infrared spectrum, as seen in ceramics such as zirconium oxide, facilitates the clear visualization of tissues for medical imaging applications [57]. Certain ceramics exhibit fluorescence capabilities, allowing real-time observation of biological processes at the cellular level, crucial for biomedical research and targeted therapies [58–60]. Ceramics also serve as photothermal agents in therapies, converting light into heat to selectively destroy cancerous cells. Additionally, they act as photosensitizers in photodynamic therapy, facilitating the production of reactive oxygen species to effectively combat malignancies [59]. These optical properties are crucial for the development of highly sensitive optical biosensors, facilitating the accurate detection of biomolecules and pathogens in biological samples, thus advancing diagnostic applications [61].

6.2 Biotechnology and nanomedicine

The significance of ceramic biomaterials in biotechnology and nanomedicine has substantially increased due to their unique properties, especially their optical characteristics.

In biotechnology, ceramic biomaterials are primarily used in implants and prosthetics due to their low thermal conductivity, which minimizes damage to surrounding tissues, and their advanced optical properties for monitoring and diagnosis. Ceramics like oxygen-deficient zirconia or titania are ideal for medical imaging applications, offering clear and precise tissue visualization [57].

In nanomedicine, the optical properties of ceramics are utilized in nanoparticles for photothermal and photodynamic therapies. In photothermal therapy, nanoparticles generate heat when exposed to light, destroying cancerous cells without harming healthy tissues. In photodynamic therapy, they act as photosensitizers, producing reactive oxygen species to destroy malignant cells. Conversely, ceramics with optical absorbance in the NIR are ideal for imaging and monitoring cancer treatment. Infrared thermal imaging and photoacoustic imaging allow simultaneous diagnosis and monitoring, offering rapid, non-invasive high-resolution sensitivity.

Furthermore, the integration of magnetic elements facilitates precise monitoring using MRI (magnetic resonance imaging) [59].

Ceramics are also essential in the manufacture of optical biosensors used for the selective detection of various neurotransmitters and biological molecules. These biosensors harness the high sensitivity and selectivity provided by the optical properties of ceramics, allowing for the detection of very low concentrations of analytes in biological samples [60]. The optical properties of ceramics play a crucial role in the development of imaging devices for medical diagnostics. Fluorescence of certain ceramic materials is used in molecular imaging techniques, allowing real-time visualization of biological processes at the cellular level [58, 62].

In the past decade, there has been significant advancement in diagnostic tools and imaging agents utilizing ceramic materials. Quantum dots enable high-resolution imaging crucial for disease detection and monitoring. In fact, silica quantum dots (Cornell dots) are used for tumor imaging and monitoring in therapeutic and diagnostic applications [59].

Mesoporous silica nanoparticles (MSNs) have attracted considerable interest in biomedical research owing to their tunable mesoporous structure, high specific surface area, large pore volume, and customizable particle size. These attributes make MSNs well suited for concurrent diagnosis and therapy by enabling the encapsulation and controlled release of therapeutic agents. MSNs have rapidly advanced in bioimaging techniques, enabling their use in optical imaging, MRI, PET (positron emission tomography), CT (computed tomography), and ultrasound, as outlined in **Figure 3**. They have shown high potential in *in vivo* studies for diagnostics and drug delivery. However, improvements in imaging efficiency and colloidal stability are needed for clinical applications. Enhancing targeting, smaller particle sizes, and better surface passivation are critical areas for future research [58, 60].

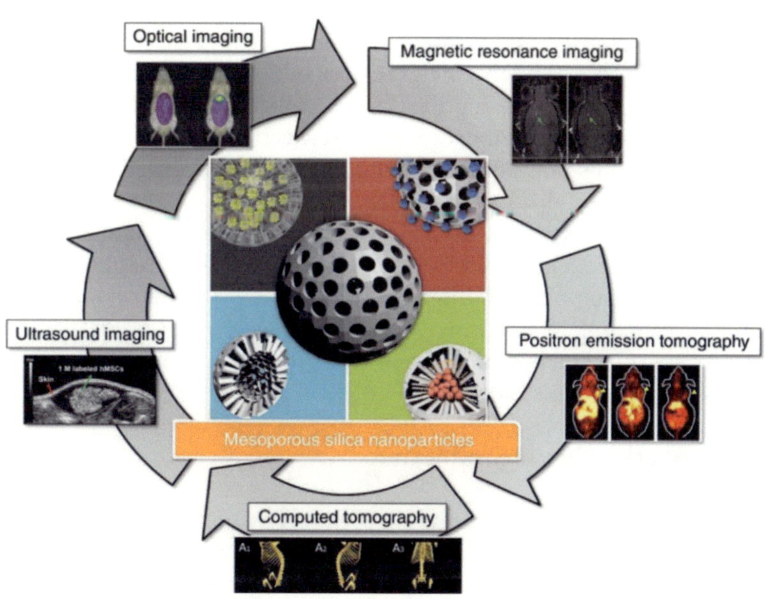

Figure 3.
Illustration of different imaging techniques using mesoporous silica nanoparticles. Reprinted with permission from reference [60].

6.3 Perspectives

Ceramics offer unique versatility for biomedical applications, distinguished by their biocompatibility, mechanical strength, and advanced optical properties, which are crucial for the development of innovative medical treatments and diagnostics [55–57, 63]. These characteristics not only enhance the effectiveness of implants and prosthetics but also enable photothermal and photodynamic therapies, as well as selective detection of various neurotransmitters and biological molecules at ultra-low levels [58–60]. Ongoing research in these areas promises to further advance the therapeutic and diagnostic capabilities of ceramics in modern medicine.

7. Sustainable and circular practices in ceramic production

An emblematic initiative within the Europe 2020 strategy is the circular economy, which aims to generate smart, sustainable, and inclusive growth. The evolution of the ceramic sector toward sustainability and the circular economy represents a significant transition from the traditional "make-use-dispose" model to a more rational and collaborative production and consumption strategy [64].

This shift has been driven by the need to reduce environmental impact, improve resource efficiency, and comply with increasingly stringent environmental regulations. Thus, the concept of a circular economy should be regarded as a novel approach for industrial ecosystems, encompassing the incorporation of new materials and improved products into the supply chain as resources. This integration results in a decreased reliance on primary resources and a reduction in waste generation [65].

The adoption of circular economy principles in the ceramics industry has led to a transformation in production and waste management processes that are much more innovative and sustainable. Sustainable practices in ceramic tile production have emerged as a response to the growing need to minimize the industry's environmental impact [66]. As environmental concerns intensify, ceramic companies are adopting innovations and strategies that not only comply with environmental regulations but also promote a circular economy and resource efficiency.

The excessive extraction of raw materials such as clay, feldspar, and silica, the exhaustive use of water and energy as well as the emission of greenhouse gases are just some of the issues that have highlighted the need for a more sustainable approach in this industry. According to the Global Material Resources Outlook to 2060, "the global consumption of materials such as biomass, fossil fuels, metals, and minerals is expected to double in the next forty years" [67].

7.1 Sustainable practices in tile production

- *Reuse of industrial waste*: One of the most notable practices is the reuse of industrial waste since this technique not only reduces the need for virgin raw materials but also minimizes the amount of waste sent to landfills. Innovative companies are developing ceramic pigments and glazes from industrial waste, achieving a dual advantage: waste reduction and the creation of sustainable products [68]. Ceramic glaze production can also significantly benefit from the reuse of ceramic waste, which includes both clay and glaze residues generated during various stages of the manufacturing process. Using ceramic waste, such as glass waste, in glaze compositions helps reduce the environmental impact by substituting traditional frits. This approach not only minimizes waste sent to landfills but also conserves raw materials and energy.

- *Industrial symbiosis:* It is a strategic approach that significantly enhances sustainability in the ceramic production process. By facilitating the exchange of materials, energy, and by-products between different industries, industrial symbiosis reduces waste and optimizes resource efficiency. This collaborative model allows for the valorization of waste products, such as using ceramic waste in the production of new materials, thereby minimizing the extraction of virgin resources and reducing environmental impact [69]. Additionally, industrial symbiosis supports circular economy principles by integrating sustainable practices within the supply chain, leading to innovative production processes that lower carbon emissions and enhance economic savings

- *Energy efficiency and use of renewable energies:* The ceramic sector, traditionally energy-intensive, relies heavily on high-temperature processes such as firing in kilns, which consume substantial amounts of fuel. Ceramic factories are gradually integrating systems that optimize energy use, thus reducing carbon emissions. Strategies to enhance energy efficiency include optimizing furnace operations, improving insulation, and recovering waste heat for reuse in the production cycle [70]. Therefore, the adoption of renewable energy sources, such as solar and wind power, further reduces reliance on fossil fuels and decreases greenhouse gas emissions.

- *Implementation of the circular economy:* The circular economy in the ceramic industry is based on the conservation and rehabilitation of resources through recycling, reuse, and redesign. Research highlights the feasibility of incorporating waste materials into new ceramic products, such as wall tiles, which can include recycled content without compromising quality [66]. The circular economy framework also fosters innovation in production methods, enhancing sustainability through efficient resource use and reduced emissions. Industry-scale applications, supported by technological advancements like artificial neural networks, optimize material properties and process efficiency, demonstrating significant environmental and economic benefits [71].

- *Use of artificial intelligence*: Artificial intelligence (AI) significantly enhances the implementation of the principles of circularity in the ceramic production process by optimizing resource use, minimizing waste, and improving overall efficiency. AI-driven technologies streamline the monitoring and analysis of production processes, leading to substantial reductions in energy consumption and material waste while increasing productivity. By leveraging predictive analytics, AI can anticipate equipment maintenance needs and optimize production schedules, further reducing downtime and enhancing operational efficiency [72]. Additionally, AI facilitates the recycling and reuse of ceramic materials by improving sorting and processing techniques, thus supporting the circular economy's goal of keeping resources in use for as long as possible. The integration of AI not only contributes to sustainable production practices but also aligns the ceramic industry with broader environmental objectives.

7.2 Perspectives

The adoption of sustainable practices in ceramic tile production is essential to mitigate the industry's environmental impact and promote a circular economy. Strategies, such as waste reuse, energy efficiency, the incorporation of renewable energy sources, and the application of artificial intelligence, not only provide environmental benefits but also yield long-term economic gains. It is imperative that that more companies within the sector embrace these practices to secure a sustainable future.

8. Conclusions

The chapter emphasizes the importance of new materials and their applications in addressing global challenges, focusing on light-mediated processes in advanced ceramics. Innovations in photocatalytic ceramics provide self-cleaning, air purification, and antibacterial functions, which enhance public health and safety. The development of nanomaterials improves energy storage and catalysis, offering efficient and eco-friendly alternatives for solar cells. Advancements in photonics have significantly boosted the properties and applications of multifunctional ceramics, with improvements in luminescence and nonlinear optics benefiting LEDs, OLEDs, laser technology, and optical signal processing. Laser processing allows for precise surface patterning in optics, electronics, and biomedicine, although further research is needed for optimization.

In biotechnology and nanomedicine, ceramics enhance medical implants, prosthetics, and innovative therapies due to their biocompatibility and mechanical strength. Sustainability is a key focus, with the circular economy in ceramic production minimizing environmental impacts and improving resource efficiency. These advancements not only enhance our current capabilities but also pave the way for future solutions, addressing both present and unforeseen challenges. The continuous exploration and development of advanced materials are essential for sustainable progress and the well-being of humanity.

Acknowledgements

The authors appreciate financial support from research projects PID2020-116719RB-C43, funded by the Ministry of Science and Innovation (reference MCIN/AEI/10.13039/501100011033) and TED2021-130963B-C22 (reference AEI/10.13039/501100011033/European Union NextGenerationEU/PRTR), sponsored by the "Agencia Estatal de Investigación."

Conflict of interest

The authors declare no conflict of interest.

DOI: http://dx.doi.org/10.5772/intechopen.1007093

Author details

Pablo Serna-Gallén*, Robinson Cadena, Samuel Porcar, Jaime González Cuadra, Abderrahim Lahlahi, Santiago Toca, Diego Fraga and Juan Carda
Department of Inorganic and Organic Chemistry, Universitat Jaume I, Castelló de la Plana, Spain

*Address all correspondence to: pserna@uji.es

References

[1] Energy, Climate change, Environment. European Comssion. n.d. Available from: https://energy.ec.europa.eu/topics/research-and-technology/strategic-energy-technology-plan_en [Accessed: July 05, 2024]

[2] International Partnerships. European Comssion. n.d. Available from: https://international-partnerships.ec.europa.eu/policies/sustainable-development-goals_en [Accessed: July 05, 2024]

[3] Zhao Y, Li L, Wang Q, Ding G, Zhang C. Electrical insulation improvement using a CeYSZ/Al_2O_3 double ceramic layer underlayer for thin-film sensors. Ceramics International. 2024;**50**:22165-22173. DOI: 10.1016/j.ceramint.2024.03.333

[4] Benchorfi H, Duclère JR, Chenu S, Messaddeq Y, Delaizir G, Rivera VAG. Exploring the optical properties of the 1.53 μm emission in Er^{3+}-doped glass, anti-glass and ceramic in TeO_2–Ta_2O_5–Bi_2O_3 system. Optical Materials (Amsterdam). 2024;**153**:115602. DOI: 10.1016/j.optmat.2024.115602

[5] Hartman T, Geitenbeek RG, Whiting GT, et al. Operando monitoring of temperature and active species at the single catalyst particle level. Nature Catalysis. 2019;**2**:986-996. DOI: 10.1038/s41929-019-0352-1

[6] Raizada P, Sudhaik A, Singh P. Photocatalytic water decontamination using graphene and ZnO coupled photocatalysts: A review. Materials Science for Energy Technologies. 2019;**2**:509-525. DOI: 10.1016/j.mset.2019.04.007

[7] Shi Y, Huang J, Zeng G, Cheng W, Hu J. Photocatalytic membrane in water purification: Is it stepping closer to be driven by visible light? Journal of Membrane Science. 2019;**584**:364-392. DOI: 10.1016/j.memsci.2019.04.078

[8] Wang X, Wang H, Cheng J, Li H, Wu X, Zhang D, et al. Initiative ROS generation of Cu-doped ZIF-8 for excellent antibacterial performance. Chemical Engineering Journal. 2023;**466**:143201. DOI: 10.1016/j.cej.2023.143201

[9] Cuadra JG, Molina-Prados S, Mínguez-Vega G, Estrada AC, Trindade T, Oliveira C, et al. Multifunctional silver-coated transparent TiO_2 thin films for photocatalytic and antimicrobial applications. Applied Surface Science. 2023;**617**:156519. DOI: 10.1016/j.apsusc.2023.156519

[10] Cui M, Wang H, Fan X, Zhang J, Xing C, Yan W. Photocatalytic degradation of four organophosphorus pesticides in aqueous solution using D-cys/Au NPs modified TiO_2 by natural sunlight. Applied Surface Science. 2024;**663**:160197. DOI: 10.1016/j.apsusc.2024.160197

[11] Santos AL, Gomes NCM, Henriques I, Almeida A, Correia A, Cunha Â. Contribution of reactive oxygen species to UV-B-induced damage in bacteria. Journal of Photochemistry and Photobiology. B. 2012;**117**:40-46. DOI: 10.1016/j.jphotobiol.2012.08.016

[12] Cuadra JG, Estrada AC, Oliveira C, Abderrahim LA, Porcar S, Fraga D, et al. Functional properties of transparent ZnO thin films synthetized by using spray pyrolysis for environmental and biomedical applications. Ceramics International. 2023;**49**:32779-32788. DOI: 10.1016/j.ceramint.2023.07.246

[13] Loo YY, Rukayadi Y, Nor-Khaizura MAR, Kuan CH, Chieng BW, Nishibuchi M, et al. In vitro antimicrobial activity of green synthesized silver nanoparticles against selected gram-negative foodborne pathogens. Frontiers in Microbiology. 2018;**9**:1555. DOI: 10.3389/fmicb.2018.01555

[14] Yin IX, Zhang J, Zhao IS, Mei ML, Li Q, Chu CH. The antibacterial mechanism of silver nanoparticles and its application in dentistry. International Journal of Nanomedicine. 2020;**15**:2555-2562. DOI: 10.2147/IJN.S246764

[15] Fayaz AM, Balaji K, Girilal M, Yadav R, Kalaichelvan PT, Venketesan R. Biogenic synthesis of silver nanoparticles and their synergistic effect with antibiotics: A study against gram-positive and gram-negative bacteria. Nanomedicine. 2010;**6**:103-109. DOI: 10.1016/j.nano.2009.04.006

[16] Aal NA, Al-Hazmi F, Al-Ghamdi AA, Al-Ghamdi AA, El-Tantawy F, Yakuphanoglu F. Novel rapid synthesis of zinc oxide nanotubes via hydrothermal technique and antibacterial properties. Spectrochimica Acta. Part A, Molecular and Biomolecular Spectroscopy. 2015;**135**:871-877. DOI: 10.1016/j.saa.2014.07.099

[17] Aymonier C, Schlotterbeck U, Antonietti L, Zacharias P, Thomann R, Tiller JC, et al. Hybrids of silver nanoparticles with amphiphilic hyperbranched macromolecules exhibiting antimicrobial properties. Chemical Communications. 2002;**24**:3018-3019. DOI: 10.1039/b208575e

[18] Supekar AT, Bhujbal PK, Salunke SA, Rathod SM, Patole SP, Pathan HM. Bismuth sulfide and antimony sulfide-based solar cells: A review. Energy and Environment. 2023;**19**:848. DOI: 10.30919/esee8c848

[19] Di Mare S, Menossi D, Salavei A, Artegiani E, Piccinelli F, Kumar A, et al. SnS thin film solar cells: Perspectives and limitations. Coatings. 2017;7(2):34. DOI: 10.3390/coatings7020034

[20] Mavlonov A, Razykov T, Raziq F, Gan J, Chantana J, Kawano Y, et al. A review of Sb_2Se_3 photovoltaic absorber materials and thin-film solar cells. Solar Energy. 2020;**201**:227-246. DOI: 10.1016/j.solener.2020.03.009

[21] Gohri S, Madan J, Mohammed MKA, Pandey R. Inherent internal p-n junction assisted single layered n-type iron pyrite solar cell. Materials Research Express. 2023;**10**:024001. DOI: 10.1088/2053-1591/acb982

[22] Wang S, Zhao Y, Che B, Li C, Chen X, Tang R, et al. A novel multi-Sulfur source collaborative chemical bath deposition technology enables 8%-efficiency Sb_2S_3 planar solar cells. Advanced Materials. 2022;**34**:2206242. DOI: 10.1002/adma.202206242

[23] Zhao Y, Wang S, Jiang C, Li C, Xiao P, Tang R, et al. Regulating energy band alignment via alkaline metal fluoride assisted solution post-treatment enabling $Sb_2(S,Se)_3$ solar cells with 10.7% efficiency. Advanced Energy Materials. 2022;**12**:2103015. DOI: 10.1002/aenm.202103015

[24] Wang X, Tang R, Jiang C, Lian W, Ju H, Jiang G, et al. Manipulating the electrical properties of $Sb_2(S,Se)_3$ film for high-efficiency solar cell. Advanced Energy Materials. 2020;**10**:2002341. DOI: 10.1002/aenm.202002341

[25] Whittaker-Brooks L, Gao J, Hailey AK, Thomas CR, Yao N, Loo YL. Bi_2S_3 nanowire networks as electron

acceptor layers in solution-processed hybrid solar cells. Journal of Materials Chemistry C. 2015;**3**:2686-2692. DOI: 10.1039/c4tc02534b

[26] Suzuki I, Kawanishi S, Omata T, Yanagi H. Current status of n-type SnS: Paving the way for SnS homojunction solar cells. Journal of Physics: Energy. 2022;**4**:042002. DOI: 10.1088/2515-7655/ac86a1

[27] Jaramillo R, Steinmann V, Yang C, Hartman K, Chakraborty R, Poindexter JR, et al. Making record-efficiency SnS solar cells by thermal evaporation and atomic layer deposition. Journal of Visualized Experiments. May 2015;**22**(99):e52705. DOI: 10.3791/52705

[28] Atsushi W. Highlighted presentations press release development and prospect of Cu_2O/Si tandem solar cells toward carbon neutrality. In: The 70th Japan Society of Applied Physics Spring Meeting. Tokyo: The Japan Society of Applied Physics; 2023

[29] Shibasaki S, Honishi Y, Nakagawa N, Yamazaki M, Mizuno Y, Nishida Y, et al. Highly transparent Cu_2O absorbing layer for thin film solar cells. Applied Physics Letters. 2021;**119**:242102. DOI: 10.1063/5.0072310

[30] Serna-Gallén P, Beltrán-Mir H, Cordoncillo E. Practical guidance for easily interpreting the emission and physicochemical parameters of Eu^{3+} in solid-state hosts. Ceramics International. 2023;**49**:41078-41089. DOI: 10.1016/j.ceramint.2023.01.141

[31] Pan H, Liu Q, Chen X, Liu X, Chen H, Xie T, et al. Fabrication and properties of Gd_2O_2S:Tb scintillation ceramics for the high-resolution neutron imaging. Optical Materials (Amsterdam). 2020;**105**:109909. DOI: 10.1016/j.optmat.2020.109909

[32] Tessitore G, Mandl GA, Maurizio SL, Kaur M, Capobianco JA. The role of lanthanide luminescence in advancing technology. RSC Advances. 2023;**13**:17787-17811. DOI: 10.1039/d3ra00991b

[33] Erol E, Vahedigharehchopogh N, Kibrisli O, Ersundu MC, Ersundu AE. Recent progress in lanthanide-doped luminescent glasses for solid-state lighting applications - A review. Journal of Physics Condensed Matter. 2021;**33**:483001. DOI: 10.1088/1361-648X/ac22d9

[34] Brites CDS, Balabhadra S, Carlos LD. Lanthanide-based thermometers: At the cutting-edge of luminescence thermometry. Advanced Optical Materials. 2019;**7**:1801239. DOI: 10.1002/adom.201801239

[35] Zhou H, Sharma M, Berezin O, Zuckerman D, Berezin MY. Nanothermometry: From microscopy to thermal treatments. ChemPhysChem. 2016;**17**:27-36. DOI: 10.1002/cphc.201500753

[36] Dos Santos LF, Martins JC, Lima KO, Gomes LFT, De Melo MT, Tedesco AC, et al. In vitro assays and nanothermometry studies of infrared-to-visible upconversion of nanocrystalline Er^{3+},Yb^{3+} co-doped Y_2O_3 nanoparticles for theranostic applications. Physica B Condens Matter. 2022;**624**:413447. DOI: 10.1016/j.physb.2021.413447

[37] Dai Y, Ma P, Cheng Z, Kang X, Zhang X, Hou Z, et al. Up-conversion cell imaging and pH-induced thermally controlled drug release from $NaYF_4$:Yb^{3+}/Er^{3+}@hydrogel core-shell hybrid microspheres. ACS Nano. 2012;**6**:3327-3338. DOI: 10.1021/nn300303q

[38] Feng X, Qiu J, Zhou S. Synergistic effect in nonlinear response in glass

ceramics. Journal of the American Ceramic Society. 2024;**107**:1936-1948. DOI: 10.1111/jace.19522

[39] Tumuluri A, Bharati MSS, Rao SV, James Raju KC. Structural, optical and femtosecond third-order nonlinear optical properties of $LiNbO_3$ thin films. Materials Research Bulletin. 2017;**94**:342-351. DOI: 10.1016/j.materresbull.2017.06.029

[40] Bermúdez V. Engineered periodic-poled lithium niobate structures doped with rare earths for multi-self-frequency conversion. Journal of Crystal Growth. 2008;**310**:1324-1330. DOI: 10.1016/j.jcrysgro.2007.10.055

[41] Wang K, Yin J, Chen X, Wang L, Xiao H, Liu X, et al. Advances on direct selective laser printing of ceramics: An overview. Journal of Alloys and Compounds. 2024;**975**:172821. DOI: 10.1016/j.jallcom.2023.172821

[42] Rethfeld B, Ivanov DS, Garcia ME, Anisimov SI. Modelling ultrafast laser ablation. Journal of Physics D: Applied Physics. 2017;**50**:193001. DOI: 10.1088/1361-6463/50/19/193001

[43] Henriques B, Fabris D, Voisiat B, Boccaccini AR, Lasagni AF. Direct laser interference patterning of zirconia using infra-red picosecond pulsed laser: Effect of laser processing parameters on the surface topography and microstructure. Advanced Functional Materials. 2024;**34**:2307894. DOI: 10.1002/adfm.202307894

[44] Bleu Y, Bourquard F, Tite T, Loir A-S, Maddi C, Donnet C, et al. Review of graphene growth from a solid carbon source by pulsed laser deposition (PLD). Frontiers in Chemistry. 2018;**6**:572. DOI: 10.3389/fchem.2018.00572

[45] Ho JY, Wong KK, Leong KC. Saturated pool boiling of FC-72 from enhanced surfaces produced by selective laser melting. International Journal of Heat and Mass Transfer. 2016;**99**:107-121. DOI: 10.1016/J.IJHEATMASSTRANSFER.2016.03.073

[46] Zhu D, Zuo P, Li F, Tian H, Liu T, Hu L, et al. Fabrication and applications of surface micro/nanostructures by femtosecond laser. Colloid and Interface Science Communications. 2024;**59**:100770. DOI: 10.1016/J.COLCOM.2024.100770

[47] Wang H, Deng D, Zhai Z, Yao Y. Laser-processed functional surface structures for multi-functional applications-a review. Journal of Manufacturing Processes. 2024;**116**:247-283. DOI: 10.1016/j.jmapro.2024.02.062

[48] Nayak BK, Iyengar VV, Gupta MC. Efficient light trapping in silicon solar cells by ultrafast-laser-induced self-assembled micro/nano structures. Progress in Photovoltaics: Research and Applications. 2011;**19**:631-639. DOI: 10.1002/pip.1067

[49] Zhao B, Wang R, Yang J. Homogeneous nanostructuring of a molybdenum surface by dual-color correlated femtosecond laser irradiation for solar absorber applications. ACS Applied Nano Materials. 2023;**6**:21092-21100. DOI: 10.1021/acsanm.3c03994

[50] Arthanari S, Park J-E, Bose S, Kang HW, Kim S, Yang M, et al. Structural color generation on transparent and flexible substrates by nanosecond laser induced periodic surface structures. Adv. Materials and Technologies. 2023;**8**:2201725. DOI: 10.1002/admt.202201725

[51] Vorobyev AY, Guo C. Colorizing metals with femtosecond laser pulses.

Applied Physics Letters. 2008;**92**:041914. DOI: 10.1063/1.2834902

[52] Li G, Li J, Hu Y, Zhang C, Li X, Chu J, et al. Femtosecond laser color marking stainless steel surface with different wavelengths. Applied Physics A. 2015;**118**:1189-1196. DOI: 10.1007/s00339-014-8868-3

[53] Liu Y, Li S, Niu S, Cao X, Han Z, Ren L. Bio-inspired micro-nano structured surface with structural color and anisotropic wettability on Cu substrate. Applied Surface Science. 2016;**379**:230-237. DOI: 10.1016/J.APSUSC.2016.03.234

[54] Rico V, López-Gascón C, Espinós JP, Lahoz R, Laguna M, González-Elipe AR, et al. Metallization of ceramic substrates by laser induced decomposition of coordination complexes. Journal of the European Ceramic Society. 2016;**36**:2831-2836. DOI: 10.1016/j.jeurceramsoc.2016.04.016

[55] Hench LL. Bioceramics. Journal of the American Ceramic Society. 1998;**81**:1705-1728. DOI: 10.1111/j.1151-2916.1998.tb02540.x

[56] Katti KS. Biomaterials in total joint replacement. Colloids and Surfaces. B, Biointerfaces. 2004;**39**:133-142. DOI: 10.1016/j.colsurfb.2003.12.002

[57] Vallet-Regí M, González-Calbet JM. Calcium phosphates as substitution of bone tissues. Progress in Solid State Chemistry. 2004;**32**:1-31. DOI: 10.1016/j.progsolidstchem.2004.07.001

[58] Baeza A, Vallet-Regí M. Mesoporous silica nanoparticles as theranostic antitumoral nanomedicines. Pharmaceutics. 2020;**12**:1-16. DOI: 10.3390/pharmaceutics12100957

[59] Bigham A, Raucci MG, Zheng K, Boccaccini AR, Ambrosio L. Oxygen-deficient bioceramics: Combination of diagnosis, therapy, and regeneration. Advanced Materials. 2023;**35**:2302858. DOI: 10.1002/adma.202302858

[60] Cha BG, Kim J. Functional mesoporous silica nanoparticles for bio-imaging applications. Wiley Interdisciplinary Reviews. Nanomedicine and Nanobiotechnology. 2019;**11**:e1515. DOI: 10.1002/wnan.1515

[61] Trewyn BG, Giri S, Slowing II, Lin VSY. Mesoporous silica nanoparticle based controlled release, drug delivery, and biosensor systems. Chemical Communications. 2007;**2007**:3236-3245. DOI: 10.1039/b701744h

[62] Colilla M, González B, Vallet-Regí M. Mesoporous silica nanoparticles for the design of smart delivery nanodevices. Biomaterials Science. 2013;**1**:114-134. DOI: 10.1039/c2bm00085g

[63] Thomas S, Harshita BSP, Mishra P, Talegaonkar S. Ceramic nanoparticles: Fabrication methods and applications in drug delivery. Current Pharmaceutical Design. 2015;**21**:6165-6188. DOI: 10.2174/1381612821666151027153246

[64] Wietschel L, Halter F, Thorenz A, Schüppel D, Koch D. Literature review on the state of the art of the circular economy of ceramic matrix composites. Open Ceramics. 2023;**14**:100357. DOI: 10.1016/j.oceram.2023.100357

[65] Okogwu C, Agho M, Adeyinka M, Odulaja B, Eyo-Udo N, Daraojimba C, et al. Exploring the integration of sustainable materials in supply chain management for environmental impact. Engineering Science & Technology Journal. 2023;**4**:49-65. DOI: 10.51594/estj.v4i3.546

[66] Castellano J, Sanz V, Cañas E, Sánchez E. Industry-scalable wall tile composition based on circular economy. Boletín de La Sociedad Española de Cerámica y Vidrio. 2022;**61**:374-382. DOI: 10.1016/j.bsecv.2022.03.003

[67] OECD. Global Material Resources Outlook to 2060. Paris: OECD; 2019. DOI: 10.1787/9789264307452-en

[68] Carneiro J, Tobaldi DM, Capela MN, Seabra MP, Labrincha JA. Waste-based pigments for application in ceramic glazes and stoneware bodies. Materials. 2019;**12**:3396. DOI: 10.3390/ma12203396

[69] Castellet-Viciano L, Hernandez-Chover V, Bellver-Domingo Á, Hernández-Sancho F. Industrial symbiosis: A mechanism to guarantee the implementation of circular economy practices. Sustainability. 2022;**14**:15872. DOI: 10.3390/su142315872

[70] Castro Oliveira M, Iten M, Cruz P, Monteiro H. Review on energy efficiency progresses, technologies and strategies in the ceramic sector focusing on waste heat recovery. Energies (Basel). 2020;**13**:6096. DOI: 10.3390/en13226096

[71] García Kerdan I, Morillón Gálvez D. Artificial neural network structure optimisation for accurately prediction of exergy, comfort and life cycle cost performance of a low energy building. Applied Energy. 2020;**280**:115862. DOI: 10.1016/j.apenergy.2020.115862

[72] Özsoy T. The role of artificial intelligence in facilitating the transition to a circular economy. Nişantaşı Üniversitesi Sosyal Bilimler Dergisi. 2023;**11**:369-389. DOI: 10.52122/nisantasisbd.1345828

Chapter 3

Synthesis of WC-Co Composite Powder by In-Situ Solid Carbothermic Reduction Method

Hamed Naderi-Samani, Reza Shoja Razavi, Hasan Abbaszadeh, Afshin Amiri-Moghaddam, Mehri Mashhadi and Ali Alizadeh

Abstract

This chapter explores in detail the process of preparing WC-Co composite powder using the in-situ solid carbo-thermic reduction method. WC-Co cermets are recognized as advanced engineering materials, valued for their unique properties such as wear resistance and high hardness. The chapter discusses various parameters essential to the preparation process, including the ratio of reaction raw materials (tungsten and cobalt oxides and the reducing agent), grinding conditions, purity and particle size of raw materials, heat treatment conditions (heating temperature, duration, and type of furnace atmosphere), density within the raw material mixture, and the type of reducing agent. Furthermore, the chapter addresses challenges and limitations related to the carbon content in the reactions. As such, this chapter proves to be a valuable resource for researchers and scientists involved in the synthesis of WC-Co composite powder through the in-situ solid carbo-thermic reduction method, emphasizing the critical role of process parameters in attaining the desired properties of the composite powder.

Keywords: WC-Co cermet, carbothermic in situ reduction, phase analysis, microstructural analysis, thermodynamic analysis

1. Introduction

Engineering ceramics, due to their strong atomic bonds, exhibit high refractoriness and thermal resistance, including hot strength and creep resistance. For instance, carbides generally do not have a melting point and begin to sublimate at temperatures above 2500°C [1]. Most ceramics are lightweight yet possess high hardness, wear resistance, oxidation resistance, and chemical stability. However, they also have drawbacks such as brittleness, thermal and electrical insulation, and poor thermal shock resistance [2]. To mitigate these drawbacks, they can be combined with metals and their alloys to form cermets [3]. One notable engineering ceramic is tungsten carbide, which is gray, has a Mohs hardness of about 9, and a density roughly twice that of steel [4].

Cermets are predominantly ceramic in their primary phase but feature a continuous metallic phase along the particle boundaries, which enhances ductility and impact resistance while reducing brittleness [5, 6]. For example, a mixture of tungsten carbides with 10% by weight of Co metal produces cutting tools with reduced brittleness, which neither metal nor ceramic alone can achieve. This composite exhibits high toughness and hardness simultaneously, making it highly wear-resistant without being brittle, which is why it is used for cutting tools for hard steels, known as hard metals [7].

Cemented carbides consist of a complex phase, such as WC (tungsten carbide), and a binder phase, like Co (cobalt). The primary constituent of these materials is tungsten carbide, which provides hardness and wear resistance to the composite. The secondary component, cobalt, imparts toughness and helps bind the tungsten carbide particles together. Cobalt is predominantly used in tungsten carbide-based cermets due to its properties, such as the good wettability of tungsten carbide particles and forming strong bonds with them.

In addition to the mentioned properties of Co, recent studies indicate that the presence of a certain amount of pure Co alongside the raw materials ($C + WO_3$) for producing WC can act as a catalyst and improve the reduction conditions. In this role, cobalt accelerates the reduction of WO_3 during the carburization process. Specifically, the presence of Co reduces the oxygen pressure of the environment required to form the intermediate phase $CoWO_4$, which lowers the phase transition temperature of WO_3 to WO_2 and subsequently to W to about 750°C [8].

The hardness of tungsten carbide makes it brittle. To increase its flexibility and toughness, cobalt is used as the secondary phase. Therefore, this composite is utilized in the manufacturing of cutting tools and wear-resistant tools that require desirable toughness. The particle size of tungsten carbide in WC-Co cermets ranges from 0.3 to 40 microns, and the cobalt content varies from 3 to 30% by weight [9]. The grain size significantly affects performance; for general industrial cutting applications, tungsten carbide particle sizes are typically 1–2 microns [10]. Recently, ultrafine WC particles less than 0.1 microns have been developed, improving wear behavior [11].

According to the Hall-Petch effect ($\sigma = \sigma_0 + k.d^{-0.5}$), as grain size decreases, hardness, strength, and toughness increase. Smaller particles also enhance the driving force for sintering, leading to better composite formation. Particle size directly impacts the pressing pressure during powder compaction; smaller particles require higher pressing pressure, resulting in higher densification [12]. Over the past 20 years, significant attention has been given to nanocomposites of WC-Co. Applications include cutting tools, drilling, machining, drilling tools, molds for metal forming, ceramic and composite part molds in powder metallurgy, aerospace industries, turbine blades, milling balls, and sandblasting nozzles [13–15].

The excellent properties of this composite depend on various factors, but grain size has the most significant impact. As the grain size decreases, the material's hardness and strength increase. Additionally, finer particles increase the driving force for sintering, resulting in better composite formation conditions. The particle size directly affects the pressing pressure during the powder compaction stage; finer particles require higher pressure, leading to higher density [16–18].

One of the measures taken to reduce particle size is pressure-assisted sintering, including methods such as SPS (Spark Plasma Sintering), MS (Microwave Sintering), UPRC (Ultra-High Pressure Rapid Consolidation), HFIHS (High-Frequency Induction Heating Sintering), and HIP (Hot Isostatic Pressing). It is worth mentioning that the carbo-thermic reduction method and the in-situ production of the WC-Co composite yield finer grains compared to the method of carburizing tungsten in the initial mixture (W + Co + C) or the powder metallurgy method for the (WC + Co) mixture [19–24].

Refractory metals are commonly used in cermets because, in addition to providing toughness, they do not significantly reduce refractoriness and mechanical properties at high temperatures. Refractory metals have melting points above 1500°C and generally belong to the transition elements of the periodic table. These metals act as barriers to crack propagation at the ceramic-metal interfaces, requiring cracks to expend more energy cutting through the metal phase. As a result, the crack loses its energy and stops, thereby increasing the toughness of the piece [13].

2. In-situ methods for composite production

In-situ synthesis refers to forming composite phases (matrix or reinforcement) during milling or thermal treatment through chemical reactions. This can occur in liquid, solid, or gas phases or their combinations:

1. Direct nitration
2. Indirect nitration
3. Pyrolysis of organic materials
4. Chemical vapor deposition
5. Plasma arc spark process
6. Sol-gel process
7. Mechanical methods
8. Atomization process
9. Combustion and self-combustion synthesis
10. Carbo-thermic reduction (solid or gaseous or their combination) [25–27].

The most important parameters influencing this method are:

1. Characteristics of raw materials
2. Milling conditions
3. Effect of heat treatment temperature and holding time in the furnace (heat treatment cycle): The reaction conditions are optimized when the temperature and holding time in the furnace are minimized, which is more economically favorable and cost-effective.
4. Effect of internal furnace pressure and its atmosphere
5. Appropriate composition and purity of hydrogen gas as an auxiliary agent for the reduction process.

3. Solid carbo-thermic reduction method

In this process, composite matrix and reinforcement phases are formed through carbon reduction, which can use solid carbon (active carbon powder) or gaseous carbon (methane, CH_4, which constitutes about 80% of natural gas) or CO [28]. The reducing agent can also combine carbonaceous materials and H_2 gas. Notably, carbon polymers can serve as a carbon source. For instance, in a research study, a combination of ethylene glycol, polyethylene glycol, and polypropylene glycol was used as a carbon source to reduce tungsten oxides for WC powder production (**Figure 1**) [29].

Advantages of carbo-thermic reduction:

1. The powders produced have high compressibility due to their high porosity.

2. Efficient for producing composites where melting the metallic part on an industrial scale is challenging.

3. Easy control over factors like temperature, atmosphere, and chemical composition.

4. Potential for producing nanocomposite powders.

5. Reduced environmental pollution and hazards.

6. Better and more uniform particle distribution, improving mechanical and physical properties.

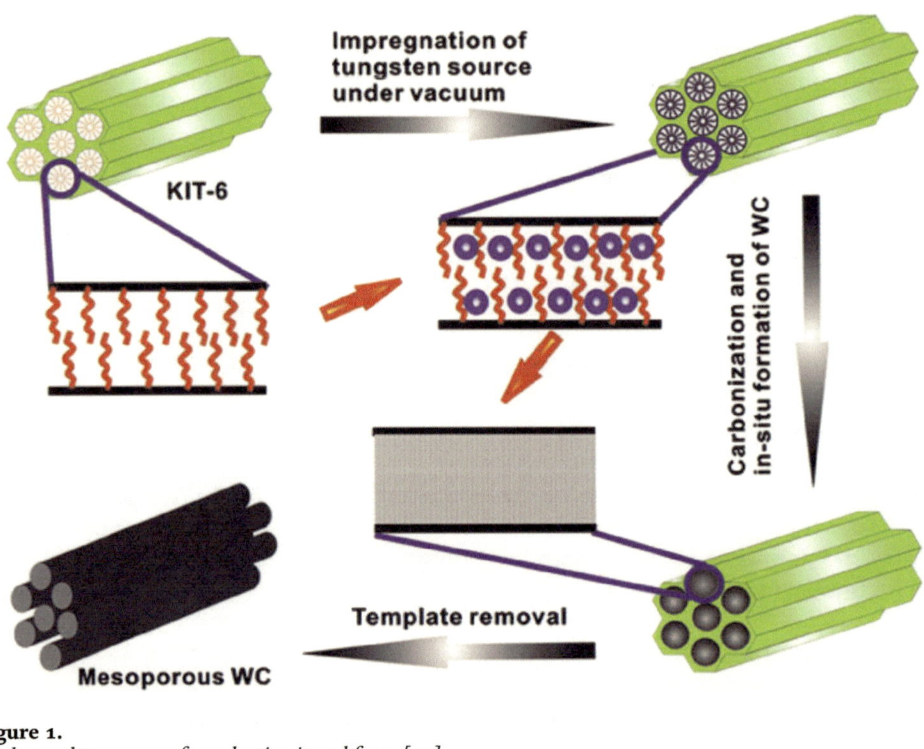

Figure 1.
Carbon polymer source for reduction in rod form [29].

7. Clean interface between soft and hard phases, enhancing interface strength.

8. Thermodynamic stability at higher temperatures.

9. Cost-effectiveness [30–33].

The main drawback of solid carbo-thermic reduction is the lengthy process due to diffusion-controlled reduction reactions and the high activation energy required for initial oxide materials.

4. WC ceramic and WC-Co cermets formed by in-situ and solid carbo-thermic reduction methods

One of the significant differences in the fabrication of WC-Co composites is the use of different raw materials, which can be categorized as follows:

1. Using tungsten and cobalt oxides as the primary materials and performing gaseous and solid reduction with various carbon sources [34, 35].

2. Using aqueous and salt compounds containing W and Co and introducing different C sources for reduction [36–39].

3. Direct use of W, Co, and C through powder metallurgy, which, although not cost-effective, is very efficient and precise for critical components [40].

In this research, the primary focus is on the first method, which offers higher economic benefits and aligns with the objectives of this study. Other methods are briefly mentioned.

For instance, in one study, $(NH_4)_6(H_2W_{12}O_4).4H_2O$ and $(CoNO_3)_2.6H_2O$ salts were used as the raw materials. These materials were mixed via spraying in an environment at 100–130°C, and then heated to 800°C for 2 hours to remove the molecular water of the salts, resulting in a powder mixture of WO_3 and $CoWO_4$. This study is mentioned here because the final raw materials are similar to those obtained via solid carbo-thermic reduction. In the next step, the calcined powder was mixed with activated carbon as the reducing agent and then heat-treated at 950°C for 4 hours under a nitrogen gas atmosphere, ultimately producing WC-Co composite powder [41].

Figures 2–4 illustrate SEM images with different magnifications of WC-Co powder. The composite powder size ranges from 100 to 300 nm, larger than the initial oxide powders.

The reactions during the carbo-thermic reduction are as follows:

$$CoWO_4 + 5C = Co + WC + 4CO_{(g)} \tag{1}$$

$$WO_3 + 4C = WC + 3CO_{(g)} \tag{2}$$

In a previous study [42], the reduction mechanism of WO_3 during carburization under a methane and hydrogen gas mixture atmosphere was examined. The results indicated that WO_3 is first converted to $W_{20}O_{58}$, then sequentially to WO_2, W, W_2C,

Figure 2.
Morphology of WC-Co composite powder [41].

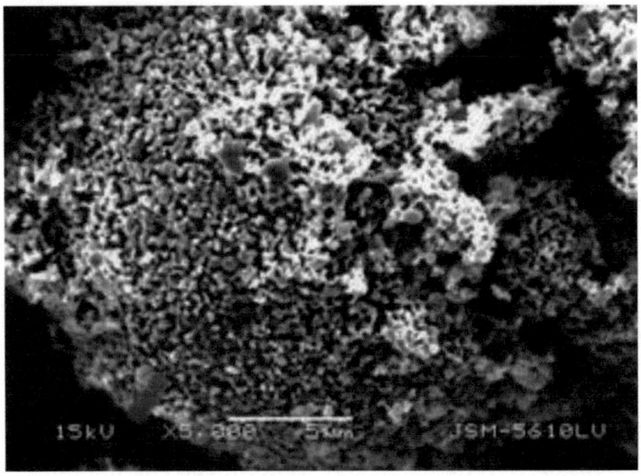

Figure 3.
Figure* 2 *at higher magnification [41].

and WC. A similar study showed that in a powder mixture of tungsten oxide, cobalt, and carbon, the $CoWO_4$ phase forms first, followed by W, W_2C, η, and WC phases. Notably, the η phase appears in two forms, Co_3W_3C and Co_6W_6C, which are intermediate and unstable phases for synthesizing WC-Co cermet [43]. It was also noted that at 900–923 K, solid-state diffusion on the surface of tungsten oxide particles intensifies and gradually penetrates the core.

For the reduction of WO_2 powder to WC using CO gas as the reducing agent, a recent study provided exciting results [44]. The conclusion was that at relatively low temperatures (813–908°C), the final product of the reduction process is WC. In contrast, at higher temperatures (1179°C), the final product is W_2C or W. Thermodynamic studies of this reduction mechanism showed that WO_2 is directly converted to

Figure 4.
***Figure** 3 at higher magnification [41].*

WC at low temperatures (813–908°C) during carburization with CO gas. However, at higher temperatures (1179°C), the reduction mechanism is two-staged, first converting WO_2 to W and then to W_2C and WC. The reason for this is the increased difficulty of carburization at higher temperatures, requiring more CO. In essence, CO loses its effective reducing capability at high temperatures.

A similar study demonstrated the two-stage synthesis of tungsten carbide powder from tungsten oxide using direct reduction and carburization by a mixture of hydrogen and methane gases. Tungsten oxide powder was first converted to tungsten powder under a hydrogen atmosphere at 800–900°C. Subsequently, with the addition of methane to the hydrogen gas at various ratios (100, 50, and 20%), the behavior of tungsten carbide powder synthesis was studied at 950°C for 60–210 minutes.

The results indicated that complete carburization of tungsten carbide was not achieved within 210 minutes, and the methane-to-hydrogen ratio was optimized at 20% methane and 80% hydrogen. Spherical tungsten carbide particles with an average size of about 1.5 microns were obtained after 210 minutes [45].

One current application of nano-structured WC powder is its use as a platinum substitute. In one study, ammonium para tungstate was used as the reducing agent and subjected to a thermal gradient of 1.1°C/min under a 20% methane flow in a hydrogen environment. Samples were withdrawn at specific temperature intervals and analyzed. The results indicated that the tungsten phase formed around 550°C, initially converting to W_2C and then to WC at higher temperatures. The heat treatment was conducted in the 350–950°C range [46, 47]. Recently, in 2018, studies showed that glucose, as a solid carbon source, is more active than carbon black, successfully reducing tungsten and cobalt oxides at 900°C with a holding time of 1 hour [48].

A literature review indicates that many studies have used tungsten and cobalt oxide materials as the initial raw materials. For example, one study used $WO_{2.9}$, Co_3O_4, and C as the raw materials. The materials were homogenized in a ball mill with ethanol, a ball-to-powder weight ratio of 3:1, and a rotation speed of 500 rpm for 5 hours. Subsequently, the powder mixture was compacted and sintered, during which the reduction reaction occurred, forming a WC-Co composite [31]. It is worth noting that compaction in the initial material mixture, especially in the solid reduction method, increases the contact surface between particles, improving the reduction reaction kinetics.

The initial reduction temperature of tungsten oxide by carbon or carbon monoxide is lower due to the more negative change in energy, the lower equilibrium temperature of the reduction reaction, the higher electronegativity difference between tungsten and carbon, and the less dense crystalline structure of tungsten oxide (BCC) compared to cobalt oxide's compact structure (HCP) [49]. DTA-TGA analysis results showed that tungsten oxide reduction reactions occurred within the 850–1000°C range, and then the cobalt oxide reduction and WC-Co cermet formation were completed simultaneously up to 1200°C. If the reduction temperature reaches the sintering temperature of cobalt (the melting point of cobalt is 1495°C), the reduction and sintering processes can be observed in a two-stage process, as reported in the study related to the source [49]. However, most studies have synthesized WC-Co cermet powder and sintered it in two separate stages, initially obtaining WC-Co cermet powder at around 1200°C using solid or gaseous reduction methods, and then performing sintering with methods such as hot pressing, spark plasma, etc. [50].

In 2020, a new two-stage reduction method was employed [51], resulting in a WC-Co composite powder with improved toughness after spark plasma sintering (SPS). As known, WC-Co composites exhibit high hardness but poor toughness. This method used pure tungsten, cobalt, and carbon powders as raw materials. The materials were combined in a ball mill at 350 rpm for 25 hours with a ball-to-powder ratio of 3:1 in an ethanol environment. The resulting mixture underwent partial reduction heat treatment in a vacuum furnace to form amorphous and unstable $Co_3W_9C_4$ and Co_2W_4C phases. The resulting products were re-milled with the necessary carbon to form a WC-10wt%Co composite for 15 hours. Finally, the combined powder was subjected to SPS, heating to 850°C for 5 minutes, holding for 4 minutes, then rapidly heating to 1050°C in 1 minute, holding for 3 minutes, cooling to 600°C in 3 minutes, and finally reaching room temperature in the furnace. During heating, the SPS pressure increased from 10 to 30 N. **Figure 5** schematically illustrates the above process.

This two-stage heat treatment modifies the interface between WC and Co phases from non-coherent to coherent, increasing interfacial strength and resistance to intergranular fracture. The coherent twin boundaries in the cobalt phase allow greater plastic deformation during crack propagation, enhancing fracture toughness.

A 2017 study [52] utilized a mixture of coke and alumina with a 50–50 weight ratio as a protective layer instead of using vacuum or hydrogen gas atmospheres to prevent reoxidation of tungsten carbide and cobalt metal and the formation of complex oxides. The results confirmed that alumina as an inert filler between coke particles prevents oxygen penetration, reduces the activity of coke, and slows down the oxidation or combustion of carbon. The discussion continues on the influential parameters in the synthesis of WC-Co composite powder by in-situ solid carbo-thermic reduction.

4.1 Milling conditions

Microscopic results in **Figure 6** indicate that the primary powder size decreases with increased milling time. The results show that the initial powder mixture, milled for 20, 30, and 40 hours, reduced the $WO_{2.9}$ and Co_3O_4 grain sizes to around 40–50 nm after 20 hours of milling, while the carbon grain size remained larger, ranging from 200 to 500 nm, indicating carbon's tendency to re-agglomerate.

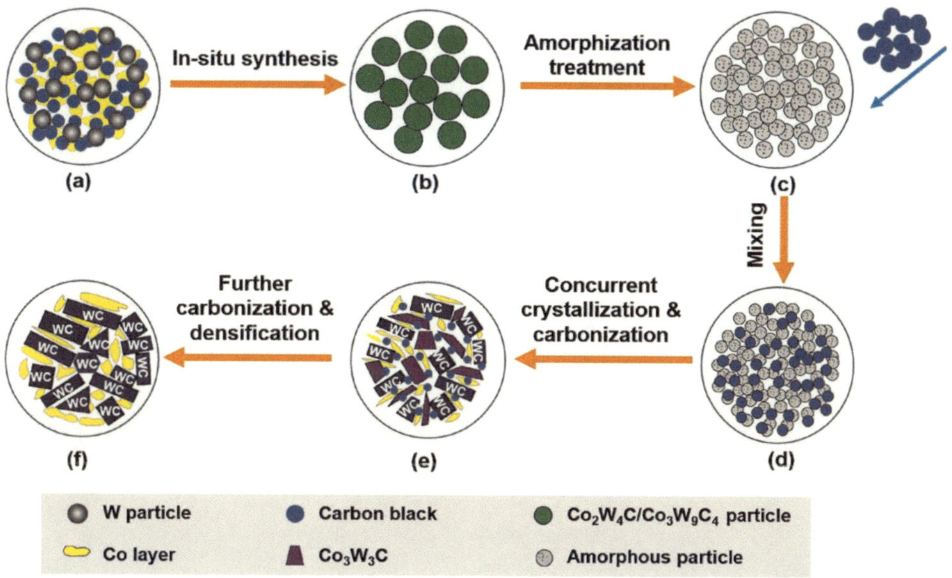

Figure 5.
Schematic of the overall process for producing WC-Co composite via two-stage reduction [51].

By increasing the milling time to 30 hours, the size of the oxide powder particles becomes finer, ranging from 20 to 40 nm, and the size of the carbon grains also becomes slightly finer. Subsequently, by increasing the milling time to 40 hours, the size of the oxide powder particles does not change much, but the distribution and dispersion of the carbon particles increase. Their grain size becomes about 250 nm, which is suitable for the reduction process due to having the highest surface energy and lower activation energy for reduction [25].

In a similar example, XRD results on the products show that the samples milled for 40 hours do not contain the η phase (the η phase includes compounds like Co_3W_3C, Co_6W_6C, and Co_2W_4C, all of which have an FCC structure but differ in the number of atoms in the crystal lattice). This phase persists if powder synthesis is done at 1100°C for 1 hour [53]. Moreover, the main phases are WC and Co. **Figure** 7 indicates that the η phase is present in the final product for shorter milling times.

Increasing the milling time before reduction not only affects the formed phases but also facilitates the production of nanocomposite particles. Results of three samples with different milling times show that the grain size of the final composite powder after 20, 30, and 40 hours of milling (**Figure 8**) is 424, 355, and 306 nm, respectively [25].

In a new study conducted in 2019 [54], it was shown that the particle size of the initial WO_3 powder can affect the size and distribution of the resulting WC powder. Four types of WO_3 powder with different grain sizes were used as raw materials, and the results indicated that WO_3 powder with the smallest particles (200 nm) had the least agglomeration. Also, its distribution and uniformity after milling were higher, which reduced the reduction temperature of tungsten oxides and increased the production efficiency of WC powder. Under thermal treatment at 1100°C for 3 hours in a vacuum atmosphere, tungsten oxides were reduced with carbon, and WC powder with a size of 100 nm was obtained. In contrast, when using WO_3 powder with the largest particle size (20–100 microns), the agglomeration and clustering of oxide

Figure 6.
SEM images of the powder: (a) $WO_{2.9}$, (b) Co_3O_4, and (c) carbon before milling, and (d), (e), and (f), respectively, from right to left after 20, 30, and 40 hours of milling [25].

particles increased, and after milling, good uniformity and homogeneity were not achieved, resulting in WC powder with a particle size of 250 nm. Even the milling temperature can affect the formation conditions of the powder [55]. A study showed that at low milling temperatures (−30°C) compared to ambient temperature milling, the particle shape disorder was less, and the final WC particle size was larger (26 vs. 21 nm). This factor can influence the sintering conditions of the WC-Co cermet powder.

Therefore, according to available sources, the raw materials for producing tungsten carbide ceramics and tungsten carbide-cobalt cermet must be well-milled;

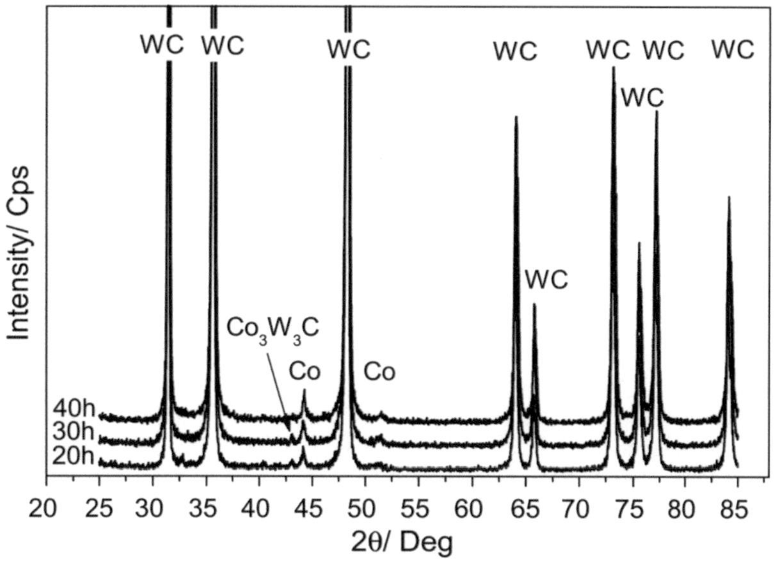

Figure 7.
XRD patterns of composite samples with different milling times [25].

Figure 8.
Three samples with different milling times (a) 20 hours, (b) 30 hours, and (c) 40 hours [25].

otherwise, the powder synthesis will not be successful. One of the main issues in industrial-scale production is that to increase production efficiency, only a mixer is used, and the use or time of ball milling is minimized, which can be a weakness during powder synthesis. (For the synthesis of WC and WC-Co, the raw materials must be milled for at least 10 hours. However, if the aim is to use an industrial mixer since little strain energy is stored in the raw materials, the time and temperature of thermal treatment must be increased).

4.2 Effect of carbon content and material ratios in initial powder mixture

If the carbon content in the initial powder mixture exceeds the permissible limit, free carbon or intermediate η phases can form in the composite powder, which can affect the properties of the WC-Co composite. The η phase includes compounds like Co_3W_3C and Co_6W_6C. Therefore, the amount of carbon added to the initial powder is crucial. The amount of carbon to be added is determined experimentally. Different amounts of carbon are examined under constant conditions for composite fabrication. In a study to determine the carbon content in the initial materials, carbon amounts ranged from 16.8 to 17.1% by weight under 40 hours of milling and 1080°C reduction for 4 hours. XRD results are shown in **Figure 9** [25].

According to the results in **Table 1**, the free carbon in the 16.8% weight sample is less than all other samples. The theoretical carbon amount under these conditions is 5.75% by weight, and since the actual amount is less, there is a carbon deficiency for composite production. For the second sample containing 16.9% carbon by weight, free carbon is still low, but intermediate η phases are present in the composite structure. However, for the third and fourth samples with 17 and 17.1% carbon by weight, only WC and Co phases are observed. Therefore, according to these results, the initial powder should contain 17% carbon by weight to achieve the best and most optimized structure in the composite. An optimal amount of the η phase can be suitable for the WC-Co composite and increase its corrosion resistance, especially in zinc-containing environments. Additionally, recent studies have shown that forming the η phase reduces the energy required for WC phase formation, thus lowering the reduction temperature and significantly impacting the final powder grain size [43].

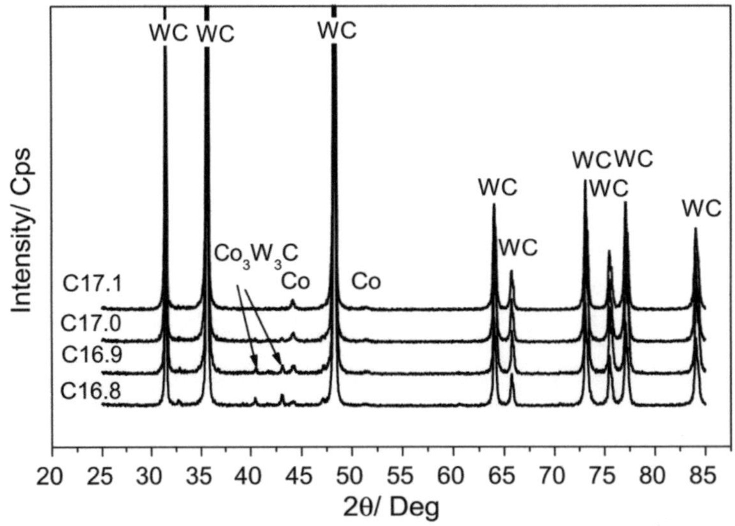

Figure 9.
XRD analysis of composite powders with different initial carbon amounts [25].

Sample	Excess carbon in raw materials (wt%)	Total actual carbon (wt%)	Free carbon (wt%)	Theoretical carbon (wt%)
C 16.8	16.8	5.71	0.05	5.66
C 16.9	16.9	5.88	0.06	5.82
C 17.0	17.0	5.97	0.20	5.77
C 17.1	17.1	6.11	0.34	5.77

Table 1.
Free carbon and total carbon amounts in four samples with different theoretical and actual initial carbon contents [25].

4.3 Effect of furnace atmosphere and pressure

The furnace atmosphere plays a crucial role in preparing tungsten carbide and tungsten carbide-cobalt. A vacuum atmosphere can reduce the reduction reaction temperature and improve the final powder phases, eliminating unwanted phases. **Figure 10** shows the XRD pattern for powder produced at 1050°C under different vacuum levels. The best powder production conditions are under gas pressure of 4×10^{-4} Pa, where no additional phases are present and high-quality powder is obtained.

4.4 Effect of thermal treatment temperature

Figure 11 shows the XRD patterns of composites produced in the 850–1050°C [25]. Initially, when the temperature rises to about 850°C, the reduction of tungsten oxide and the release of tungsten become possible, forming the WC phase. At temperatures above 850°C, the formation of η and W phases increases. With further temperature

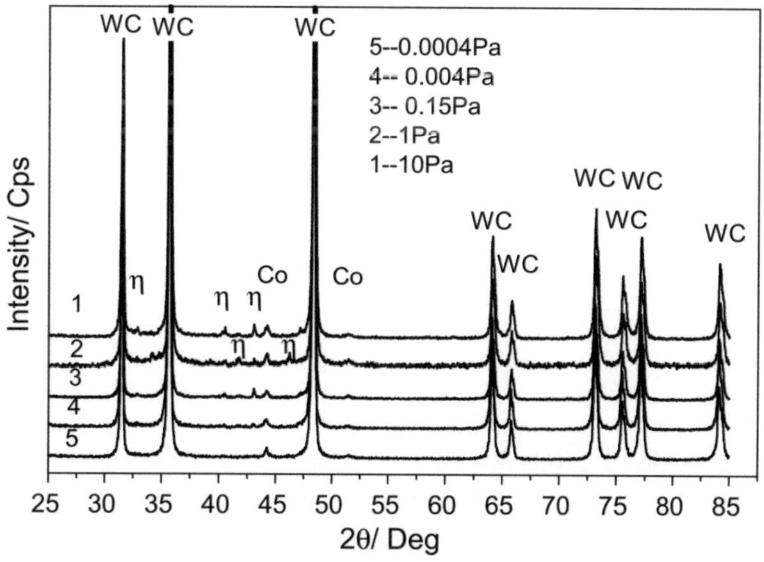

Figure 10.
XRD patterns for powders produced at 1050°C under different vacuum levels [25].

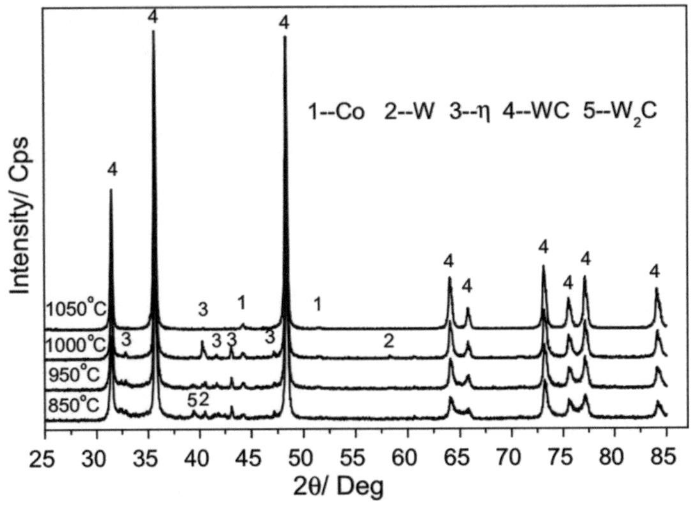

Figure 11.
XRD patterns of composites produced in the temperature range of 850–1050°C [25].

increase to 1050°C, the formation of η phases decreases, and cobalt oxide reduction completes, resulting only in WC and Co phases.

Besides the effects of furnace temperature on formed phases, it can also influence the grain size of the produced composite powder (**Figures 12** and **13**). Results show that the initial powder grain size was around 115 nm, indicating fine raw materials. With thermal treatment in the temperature range of 900–950°C, grain growth occurs, possibly due to the high activation energy generated during milling, causing particles to reduce their surface area by increasing size significantly.

Therefore, a review of the literature shows that the temperature range for the reduction and formation of tungsten carbide and tungsten carbide-cobalt varies depending on the type of raw materials used, milling conditions, and the synthesis equipment and method. In most references, an approximate temperature range of 1000–1200°C has been chosen to synthesize tungsten carbide and tungsten carbide-cobalt powders.

4.5 Effect of heat treatment time

There is an optimal heat treatment time; if it is less than the required amount, the reduction and carburization reactions for synthesizing the products may not be completed. If it is too long, grain coarsening occurs in the composite structure, and issues of reoxidation and the formation of unwanted phase compounds arise. Evidence shows that during the reduction process at 1050°C, for different times ranging from 6 to 10 hours, the grain size changes from 350 to 545 nm, and powder particle coarsening occurs [25].

4.6 Thermodynamic analysis for the formation of WC-Co composite powder (WC powder) from metal oxides using in-situ and carbo-thermic reduction methods

When discussing chemical reactions, thermodynamic laws come into play. Specifically, the first law of thermodynamics deals with the energy of reactions, and the

Figure 12.
SEM images of the morphology of powder grains produced at temperatures of 900, 950, 1000, 1050, 1100, and 1150°C, respectively, shown from right to left in images (a), (b), (c), (d), (e), and (f) [25].

second law addresses the transfer of this energy, guiding the direction of reactions. For instance, if $\Delta G = 0$, the reaction is at equilibrium and can proceed in both directions. If $\Delta G < 0$, the reaction is spontaneous and moves forward, while if $\Delta G > 0$, the reaction is non-spontaneous and requires energy input. ΔG can be calculated using the following formula:

$$\Delta G = -\Delta A_1 T \operatorname{Ln} T - \frac{1}{2} \times 10^{-3} \Delta A_2 T^2 - \frac{1}{2} \times 10^{5} \Delta A_3 T^{-1} - \frac{1}{6} \times 10^{-6} \Delta A_4 T^3 - \frac{1}{6} \times 10^{8} \Delta A_5 T^{-2} + A_6 T + A_7 \quad (3)$$

Where ΔA_i and $i = 1 - 5$ pertain to the thermodynamic data of reactants and products, and A_7 and A_6 are constants related to stability. The reaction's starting temperature can be found under conditions where $\Delta G = 0$.

When tungsten oxide (WO_3) and cobalt oxide are combined and heated in a carbon environment at specific temperatures, numerous reactions occur, including forming intermediate phases before yielding the final product. Thus, controlling the conditions of these reactions is complex [30].

The following reactions occur during the carbo-thermic reduction of WO_3:

$$WO_3 + 0.05C = WO_{2.9} + 0.05CO_2 \quad (4)$$

$$WO_3 + 0.1\,C = WO_{2.9} + 0.1CO \quad (5)$$

$$WO_{2.9} + 0.09C = WO_{2.72} + 0.09CO_2 \quad (6)$$

$$WO_{2.9} + 0.18C = WO_{2.72} + 0.18CO \quad (7)$$

$$WO_{2.72} + 0.36C = WO_2 + 0.36CO_2 \quad (8)$$

$$WO_{2.72} + 0.72C = WO_2 + 0.72CO \quad (9)$$

$$WO_2 + C = W + CO_2 \quad (10)$$

$$WO_2 + 2C = W + 2CO \quad (11)$$

The ΔG values for these reactions can be obtained using formula 3 and thermodynamic tables. The free energy changes with temperature for oxides WO_2, $WO_{2.72}$, $WO_{2.9}$, and WO_3 with gaseous products CO_2 and CO are illustrated in **Figure 14**.

In **Figure 14(a)**, reactions (4) and (5) are depicted. When the temperature exceeds 390 K and 705 K, respectively, the free energy for both reactions is less than

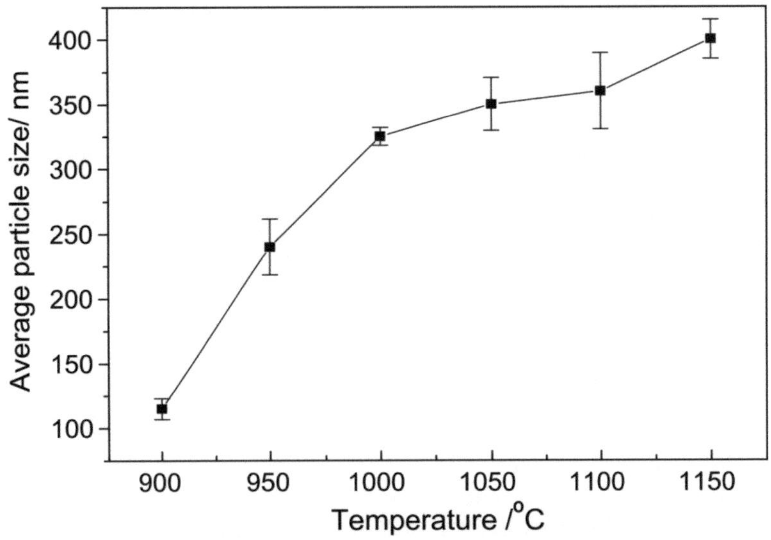

Figure 13.
Schematic relationship between temperature and particle size of the composite powder [25].

zero, indicating they can occur under these conditions. Below 970 K, the free energy for reaction (4) is lower than for reaction (5), making CO_2 stable at temperatures below 970 K and CO at higher temperatures.

Figure 14(b) concerns reactions (6) and (7), showing that at room temperature, the free energy for both reactions is below zero, indicating $WO_{2.9}$ reduces well at lower temperatures. Above room temperature, the free energy for reaction (6) is lower than for reaction (7), thus CO_2 is produced in this case.

Figure 14(c) indicates that reactions (8) and (9) have free energies below zero above 905 K and 940 K, meaning they can occur under these conditions. Below 970 K, the free energy for reaction (8) is lower than for reaction (9), making CO_2 stable, while above 970 K, reaction (9) has a lower free energy, making CO stable.

In **Figure 14(d)**, reactions (10) and (11) have free energies below zero above 1015 K and 1055 K, meaning they can proceed under these conditions. Below 970 K, reaction (10) is more stable, but since $\Delta G > 0$, the reaction does not occur and CO_2 is not produced. Above 970 K, CO is produced.

In summary, the gas produced for the reduction of WO_3 and $WO_{2.27}$ can be CO and CO_2 together, but for WO_2 and $WO_{2.9}$, only one type of gas, either CO or CO2, exists, with CO being more stable at higher temperatures.

Next, the thermodynamic analysis of the formation of tungsten carbide (WC) is presented. W_2C and WC are two different chemical compounds of tungsten carbide formed during the carburization of tungsten, as shown in reactions (12)–(14):

$$W + 0.5C = 0.5W_2C \tag{12}$$

$$W + C = WC \tag{13}$$

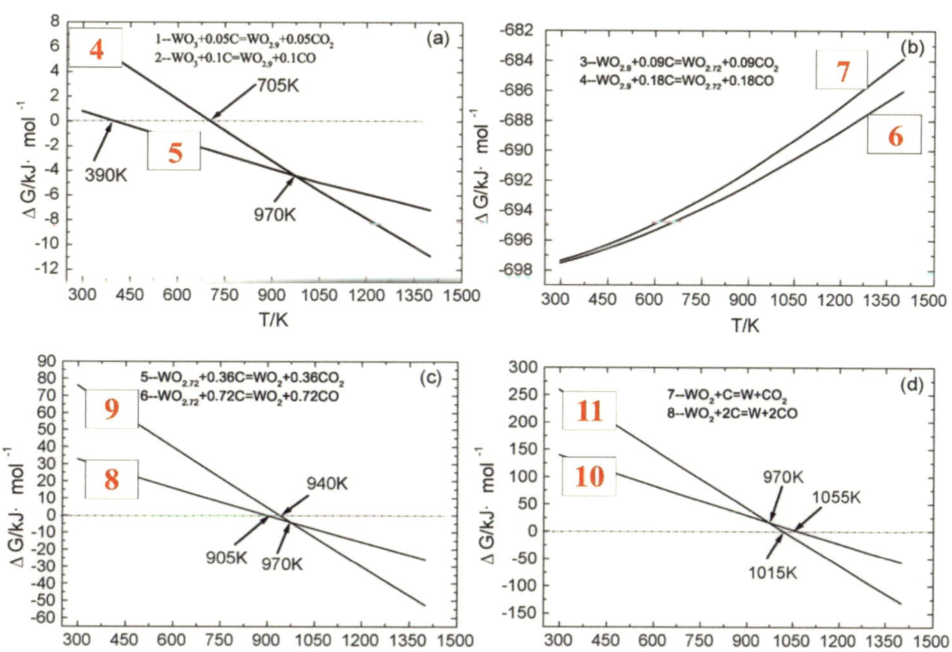

Figure 14.
Changes in ΔG with temperature for the carbo-thermic reduction of tungsten oxides according to reactions (4)–(11) [30].

$$0.5W_2C + 0.5C = WC \tag{14}$$

Figure 15 shows the curves of free energy changes with temperature for these reactions. The ΔG for reaction (13) is significantly lower than for reactions (12) and (14). Below 1500 K, WC is more likely to form and is more stable than W_2C. However, above 1500 K, the ΔG for reaction (12) becomes lower than for reaction (13), making W2C stable, and even increasing carbon content at higher temperatures cannot convert W_2C to WC.

This section continues with the thermodynamic analysis of the carbo-thermic reduction reactions of Co_3O_4 by carbon. The carbo-thermic reduction reactions of Co_3O_4 producing CoO and Co are listed below:

$$Co_3O_4 + 2C = 3Co + 2CO_2 \tag{15}$$

$$Co_3O_4 + 4C = 3Co + 2CO \tag{16}$$

$$Co_3O_4 + 0.5C = 3CoO + 0.5CO_2 \tag{17}$$

$$Co_3O_4 + C = 3CoO + CO \tag{18}$$

Figure 16 shows the curves of free energy changes with temperature for these reactions. All these reactions can occur at low temperatures. The starting temperature for reaction (16) is around 625 K, and for other reactions, it is below 360 K. At this low-temperature range, reactions (15) and (17) start sooner, both producing CO_2 gas. Below 515 K, reaction (17) has a lower ΔG and occurs more quickly, forming CoO. Above 515 K, reaction (18) has a lower ΔG and proceeds more easily, forming Co. At 970 K, reaction (16) has the lowest ΔG and is thermodynamically more stable, with the final product being cobalt and CO gas.

It is worth noting that one of the differences in the synthesis of tungsten carbide and tungsten carbide-cobalt is the presence of cobalt oxide in the initial material composition. This combination leads to the formation of the intermediate compound "$CoWO_4$" in the structure, which can improve the synthesis process of tungsten carbide–-. Therefore, it has been stated that it can act as a catalyst in synthesizing

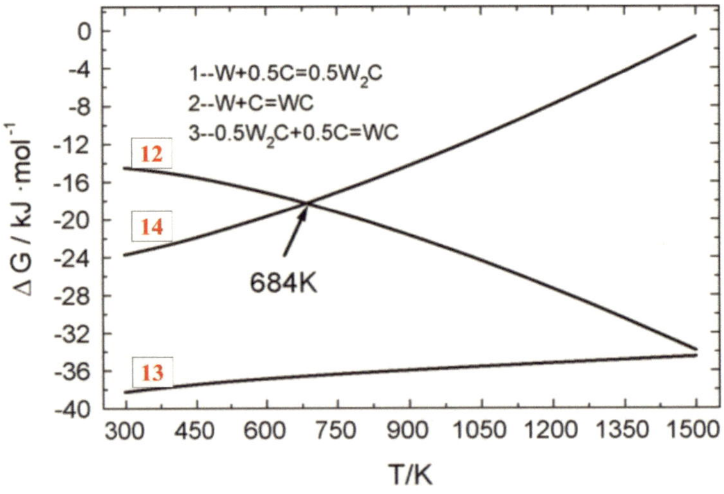

Figure 15.
Changes in free energy with temperature for the formation reactions of tungsten carbide [30].

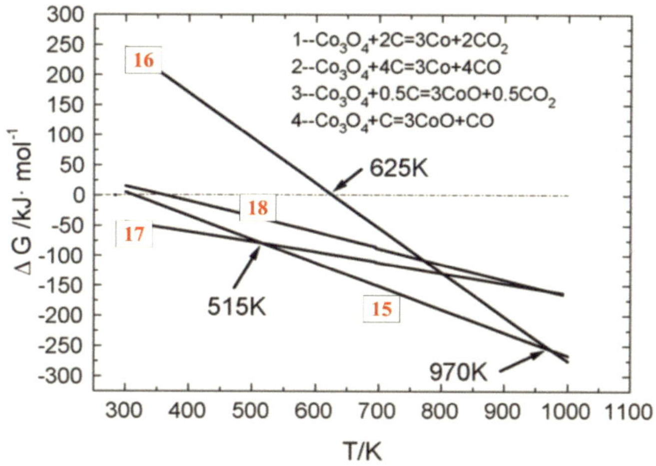

Figure 16.
Changes in free energy with temperature for the carbo-thermic reactions of Co_3O_4 [30].

tungsten carbide alongside cobalt. The thermodynamic analysis of the intermediate compound $CoWO_4$ is provided below. The reaction between WO_3, Co_3O_4, and carbon at lower temperatures results in forming the intermediate and unstable compound $CoWO_4$, which can decompose at higher temperatures. The formation and decomposition reactions of $CoWO_4$ are presented below.

$$Co_3O_4 + 0.5C + 3WO_3 = 3CoWO_4 + 0.5CO_2 \quad (19)$$

$$Co_3O_4 + C + 3WO_3 = 3CoWO_4 + CO \quad (20)$$

$$CoO + 3WO_3 = CoWO_4 \quad (21)$$

$$CoWO_4 + 2.5C = 0.5W_2C + Co + 2CO_2 \quad (22)$$

$$CoWO_4 + 4.5C = 0.5W_2C + Co + 4CO \quad (23)$$

$$CoWO_4 + 2C = W + Co + 2CO_2 \quad (24)$$

$$CoWO_4 + 4C = W + Co + 4CO \quad (25)$$

$$CoWO_4 + 3C = WC + Co + 2CO_2 \quad (26)$$

$$CoWO_4 + 5C = WC + Co + 4CO \quad (27)$$

Figure 17, parts (a) and (b), corresponding to the curves of Gibbs free energy changes with temperature related to reactions 19–27. Part (a) in **Figure 17** depicts the formation reactions of $CoWO_4$, which occur at low temperatures around 300 Kelvin. Under these conditions, the Gibbs free energy change (ΔG) for reactions 19–21 is negative. Reaction 17 has a more negative ΔG than reactions 20 and 21, indicating the stability of the phases $CoWO_4$ and CO_2 under these conditions. It is noteworthy that reaction 21 indicates that $CoWO_4$ can be produced from the combination of CoO and WO_3, and the temperature does not affect the ΔG of the reaction.

In part (b) of **Figure 17**, the decomposition reactions of $CoWO_4$ are shown, which occur at relatively high temperatures. The onset temperature for transformations corresponds to reaction 26 (945 Kelvin). Under these conditions, the products of

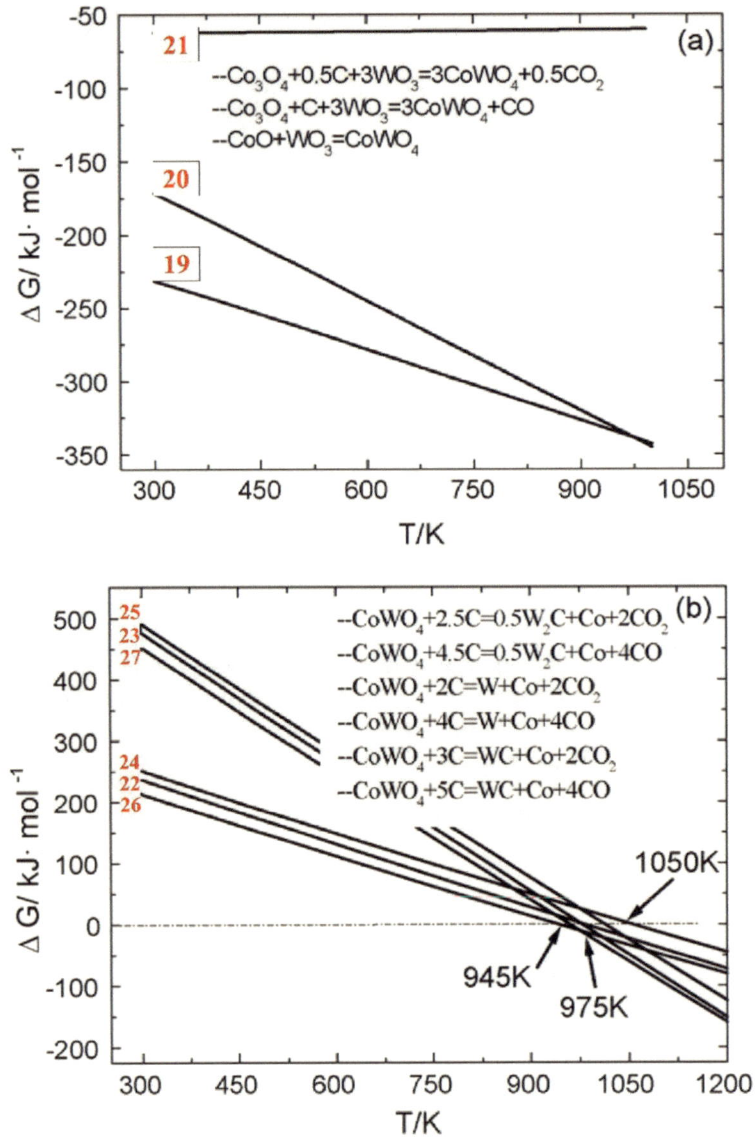

Figure 17.
Changes in Gibbs free energy with temperature for (a) the formation and (b) the decomposition of $CoWO_4$ [30].

reaction 26, Co, WC, and CO_2, are stable, and the highest formation temperature of the products is 1050 Kelvin, corresponding to reaction 24. Within this temperature range, when the temperature reaches 975 Kelvin, the ΔG of reaction 27 is at its minimum, and WC, CO gas, and Co(s) are formed.

Therefore, when $CoWO_4$ reacts with carbon, the gaseous reaction products at low temperatures are CO_2 and CO, and the solid components in the final products depend on the reaction temperature and the amount of carbon. At a constant temperature, with an increase in the amount of carbon, the products change from W to W_2C and WC.

5. Conclusion

The literature review indicates the optimal conditions for various factors influencing the carbo-thermic reduction process and the synthesis of tungsten carbide (tungsten carbide-cobalt). For the milling of raw materials in the solid-state method, a duration of 5–20 hours with a ball-to-powder ratio of 3 (up to 10) to 1 and a speed of 300 rpm is suitable. Additionally, the best atmosphere for solid-state reduction is a vacuum, while an oxidizing atmosphere or argon protective gas does not perform well. The optimal time and temperature for the heat treatment process (thermal cycle) for the solid-state reduction method are within the temperature range of 1000–1200° C and a holding time of at least 3 hours. Information on the amount of carbon in the initial material composition is incomplete and requires further investigation.

Conflict of interest

The authors declare no conflict of interest.

Author details

Hamed Naderi-Samani[1*], Reza Shoja Razavi[2], Hasan Abbaszadeh[1], Afshin Amiri-Moghaddam[3], Mehri Mashhadi[1] and Ali Alizadeh[1]

1 Faculty of Materials and Manufacturing Technologies, Malek Ashtar University of Technology, Iran

2 Materials and Energy Research Centre Iran, Karaj, Iran

3 Materials Engineering Group, Department of Mining and Metallurgical Engineering, Yazd University, Yazd, Iran

*Address all correspondence to: hamednaderi771@gmail.com

References

[1] Zhao S, Song X, Wang M, Liu X, Zhang G, Zhang J. Preparation of ultrafine WC-Co cermets by pretreatment with spark plasma sintering. Rare Metals. 2009;**28**:391-395

[2] Shengda G, Tao S, Rui B, Jiangao Y, Jianhong Y. Synthesis and characterization of WC-6Co nanocrystalline composite powder. Rare Metal Materials and Engineering. 2018; **47**:1986-1992

[3] Pervikov AV, Krinitcyn MG, Glazkova EA, Rodkevich NG, Lerner MI. Synthesis of tungsten carbide from bimodal tungsten powder produced by electrical explosion of wire. International Journal of Refractory Metals and Hard Materials. 2022;**103**:105733

[4] Schwanekamp T, Marginean G, Reuber M, Ostendorf A. Impact of cobalt content and grain growth inhibitors in laser-based powder bed fusion of WC-Co. International Journal of Refractory Metals and Hard Materials. 2022;**105**:105814

[5] Lopez Ezquerra B, Gonzalez Ojeda R, Iparraguirre I, Rodriguez N, Sanchez-Moreno JM. Presintering of TiCN-TiC-WC-Cr_3C_2-Ni cermets under N_2-H_2 atmospheres. International Journal of Refractory Metals and Hard Materials. 2017;**62**(Part A):21-28

[6] Lin T, Li Q, Han Y, Song K, Wang X, Shao H, et al. Effects of Nb and NbC additives on microstructure and properties of WC-Co-Ni cemented carbides. International Journal of Refractory Metals and Hard Materials. 2022;**103**:105782

[7] Mégret A, Vitry V, Delaunois F. High-energy ball milling of WC-10Co: Effect of the milling medium and speed on the mechanical properties. International Journal of Refractory Metals and Hard Materials. 2022;**104**:105774

[8] Park C, Kim J, Kang S. Effect of cobalt on the synthesis and sintering of WC-Co composite powders. Journal of Alloys and Compounds. 2018;**766**:564-571

[9] Jianxin D, Hui Z, Ze W, Yansongand L, Jun Z. Friction and wear behaviors of WC/Co cemented carbide tool materials with different WC grain sizes at temperatures up to 600°C. Journal of Refractory Metals and Hard Materials. 2012;**31**:196-204

[10] Khan MA, Gupta K. A study on machinability of nickel based superalloy using micro-textured tungsten carbide cutting tools. Materials Research Express. 2020;7(1):016537

[11] Xu L, Song W, Ma S, Zhou Y, Pan K, Wei S. Effect of slippage rate on frictional wear behaviors of high-speed steel with dual-scale tungsten carbides (M6C) under high-pressure sliding-rolling condition. Tribology International. 2021;**154**:106719

[12] Zhao Z, Liu F, Cao L, Du Y, Li B, Li J, et al. Investigation of indentation response, scratch resistance, and wear behavior of tungsten carbide coatings fabricated by two-step interstitial carburization on tungsten. Ceramics International. 2021;**47**(21):30636-30647

[13] Basu B, Balani K. Advanced Structural Ceramics. 1st ed. Wiley-American Ceramic Society; October 11, 2011

[14] Shahien M, Suzuki M, Shinoda K, Akedo J. Ceramic coatings deposited from fine particles by different spraying processes. Journal of Thermal Spray Technology. 2020;**29**(8):2033-2047

[15] Angelo P, Subramanian R, Ravisankar B. Powder Metallurgy: Science, Technology and Applications. 2nd ed. Delhi: PHI Learning Pvt. Ltd.; November 1, 2022

[16] Medeirs FFP, De Oliveira SA, De Souza CP, Da Silva AGP, Gomes UU, De Souza JF. Synthesis of tungsten carbide through gas–solid reaction at low temperatures. Materials Science and Engineering A. 2001;**315**:58-62

[17] Guo LT, Wu XJ, Li ZQ. Carbon nanotubes strengthened nanophase WC–Co hard alloys. Advanced Engineering Materials. 2006;**8**(1–2):62-72

[18] Zawrah MF. Synthesis and characterization of WC–Co nanocomposites by novel chemical method. Ceramics International. 2007; **33**(2):155-161

[19] Zhang ZY, Sverker W, Wang MS, Muhammed M. Processing of nanostructured WC–Co powder from precursor obtained by co-precipitation. Nanostructured Materials. 1999;**12**(1–4): 163-166

[20] Roy SK, Dey R, Mitra A, Mukherjee S, Mitra MK, Das GC. Optimization of process parameters for the synthesis of silica gel-WC nanocomposite by design of experiment. Materials Science and Engineering: C. 2007;**27**(4):725-728

[21] Kim BK, Ha GH, Lee DW, Lee GG. Chemical processing of nanostructured cemented carbide. Advanced Performance Materials. 1998;**5**:341-352

[22] Armstrong RW. The hardness and strength properties of WC-Co composites. Materials. 2011;**4**:1287-1308

[23] Yi Z, Shao G, Duan X, Sun P, Shi X, Xiong Z, et al. Preparation of WC-Co powder by direct reduction and carbonization. China Particuology. 2005; **3**:286-288

[24] Lee G-H, Kang S. Sintering of nano-sized WC–Co powders produced by a gas reduction–carburization process. Journal of Alloys and Compounds. 2006; **419**:281-289

[25] Liu W, Song X, Zhang J, Zhang G, Liu X. Preparation of ultrafine WC–Co composite powder by in situ reduction and carbonization reactions. International Journal of Refractory Metals and Hard Materials. 2009;**27**: 115-120

[26] Fernique RMT, Savoie S, Gariépy M, Braidy N, Schulz R. A simple route to produce tungsten carbide powders by high-energy ball milling and annealing. Ceramics International. 2020;**46**: 1736-1742

[27] Ma J, Zhu SG. Direct solid-state synthesis of tungsten carbide nanoparticles from mechanically activated tungsten oxide and graphite. International Journal of Refractory Metals and Hard Materials. 2010;**28**: 623-627

[28] Yang Q, Yang J, Wen Y, Zhang Q, Chen L, Chen H. A novel route for the synthesis of ultrafine WC-15 wt %Co cemented carbides. Journal of Alloys and Compounds. 2018;**748**:577-582

[29] Cui X, Zhou X, Chen H, Hua Z, Huixia W, He Q, et al. In-situ carbonization synthesis and ethylene hydrogenation activity of ordered mesoporous tungsten carbide. International Journal of Hydrogen Energy. 2011;**36**:10513-10521

[30] Liu W, Song X, Zhang J, Zhang G, Liu X. Thermodynamic analysis for in situ synthesis of WC–Co composite

powder from metal oxides. Materials Chemistry and Physics. 2008;**109**: 235-240

[31] Wei C, Song X, Zhao S, Zhang L, Liu W. In-situ synthesis of WC–Co composite powder and densification by sinter-HIP. International Journal of Refractory Metals and Hard Materials. 2010;**28**:567-571

[32] Andersson K, M. Aqueous processing of WC-Co powders [doctoral thesis]. Kollegiesalen, KTH, Valhallavägen 79, Stockholm: Royal Institute of Technology, Department of Chemistry, Surface Chemistry; April 16, 2004

[33] Richerson DW, Lee WE. Modern Ceramic Engineering: Properties, Processing, and Use in Design. 4th ed. CRC Press; May 10, 2018

[34] Liu X, Song X, Wang H, Liu X, Tang F, Hao L. Complexions in WC-Co cemented carbides. Acta Materialia. 2018;**149**:164-178

[35] Petukhov AS, Uvarova IV, Bondarenko VP, Pavlotskaya EG, Martynova LM, Konchakovskaya LD, et al. Mechanism of low temperature synthesis of tungsten carbide and WC-Co mixtures in a methane-hydrogen atmosphere. Poroshkovaya Metallurgiya. 1992;**10**:31-35

[36] Adorjan C, Bock A, Myllymäki S, Schubert W-D, Kontturi K. WC/Co-composite powders via hydrothermal reduction of Co_3O_4 suspensions. International Journal of Refractory Metals and Hard Materials. 2008;**26**: 569-574

[37] García T, Luis A, Li Q, Jensen JO, Bjerrum NJ. High surface area tungsten carbides: Synthesis, characterization and catalytic activity towards the hydrogen evolution reaction in phosphoric acid at elevated temperatures. International Journal of Electrochemical Science. 2014; **9**:1016-1032

[38] Lin H, Tao B, Li Q, Li Y. In situ synthesis of WC–Co nanocomposite powder via core–shell structure formation. Materials Research Bulletin. 2012;**47**:3283-3286

[39] Danny Xiao T, Tan X, Yi M, Peng S, Peng F, Yang J, et al. Synthesis of commercial-scale tungsten carbide-cobalt (WC/Co) nanocomposite using aqueous solutions of tungsten (W), cobalt (Co), and carbon (C) precursors. Journal of Materials Science and Chemical Engineering. 2014;**2**:1-15

[40] Guo S, Fei Y, Zhou Y, Yang J, Chen H. Investigation on reduction and carbonization process of WC-Co composite powder obtained by In situ synthesis. Journal of Alloys and Compounds. 2019;**775**:1086-1093

[41] Yi Z, Shao G, Duan X, Sun P, Shi X, Xiong Z, et al. Preparation of WC-Co powder by direct reduction and carbonization. China Particuology. 2005; **3**(05):286-288

[42] Lofberg A, Frennet A, Leclercq G, Leclercq L, Giraudony JM. Mechanism of WO3 reduction and carburization in CH_4/H_2 mixtures leading to bulk tungsten carbide powder catalysts. Journal of Catalysis. 2000;**189**:170-183

[43] Wang H, Hou C, Liu X, Liu X, Song X. Phase evolution in synthesis of nanocrystalline WC-η composite powder by solid-state in situ reactions. International Journal of Refractory Metals and Hard Materials. 2018;**71**:21-27

[44] Dang J, Yijie W, Lv Z, Lv X. Preparation of tungsten carbides by reducing and carbonizing WO2 with CO.

Journal of Alloys and Compounds. 2018; **745**:421-429

[45] Habibi AH, Shoja Razavi R, Borhani GH, Erfanmanesh M. Effect of argon shroud protection on the laser cladding of nanostructured WC-12Co powder. Journal of Materials Engineering and Performance. 2021;**30**: 3313-3320

[46] Wang K-F, Chou K-C, Zhang G-H. Preparation of high-purity and ultrafine WC-Co composite powder by a simple two-step process. Advanced Powder Technology. 2020;**31**:1940-1945

[47] Wu Y-C, Yang Y, Tan X-Y, Luo L, Zan X, Zhu X-Y, et al. Preparation technology of ultra-fine tungsten carbide powders: An overview. Frontiers in Materials. 2020;7:94

[48] Sheng-da GUO, Rui BAO, Ping YANG, Liang LIU, Jian-hong YI. Morphology and carbon content of WC−6%Co nanosized composite powders prepared using glucose as carbon source. Transactions of the Nonferrous Metals Society of China. 2018;**28**:722-728

[49] Amiri-Moghaddam A, Shoja-Razavi R, Abbaszadeh H, Naderi-Samani H. In-situ synthesis of WC and WC-17Co powder under hydrogen atmosphere by solid carbothermic reduction method. Ceramics International. 2023;**49**(6): 9806-9818

[50] Liu W, Song X, Wang K, Zhang J, Zhang G, Liu X. A novel rapid route for synthesizing WC–Co bulk by in situ reactions in spark plasma sintering. Materials Science and Engineering A. 2009;**499**:476-481

[51] Peng Y, Wang H, Zhao C, Huaxin H, Liu X, Song X. Nanocrystalline WC-Co composite with ultrahigh hardness and toughness. Composites Part B Engineering. 2020;**197**:108161

[52] Amiri-Moghaddam A, Kalantar M. In-situ synthesis of WC–X%Co composite in the WO3–Co3O4–C system by carbothermal reduction method. Journal of the Australian Ceramic Society. 2017;**53**:839-845

[53] Pan Y, Xiong H, Li Z, Long X. Synthesis of WC-Co composite powders with two-step carbonization and sintering performance study. International Journal of Refractory Metals and Hard Materials. 2019;**81**: 127-136

[54] Haoyang W, Wang Q, Qin M, Yin R, Zhang Z, Jia B, et al. Synthesis of tungsten carbide nanopowders by direct carbonization of tungsten oxide and carbon: Effects of tungsten oxide source on phase structure and morphology evolution. Ceramics International, Ceramics International. 2020;**46**: 8787-8795

[55] Hewitt SA, Laoui T, Kibble KK. Effect of milling temperature on the synthesis and consolidation of nanocomposite WC–10Co powders. International Journal of Refractory Metals and Hard Materials. 2009;**27**: 66-73

Chapter 4

Conduction Mechanisms and Permittivity: Conductivity Correlation in the Gd-Based Perovskite Structure

Khouloud Moualhi and Mouldi Zouaoui

Abstract

This chapter investigates the dielectric properties of $GdCa_2Cu_3O_8$, a recently discovered material showcasing colossal permittivity. Through a detailed examination, we explore the original mechanisms behind its remarkable dielectric behavior, crystal structure, and microstructure properties. The temperature-dependent dielectric measurements revealed a constant permittivity value at room temperature (10^4), indicative of colossal permittivity behavior. We discuss the origin of this phenomenon in terms of intrinsic and extrinsic mechanisms, including defect-related effects, charge carrier hopping, and Maxwell-Wagner polarization. By highlighting the latest research advancements, this chapter aims to provide readers with a thorough understanding of the possibilities presented by this innovative material in the ever-evolving field of functional materials and their applications. Moreover, it explores the diverse range of applications enabled by $GdCa_2Cu_3O_8$, from high-capacitance devices to energy storage systems. Our findings provide valuable insights into the understanding and utilization of colossal permittivity in $GdCa_2Cu_3O_8$ perovskite for advanced electronic applications (capacitors Y5R and Y6R).

Keywords: ceramic capacitor, colossal dielectric constant, dielectric model, interfacial polarization, conduction mechanisms

1. Introduction

Advanced ceramics represent a fascinating material class that has developed various industries due to their unique properties and applications [1–4]. These ceramics, often engineered at the molecular or nanostructure level, offer exceptional mechanical strength, thermal stability, chemical inertness, and electrical insulation properties. Advanced ceramics are typically based on oxides, carbides, nitrides, and borides of metals. The development of advanced ceramics has been driven by the need for materials that can withstand extreme conditions in the aerospace, automotive, electronics, biomedical, and energy sectors [5]. Manufacturing advanced ceramics often involves sophisticated techniques such as sintering, hot pressing, and chemical vapor deposition to achieve the desired microstructure and properties. Modern technology and industry rely heavily on advanced ceramics, which offer solutions to challenges that traditional materials cannot meet. The ongoing research and development in this field promise further innovations and applications in the years to come, contributing to advancements in areas ranging from electronics to healthcare and beyond [6].

Materials exhibiting colossal permittivity have garnered significant attention in recent years due to their exceptional electrical properties and potential applications in various technological fields. Colossal permittivity refers to a phenomenon where materials exhibit extremely high dielectric constants, often several orders of magnitude greater than those of conventional dielectrics. This remarkable characteristic allows them to store significant amounts of electrical charge under an applied electric field, making them ideal candidates for capacitive devices and energy storage applications [7]. Studying and exploring materials with colossal permittivity have opened new avenues for research in materials science, solid-state physics, and engineering. The exploration of materials with colossal permittivity has become a new frontier in materials science and engineering in recent decades. These materials, characterized by their extraordinary ability to store electrical charge under an applied electric field, hold immense promise for revolutionizing various technological applications. Among the notable examples of such materials are $BaTiO_3$ (ABO_3), Gd_2CuO_4(ABO_4), and $CaCu_3Ti_4O_{12}$ ($AA'B_4O_{12}$) (**Figure 1**) [8–10]. Due to its high relative permittivity, barium titanate ($BaTiO_3$) is used in dielectric composites [3]. $BaTiO_3$'s reported results indicate that the ε_r' is high but is affected by temperature. The presence of the transition temperature event is generally a problem when discussing these materials' applications. Recently, calcium copper titanate (CCTO) has been receiving a lot of attention because of its high and stable dielectric permittivity [10].

The exceptional physical characteristics of $CaCu_3Ti_4O_{12}$ perovskite, such as its elevated dielectric properties, have made it a significant consideration for functional materials. The electrical and dielectric properties of ceramic compounds are influenced by dipolar species and localized charges, which are the parameters that determine strength, kinematics, and charge carrier interactions [10]. The process of

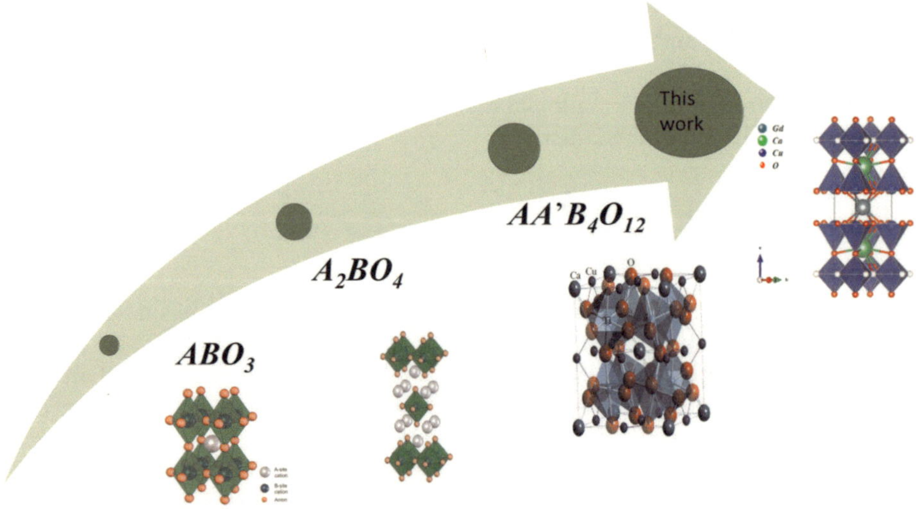

Figure 1.
Evolution of materials with colossal permittivity.

coordinating the structural, dielectric, and electrical properties of ceramics to create multifunctional compounds and utilize them in future applications is challenging. The recent proposal of fabrication of ceramic capacitors using $CaCu_3Ti_4O_{12}$ complex ceramics for their high permittivity (ε'), excellent performance, low dielectric loss, and excellent temperature stability over a large temperature domain have been made due to its motivation [6, 7]. The evolution of materials with colossal permittivity has been marked by continuous advancements in synthesis techniques, characterization methods, and theoretical understanding.

This chapter would open up an opportunity to develop new high dielectric permittivity material ($GdCa_2Cu_3O_8$). These efforts to optimize other critical properties such as conductivity, stability, and environmental sustainability [11, 12]. The crystal structure of $GdCa_2Cu_3O_8$ was discovered recently [11, 12]. $GdCa_2Cu_3O_8$ is a ternary oxide compound with exceptional dielectric permittivity values, which has been considered for capacitor applications.

2. Ceramics synthesis

The ceramic sample was synthesized using a conventional solid-state reaction (SSR) method. The compositions $GdCa_2Cu_3O_8$ dictated the weight of the reagent grades Gd_2O_3, $CaCO_3$, and CuO, which were determined by weighing them. The raw materials were weighed according to the composition $GdCa_2Cu_3O_8$ and mixed thoroughly in an agate mortar. The powders were pressed into pellets and calcined in the air in a box furnace for 24 h at 800°C. After the pellets were sintered at 950°C in the air for 24 h (**Figure 2**).

The first step in sintering is the rapid formation of bridges between the particles in contact. Several mechanisms described in **Figure 2** explain bridge formation. There is a transport of material from the surface of the grain or the center of the grain joint to the bridges. The diffusion of material can occur through surface, volume, vapor phase, or grain boundaries [13]. Obtaining thermodynamically stable phases at high temperatures through solid-state diffusion is achievable using the solid-state reaction method, also known as the ceramic method [14]. Ceramics are a diverse class of compounds such as monolithic or composite ceramics [15] (**Figure 3**).

3. Dielectric properties of ceramics

Insulating materials are known for not conducting current in an electric field. However, a microscopic approach to these phenomena allows us to distinguish more subtly between the dielectric aspect.

The nineteenth-century English Michael Faraday studied the ability of a system to store an electrical charge, he is honored with the unit "Farad" (symbol F) [16]. The capacitance of a system of two charged parallel plates with electrodes of area s separated by a distance e and filled with a material of relative permittivity ε'_r is given by:

$$C = \varepsilon_r \varepsilon_0 \frac{s}{e} \tag{1}$$

With ε_0 being the dielectric constant of vacuum.

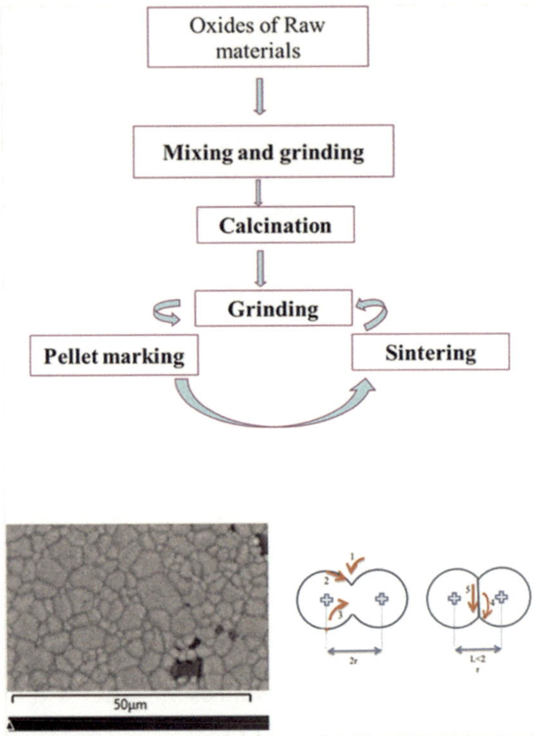

Figure 2.
A flow chart provided for developing $GdCa_2Cu_3O_\delta$ using a solid-state reaction method and SEM micrograph of ceramic [11, 12].

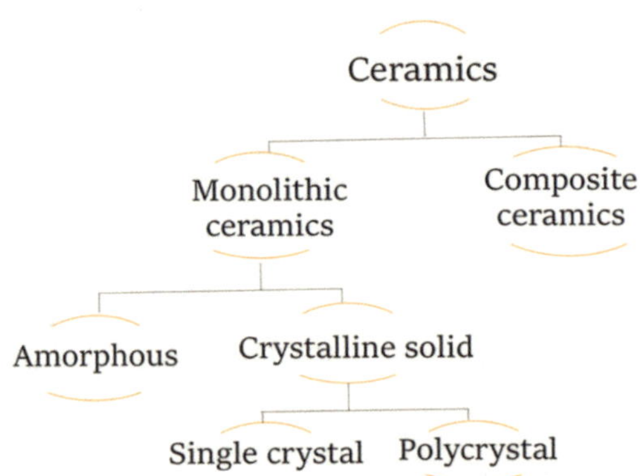

Figure 3.
Overview of the classification system of ceramics [15].

3.1 Different types of polarization mechanism in dielectrics materials

The dielectric constant (ε_r) is the property of a material that can polarize in an electric field with an alternating current. Its expression is as follows (Eq. 2):

$$\varepsilon^* (f) = \varepsilon'(f) - i\, \varepsilon''(f) \quad (2)$$

The polarization response of the dielectric to an applied electric field can be represented by the real part (ε'_r) of the dielectric constant, which is related to the capacitance of the material. Energy dissipation such as heat or other forms of energy loss is represented by the imaginary part of the dielectric constant (ε''_r). The imaginary part is a result of the material's inability to produce real-time polarization when changing the oscillation frequency [17].

The total polarization of a dielectric is the sum of four sources of charge displacement: space charge (or interfacial) displacement $(\vec{P}_i)$, the orientation of permanent dipoles $\overrightarrow{(P_o)}$, ionic (atomic) displacement $\overrightarrow{(P_a)}$, and electronic displacement $\left(\vec{P}_e\right)$ [18]:

$$\vec{P} = \overrightarrow{P_i} + \vec{P}_o + \vec{P}_a + \vec{P}_e \qquad (3)$$

Figure 4 shows the contribution of space charge, dipole orientation, ions, and electrons to the overall polarization mechanism in dielectric materials.

The frequency dependence of polarization means that the dielectric constant must also depend on the frequency, as shown in **Figure 4**. Each polarization mechanism has a characteristic frequency that is limited. The frequency range for space-charge (-interface-charge) polarization is generally between 0 and 10^2 Hz. The accumulation of charge carriers is the name for interfacial polarization, which is a term used to describe Maxwell-Wagner Sillars interfacial polarization [19]. Dipolar polarization occurs when a material initially possesses dipoles that are randomly oriented with an overall dipole moment of zero. Under the effect of an applied electric field, the dipoles orient according to the field. This results in the appearance of a non-zero dipole moment. This polarization is strongly influenced by temperature and frequency [20].

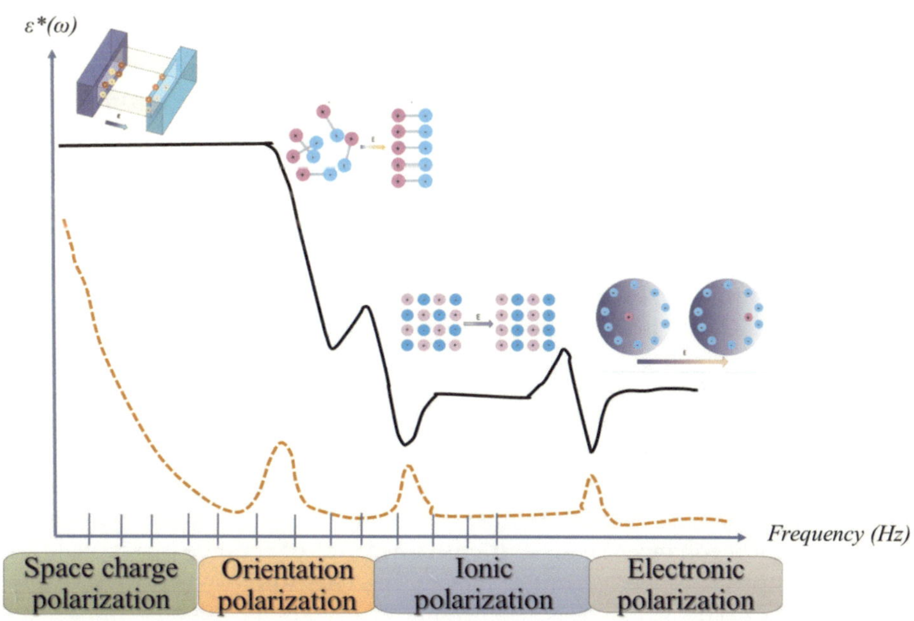

Figure 4.
Complex dielectric constant as a function of frequency and types of polarization mechanism in dielectric materials [19].

Ionic polarization is the physical transport of ions (positive and negative) in materials. The cations and anions become attracted to opposite directions when an electric field is applied. A relatively large ionic displacement is created, which can result in high dielectric constants in ceramics commonly used in capacitors [21]. The electronic polarization occurs in all atoms when an electric field is applied. The occurrence of electronic polarizations is caused by the distorted negative electrons and positive nuclei in an atom, which are in opposition to the external electric field, resulting in electric dipole moments, which occur at infrared frequencies. Electrons can follow high-frequency fields up through the optical range because of their small mass [22].

3.2 Polarization and relaxation mechanisms in dielectric materials $GdCa_2Cu_3O_\delta$

By measuring the real and imaginary parts of the impedance, we can calculate the frequency-dependent permittivity:

$$\varepsilon'(f) = \frac{e}{2\pi f \varepsilon_0 s} \frac{-Z''(f)}{Z^{*2}} \tag{4}$$

$$\varepsilon''(f) = \frac{e}{2\pi f \varepsilon_0 s} \frac{Z'(f)}{Z^{*2}} \tag{5}$$

The relative permittivity and capacitance of $GdCa_2Cu_3O_\delta$ were analyzed by measuring the frequency (100–1.3 MHz) and the results are shown in **Figure 5(a, b)**.

Figure 5(a) shows that the low-frequency range in the dielectric $GdCa_2Cu_3O_\delta$ ceramic has an elevated permittivity plateau that is a result of the grain boundary (GB) contribution. The intrinsic bulk (grain) response of the studied compound is represented by the high-frequency plateau of the permittivity. At low frequencies, the data indicate that the permittivity from the grain boundary remains constant ($\varepsilon'_r = 10^4$) over a wide frequency range, but drops in the MHz region. Below 270 K, it is seen to have a signature of a second high-frequency plateau that $\varepsilon'_r = 10^2$) is visible in the high-frequency range. In the framework, the grain boundary and grain capacitance plateaus shown in the capacitance plot versus frequency (**Figure 5(b)**) are both only an approximation for C_{gb} ($C_{gb} = 10^{-11}$ F) and C_g ($C_g = 10^{-9}$ F) respectively. The dielectric behavior of the investigated ceramic is comparable to those reported in the CCTO compound [23, 24].

The reported results in **Figure 6** indicate that the dielectric characteristics of ceramic $GdCa_2Cu_3O_\delta$ are highly significant, with a low dielectric loss $tan\ \delta = 0.2$ at 1 KHz. Furthermore, we discovered the existence of $tan\ \delta$ relaxation moves to high-frequency values but is suppressed after 415 K. The density of the material and the grain boundary resistance are two factors that affect the dielectric loss value [25]. The ceramic's low dielectric loss, elevated dielectric constant (10^4), and stability over a wide temperature range make it a suitable candidate for various capacitor types [26]. Various parameters, like annealing temperature, substitution, and nanoparticle addition, can be used to enhance the dielectric performance of ceramic materials [27, 28].

At room temperature, both samples showed high permittivity (10^4) with low losses ($\tan\delta = 0.2$). This behavior is also present in a renowned high permittivity material, CaCuTiO [29]. The activation of dielectric relaxation is indicated by the dielectric loss peak which accompanies the decrease in relative permittivity. The frequency decreases result in the appearance of dielectric loss peaks (**Figure 7(a)**).

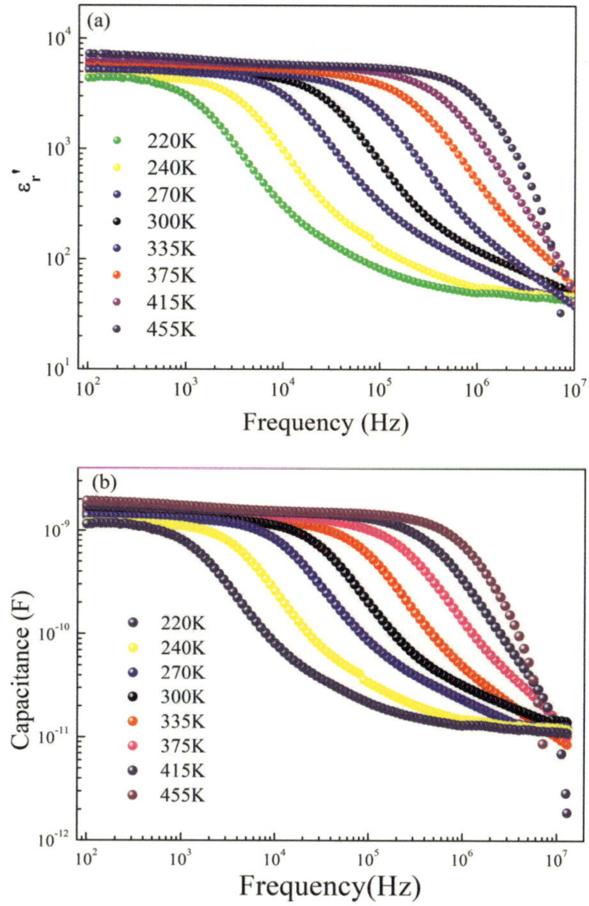

Figure 5.
Evolution of the, (a) dielectric permittivity (ε'), and capacitance (C) as a function of the frequency for the studied compound at various temperature values [11].

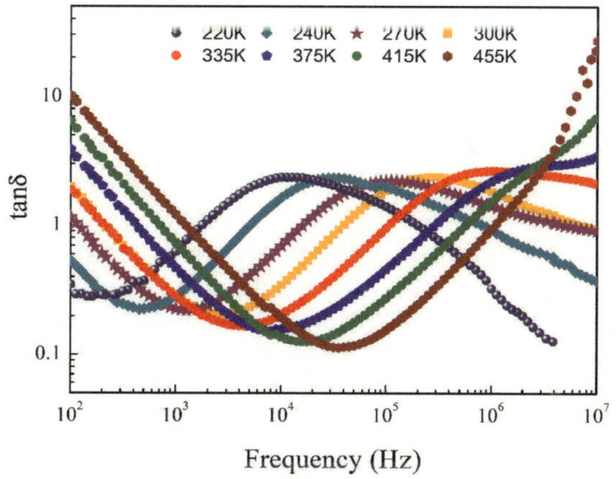

Figure 6.
Evolution of the loss tangent (tan δ) as a function of the frequency for the studied compound at various temperature values [11].

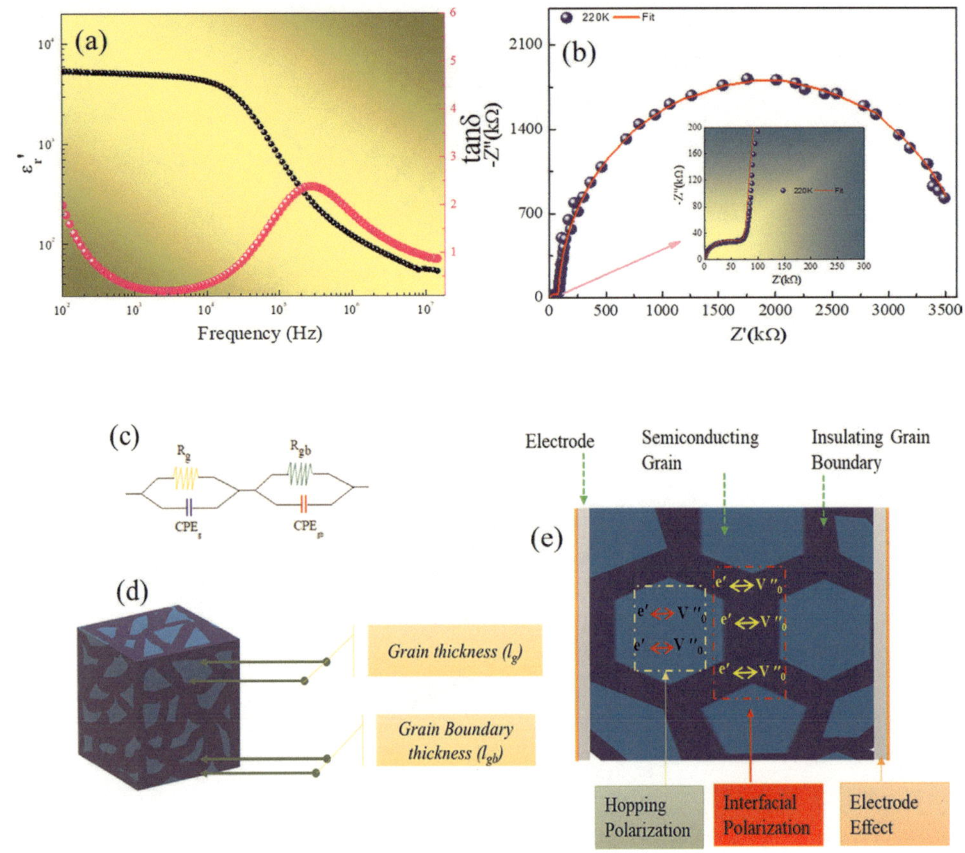

Figure 7.
(a) Frequency dependence of a dielectric constant and dielectric loss at room temperature, (b) complex impedance spectroscopy, (c) Idealized equivalent circuit model to account for GB and bulk, and (d, e) Internal barrier layer capacitor (IBLC) structure for $GdCa_2Cu_3O_\delta$ [12].

A heterogeneous microstructural system is responsible for these behaviors, which can be represented by a sequence of semiconductor grains and insulating grain boundaries. Impedance spectroscopy can define and characterize each of these contributions. According to the literature, the ceramic has an electrically heterogeneous microstructure that is caused by the significant variance in grain and boundary resistances (**Figure 7(b)**) [30]. The $GdCa_2Cu_3O_\delta$ ceramic's dielectric constant stability, order, and losses are controlled by the formation of internal barrier layer capacitance, which is the main responsible factor. It has been established that impedance spectroscopy detects different dielectric relaxation processes, including those found at the GB and bulk regions. In the simplest case, the brickwork layer model describes each grain's boundary or grain type contribution as a single RC circuit element that consists of a resistor and capacitor in parallel. The RC model works particularly well for insulators like dielectrics, where the capacitor describes the material's charge-storage capability and the parallel resistor describes the leakage current due to some untrapped carriers bypassing the ideal charge-storage element [31]. In reality, the dielectric relaxation processes in electrical ceramics are usually not ideal, and, therefore, experimental data cannot usually be modeled by ideal RC elements. To fit the data to an adequate equivalent circuit model, the non-ideality in a particular dielectric relaxation can be accounted for in the element by connecting a constant phase element (CPE) by replacing the ideal capacitor (**Figure 7(c)**) [32].

Generally, the variation of Z* as a function of frequency is given by the following relationship:

$$Z^*{}_{CPE} = \frac{1}{Q\ (j\omega)^{\alpha}} \tag{6}$$

The complex impedance depends strongly on the pseudo-capacitance Q and the exponent α (0 < α <1) [33]. For a pure capacitance α = 1 and a pure resistance α = 0 (**Figure 8**). It is possible to determine the phase angle (angle of depression) and capacitance (C) by analyzing the value of the exponent α.

$$\beta = (1 - \alpha)\frac{\pi}{2} \tag{7}$$

$$C = \left(Q.R^{1-\alpha}\right)^{\frac{1}{\alpha}} \tag{8}$$

The high relative permittivity of this material is associated with an effective internal insulating barrier or IBLC model, resulting in strong interfacial polarization [34]. Therefore, the high permittivity levels in the tested compound are related to the average area of the GB layer, the thickness of the GB layer, and the charge carriers that have accumulated at the GBs. Based on the IBLC dielectric model, conducting crystalline regions, insulating barriers, and grains/sub-grains effects are all factors that affect the dielectric loss of the ceramic being studied. This phenomenon is primarily attributed to various mechanisms of polarization and hopping (**Figure 7(d, e)**). The ceramic has a complex microstructure where the grain boundaries play a crucial role. The interfaces between grains can form capacitor-like structures due to the accumulation of charges, contributing to the overall dielectric constant. Based on the IBLC model, total relative permittivity (ε_r) is related to grain boundary permittivity εjg, grain thickness (l_g), and grain boundary thickness (l_{gb}) by the following relationship [35]:

$$\varepsilon_r = \varepsilon_{gb}\frac{l_g}{l_{gb}} \tag{9}$$

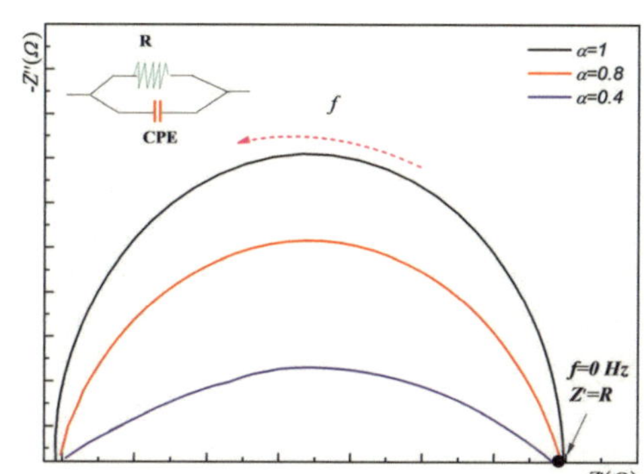

Figure 8.
Nyquist plot with different values of the exponent α (0 < α <1) [33].

According to the above relationship, the relative permittivity of a ceramic can be improved by reducing the thickness of grain boundaries and increasing grain size. In this case, Maxwell-Wagner polarization can significantly enhance the dielectric response (the high dielectric constant and low loss tanδ).

Other research groups have attributed the colossal dielectric response to the effect of hopping polarization. In this case, free electrons cause a distribution of space charges (oxygen vacancies $V_0^{\circ\circ}$ and electrons) at the grain boundaries [36]. In our case, such conduction may be governed by electron transport based on mixed valence Cu^{2+}/Cu^{3+} (p-type) [37]. The following relationships can express the process of Cu^{3+} reduction to Cu^{2+} and the appearance of oxygen vacancies in the compound $GdCa_2Cu_3O_\delta$ [38, 39].

$$O_0^x \to V_0^{\circ\circ} + e' + 1/2\ O_2\uparrow \tag{10}$$

$$Cu^{3+} + 1e' \to Cu^{2+} \tag{11}$$

The dielectric constant of ceramic can fluctuate due to changes in Cu density, different oxidation states because of the Cu element, and particle size variations.

3.3 Analytical models of uniform dielectrics

To characterize the properties of dielectric materials, the Debye equation is one of the most classical models, which explains relaxation phenomena in ceramics well. In the ideal case, the dipoles have the same relaxation time τ. These dipoles do not interact with each other.

The complex permittivity (ε*) of the Debye function can be expressed as follows [40].

$$\varepsilon^* = \varepsilon_\infty + \frac{\varepsilon_s - \varepsilon_\infty}{1 + i\omega\tau} \tag{12}$$

where ε_∞ is the dielectric constant at high frequencies, ε_s is the static dielectric constant at low frequencies, ω is the angular frequency, and τ is the relaxation time.

For the Debye model case, the Nyquist diagram corresponds to a semicircle with a maximum loss factor ε'_{max} at the top of the circle obtained for $\omega\tau = 1$. In this case, ε'_{max} is related to ε_s and ε_∞ by the expression as follows:

$$\varepsilon''_{max} = \frac{\varepsilon_s - \varepsilon_\infty}{2} \tag{13}$$

ε_s and ε_∞ represent the intersections of the circle with the x-axis (**Figure 9(a)**). The frequency spectrum obtains the dielectric constant results according to the Debye equation (**Table 1**), as demonstrated in **Figure 9(b)**. A Debye-type dielectric relaxation is characterized by a steady ε'_r value that then drops rapidly, while ε''_r shows a broad peak that passes through a maximum at relaxation frequency f_r (**Figure 9(b)**).

Generally, in a Debye system, while in Maxwell- Wagner system, at low-frequency region. The origin of heterogeneity-generated Maxwell-Wagner relaxations is likely due to the surface's roughness in a Schottky-diode mechanism or from the distribution of grain and grain boundary sizes in ceramic compounds [41]. Furthermore, the omnipresent hopping conductivity commonly leads to the typical power-law

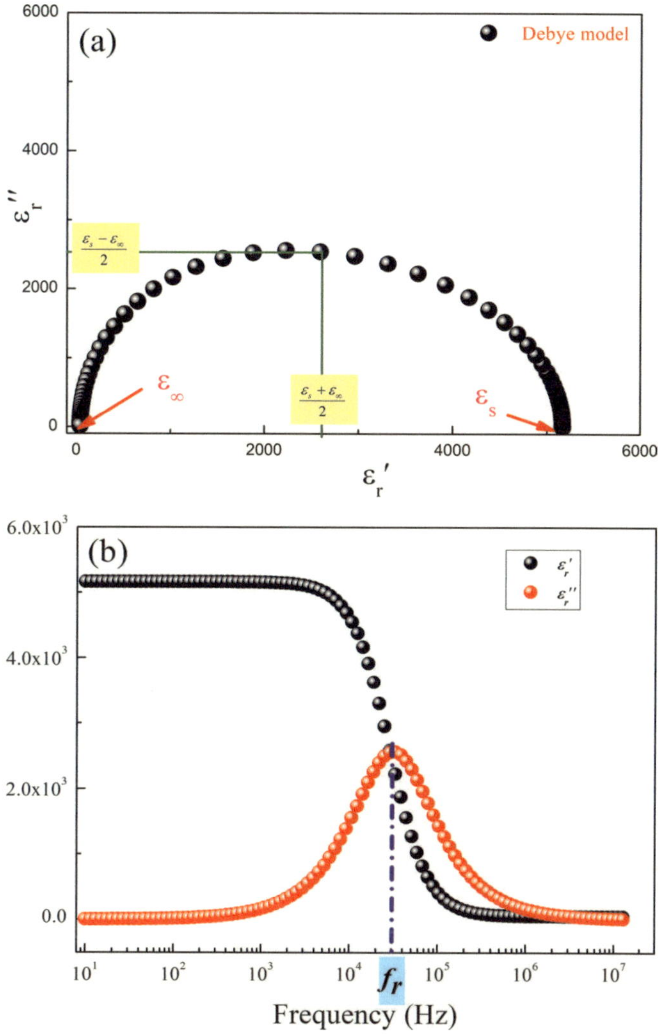

Figure 9.
(a) Evolution of ε'' as a function of ε', and (b) variation of ε' and ε'' versus frequency at room temperature [40].

Modèle	ε'	ε''
Modèle de Debye	$\varepsilon' = \varepsilon_\infty + \frac{(\varepsilon_s - \varepsilon_\infty)}{1+\omega^2\tau^2}$	$\varepsilon'' = \frac{(\varepsilon_s - \varepsilon_\infty)\omega\tau}{1+\omega^2\tau^2}$
Modèle Cole-Cole	$\varepsilon' = \varepsilon_\infty + (\varepsilon_s - \varepsilon_\infty)\frac{1+(\omega\tau)^{1-\alpha} sin\left(\frac{\pi\alpha}{2}\right)}{1+2(\omega\tau)^{1-\alpha} sin\left(\frac{\pi\alpha}{2}\right)+(\omega\tau)^{2-2\alpha}}$	$\varepsilon'' = (\varepsilon_s - \varepsilon_\infty)\frac{1+(\omega\tau)^{1-\alpha} cos\left(\frac{\pi\alpha}{2}\right)}{1+2(\omega\tau)^{1-\alpha} sin\left(\frac{\pi\alpha}{2}\right)+(\omega\tau)^{2-2\alpha}}$
Modèle de Havriliak Negami	$\varepsilon' = \varepsilon_\infty + (\varepsilon_s - \varepsilon_\infty)\frac{cos\,(\beta\varphi)}{\left[1+2(\omega\tau)^{\alpha} cos\left(\frac{\pi\alpha}{2}\right)+(\omega\tau)^{2\alpha}\right]^{\frac{\beta}{2}}}$	$\varepsilon'' = (\varepsilon_s - \varepsilon_\infty)\frac{sin\,(-\beta\varphi)}{\left[1+2(\omega\tau)^{\alpha} cos\left(\frac{\pi\alpha}{2}\right)+(\omega\tau)^{2\alpha}\right]^{\frac{\beta}{2}}}$

Table 1.
Variation of ε' and ε'' as a function of temperature using different models.

frequency dependences of the UDR. The Debye equation describes the phenomenon of relaxation for a pure dielectric. To reflect this mixed behavior, the contribution of the static conductivity σ_{dc} can then be inserted into the Debye equation, which becomes the modified Debye equation:

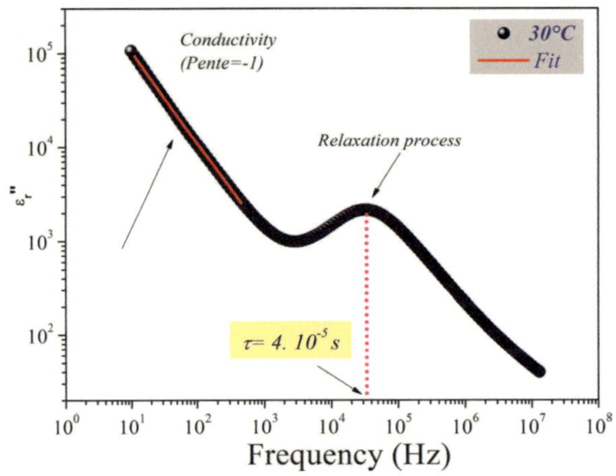

Figure 10.
The imaginary part of the permittivity as a function of frequency at room temperature [12].

$$\varepsilon^* = \varepsilon_\infty + \frac{\varepsilon_s - \varepsilon_\infty}{1 + i\omega\tau} - i\frac{\sigma_{dc}}{\omega\varepsilon_0} \tag{14}$$

The $\sigma_{dc}/(\varepsilon_0\omega)$ term in the loss expression corresponds to the contribution of DC conductivity (σ_{dc}).

The imaginary part of the permittivity versus frequency at room temperature as shown in **Figure 10**. In this system, the dielectric loss $\left(\varepsilon_r'' \ \alpha \ \frac{\sigma_{dc}}{f}\right)$ decreases linearly with increasing frequency. The movement of charge carriers within the sample becomes significant and results in an increase in dielectric losses.

The Debye relaxation model accurately defines the dielectric behavior of various gaseous and liquid materials that contain dipolar molecules [42]. However, most solids have a loss peak much broader and cannot be expressed by a single relaxation time, but instead by a distribution of relaxation times. Since the Debye model cannot correctly predict the dielectric response of numerous materials, several relaxation functions have been developed from the Debye model. For this reason, Cole-Cole, Davidson-Cole, and Havriliak Negami have proposed models to account for this relaxation time distribution [43]. Furthermore, empirical exponents have been used to increase the number of relaxation times. To fit the equivalent circuit to the ideal Debye behavior, a constant phase element (CPE) is suggested.

According to the fitting result in **Figure 11**, the dielectric response of our sample cannot be explained by Debye's model. The classical Debye equation does not fully explain the relaxation process in materials with multiple dispersions or possibly multiple defects. According to the Cole-Cole and Havriliak Negami models, the complex dielectric permittivity is given by the following relationship [44]:

$$\varepsilon* = \varepsilon_\infty + \frac{\varepsilon_s - \varepsilon_\infty}{\left(1 + (i\omega\tau)^\alpha\right)^\beta} \tag{15}$$

The factors α ($0 \leq \alpha \leq 1$) and β ($0 \leq \beta \leq 1$) control the distribution width of the relaxation time and the asymmetry of the loss peak, respectively. The case where $\beta = 1$ corresponds to the Cole-Cole model. The expression of ε' and ε'' as a function of temperature using different models is summarized in **Table 1**.

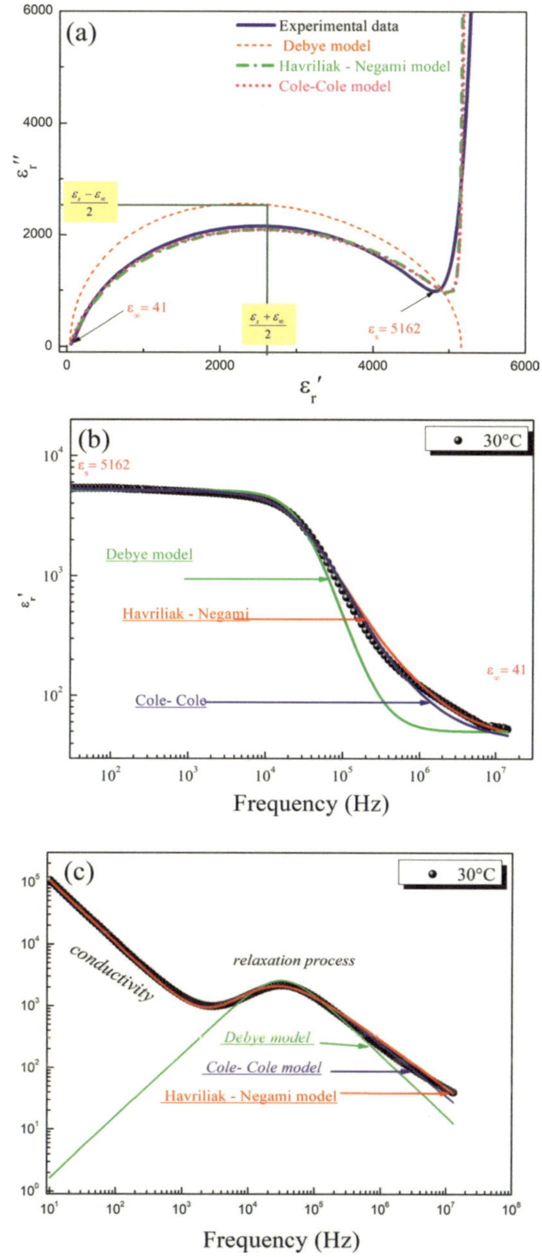

Figure 11.
(a) Nyquist plot of permittivity spectra, variation of (b) real part, and (c) imaginary part of permittivity at room temperature fitted with different models [11, 12].

The experimental and fitting (Debye, Cole-Cole, and Havriliak Negami) variation of ε'' as a function of ε', ε', and ε'' versus frequency is shown in **Figure 11(a)**. From **Figure 11(b, c)**, the fitting of the Cole-Cole model and the Havriliak Negami model is significantly better than the Debye model.

The fitting value of the characteristic parameters ε_s, ε_∞, τ, σ_{dc}, α, and β of the three models are indicated in **Table 2**. Therefore, the best model to characterize the

Model	ε_s	ε_∞	τ (s)	α	β	σ_{dc} (S.cm^{-1})
Debye	5171	41	3.10^{-5}	—	—	—
Cole-Cole	5165	41	3.10^{-5}	0.86	1	9.10^{-6}
Havriliak Negami	5162	41	4.10^{-5}	0.9	0.85	9.10^{-6}

Table 2.
Parameters ε_s, ε_∞, τ, σ_{DC}, α, and β of dielectric models.

dielectric relaxation of GdCaCuO is the Havriliak Negami model with characteristic parameters α = 0.9 and β = 0.85.

The ε_∞ generated by the dipole polarization is due to the fast oscillation of the alternating electric field, which results in the dipole following and the dipole remaining essentially in motion. The β factor is caused by one or more types of polarization.

4. Conduction mechanisms in dielectric materials

The microstructure of oxides significantly impacts their electrical transport properties and relaxation phenomena. The microstructure of perovskites frequently includes grains separated by grain boundaries. This morphology greatly influences the conductivity of perovskites and the activation of most conduction mechanisms. Ceramic structures have temperature-dependent conductivity variations influenced by various factors, including percolation thresholds, inter-grain or inter-particle distance, and the insulating layer or energy barrier. Conductive grains are arranged in particle chains separated by an insulating layer or barrier to ensure the transfer of charge carriers between conduction sites. Both factors that are characteristic of the insulating barrier have a profound effect on the electrical transport mechanisms. For perovskites, transport phenomena are also explained in terms of the resistance of grains (Rg) and grain boundary (Rgb).

Figure 12 depicts the variation of resistance (R) with temperature for grain and grain boundary. The graph shows that the resistance (R) for the sample decreases with temperature indicating the thermal activation of electrical conduction. The semiconductor nature of the synthesized ceramic was bolstered by the thermal decrease in electrical resistance.

Figure 12 indicates that the Rg resistance is usually lower than the R_{gb} resistance, which means that the conduction phenomena in the studied compound are mostly dominated by the intergranular domains [45].

Electrical conductivity is highly influenced by frequency and temperature variation in many materials. The double Jonscher law describes each conductivity spectrum in the present work [46].

$$\sigma(\omega, T) = \sigma_{dc}(T) + A(T)\ \omega^{s_1} + B(T)\ \omega^{s_2} \tag{16}$$

The term σ_{dc} explains the variation of DC electrical conductivity, A and B two constants. Two frequency exponents, s_1 and s_2 (in the range 0–1), describe the degree of interaction between mobile charges and provide information about the origin of electrical conduction.

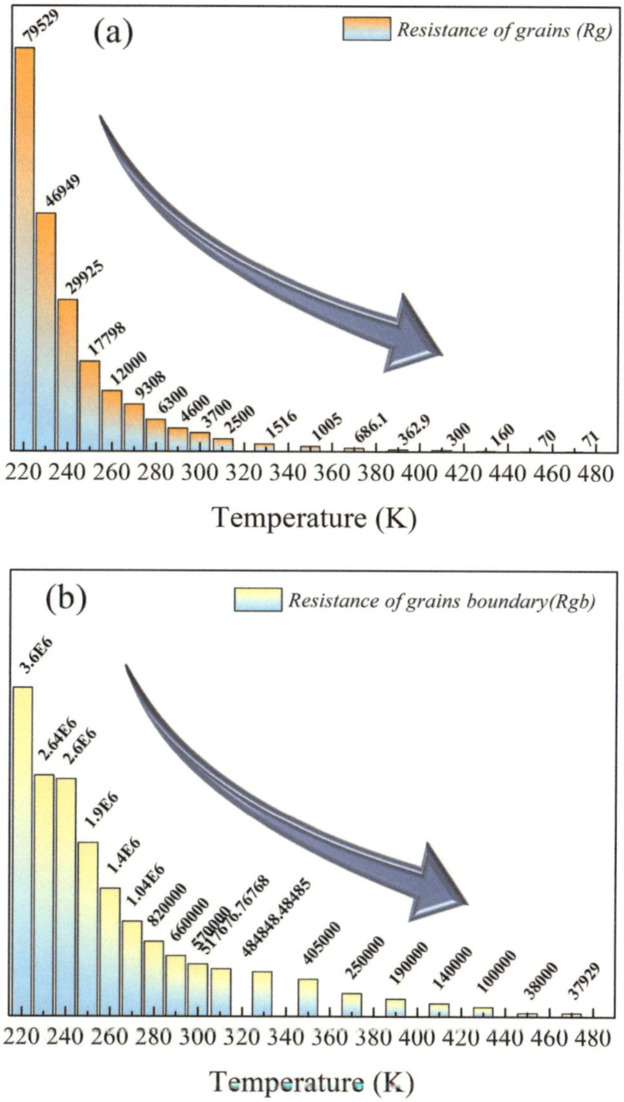

Figure 12.
Resistance of (a) grains (R_g), and (b) and grain boundaries (R_{gb}) as a function of temperature deduced from the equivalent circuit [12].

The conductivity (DC) and temperature dependences of the frequency exponents (s_1 and s_2) for the sample at the selected temperatures are depicted in **Figure 13**.

In terms of DC conductivity, Mott considered that the transport properties in dielectric materials are achieved by thermal activation of the self-caught small polarons into intermediate states [47]. The Small Polaron Hopping (SPH) mechanism is the appropriate model for describing DC conductance at high temperatures. In **Figure 14** (bottom and left), the evolution of ln ($\sigma_{dc} \times T$) is shown in relation to the inverse of the temperature (1000/T) for the material. The linear curve is able to approve the thermal activation of the SPH mechanism at elevated temperatures. The temperature dependence of DC conductivity can be described by the following relation at this temperature range.

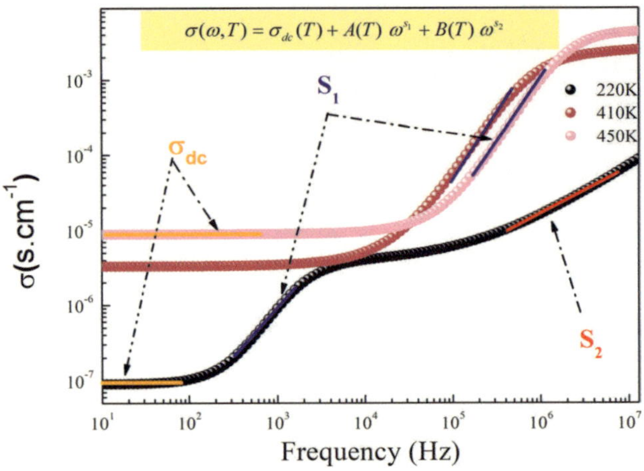

Figure 13.
Frequency dependences of the electrical conductivity (σ_{ac}) at selected temperatures [12].

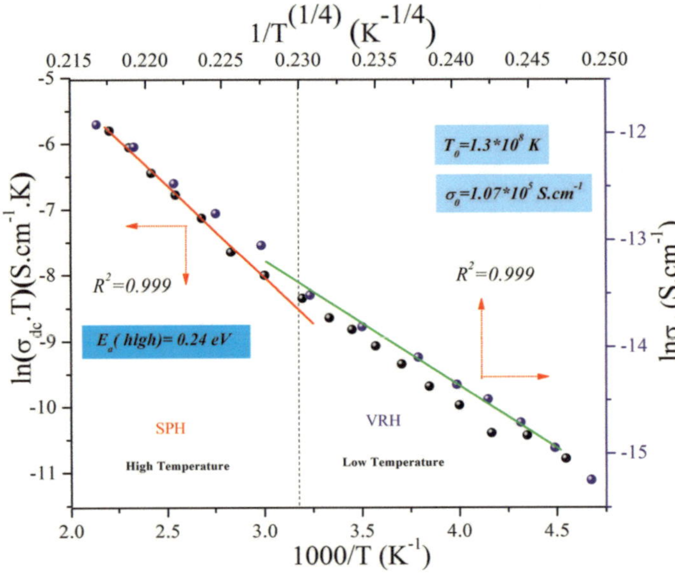

Figure 14.
The variation of $\ln(\sigma_{DC}.T)$ *as a function of the inverse of temperature* $(1000/T)$ *(bottom and left) and* $\ln(\sigma_{DC})$ *versus* $1/T^{(1/4)}$ *(top and right) [12].*

$$\sigma_{dc}(T) = \frac{\sigma_0}{T}.\exp\left(\frac{-(E_a)}{k_B.T}\right) \tag{17}$$

The parameter σ_0 is a pre-exponential factor that is independent of temperature. The activation energy $E_a = 0.24$ eV is required to move the charge carrier between two conducting sites. Dielectric relaxation occurs because oxygen vacancies have a lower activation energy than dipoles due to the same reorientation process [48].

At lower temperatures, the conduction mechanisms of perovskite are typically impacted. The linear variation shows that the VRH conduction mechanism dominates in the lower temperature range, as confirmed by the evolution of ln (σ_{dc}) as a function of $T^{-1/4}$ (**Figure 14** (top and right)). The VRH model defines the temperature dependence of electrical conductivity based on the Mott-VRH law [49].

$$\sigma_{dc} = \sigma_0 e^{-(T_0/T)^{1/4}} \tag{18}$$

The parameters $\sigma 0$ = be 1.07×10^7 $S.cm^{-1}$ and T_0 = 1.3×10^8 K have been determined.

Jonscher's power law can be used to describe the frequency-dependent dielectric response of localized carriers. The conductivity spectra for the studied compound were deduced from the dielectric plots using the universal dielectric response UDR model [50]:

$$\varepsilon' = \frac{\tan(s\pi/2)\sigma_0(T)}{\varepsilon_0} f^{s(T)-1} = A(T) f^{s(T)-1} \tag{19}$$

Therefore, plotting $f.\varepsilon r'$ as a function of frequency at a given temperature should result in a straight line with a slope of s_1 and s_2. In **Figure 15**, the lines show the fitting curves of experiments based on Eq. 19.

The thermal activation of the electrical transport mechanism is confirmed by the observed variation in the's1 and s2 values in **Figure 16**, which shows that the temperature activates it. It also proves that hopping and tunneling are the potential mechanisms that govern transport properties [51]. According to the linear fit, the values of exponents s_1 and s_2 are depicted in **Figure 16**.

The value of s_1 confirms that the Quantum Mechanical Tunneling (QMT) conduction process governs electrical conductivity [52]. In the QMT model, s_1 is independent of temperature, but rather dependent on frequency. The QMT model is founded on particle transport through a network of interconnected chains of particles or sites. The particles are separated by a thin layer that prevents them from having direct contact with each other (inset **Figure 16(a)**). The insulating layer's thinness allows for electron transfer from one conducting particle to another through tunneling.

The second frequency exponent (s_2) temperature dependence validates the influence of the correlated barrier hopping mechanism on the transport properties of the investigated compound. When the temperature increases, the frequency exponent

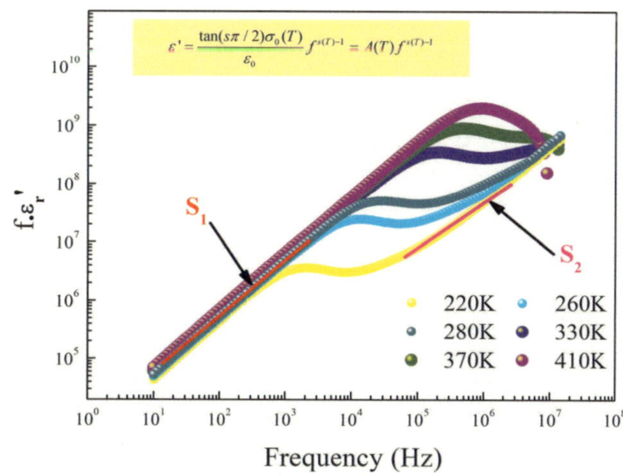

Figure 15.
Plot of ($f.\varepsilon_r'$) as a function of frequency at a given temperature for the $GdCa_2Cu_3O_8$ ceramic [12].

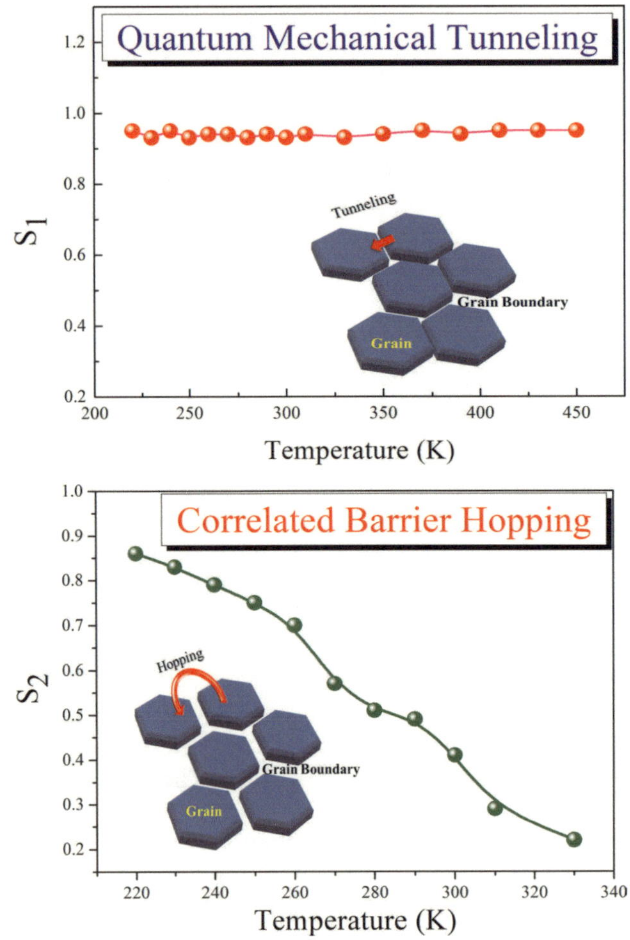

Figure 16.
Variation of the frequency exponents s_1 and s_2 with temperature for the ceramic of $GdCa_2Cu_3O_8$ [12].

decreases in the CBH model. This transport mechanism is influenced by frequency and temperature. The conduction is also supported by chains of particles or grains (inset **Figure 16(b)**). Specifically, electrons move between two conductive sites that are separated by distances, called inter-site or inter-particle distances, which are much larger than the tunnel barrier (tunnel characteristic distance).

5. Advanced ceramics for energy applications

Advanced ceramics play a crucial role in various energy applications due to their unique properties that can withstand extreme conditions such as high temperatures, corrosion, and wear. Advanced ceramics with additional functionality can greatly increase their potential impact on energy and environmental technologies. They are utilized in energy-related applications, including energy conversion, photovoltaics, energy storage, supercapacitors, lithium-ion, and capacitors (**Figure 17**) [53].

The discovery of a new ceramic ($GdCa_2Cu_3O_8$) has recently drawn much attention due to its extremely high dielectric constant ($\varepsilon' = 10^4$) and low loss ($\tan\delta = 0.2$).

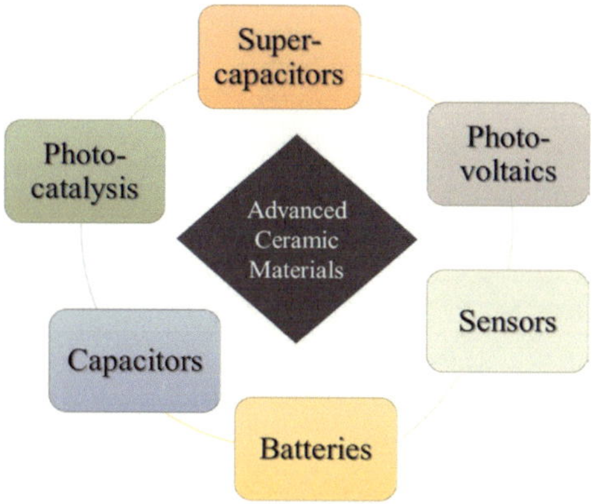

Figure 17.
Advanced ceramics materials application [53].

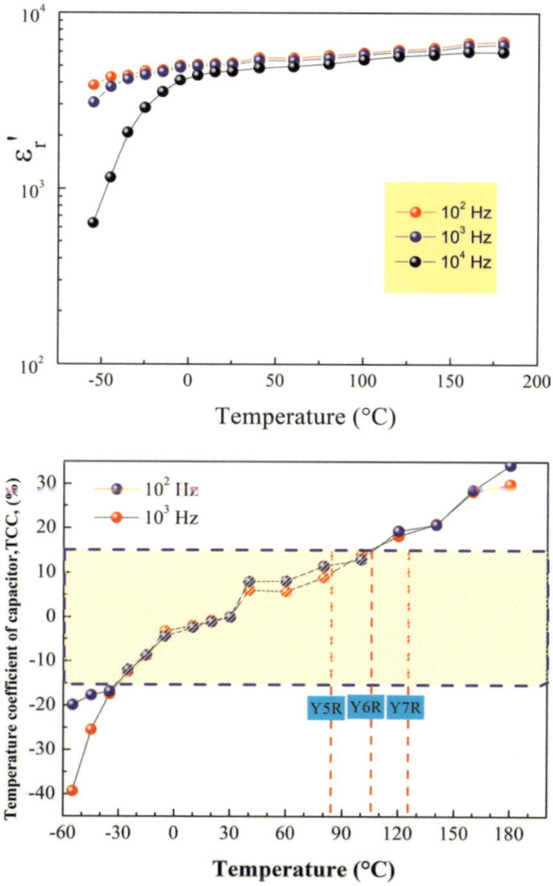

Figure 18.
Variation of ε′ and Δε′ as a function of temperature for $GdCa_2Cu_3O_8$ ceramic under various measuring frequencies [11].

Letter code **Low temperature**	**Number code** **Upper temperature**	**Letter code** **Change of capacitance** **Over the temperature range**
X = −55°C(−67°F)	4 = +65°C (+149°F)	P = ± 10%
Y = −30°C (−22°F)	5 = +85°C (185°F)	R = ± 15%
Z = +10°C (+50°F)	6 = +105°C (+221°F)	S = ± 22%
	7 = +125°C (257°F)	T = + 22/ −30%
	8 = 150°C (+302°F)	U = + 22/ −56%
	9 = +200°C (+392°F)	V = + 22/ −82%

Table 3.
The American Electronic Industries Alliance (EIA) Standards [55].

This material can potentially be utilized in various microelectronics device applications, including multilayer capacitors (MLCC) [54]. Furthermore, for each frequency value examined, the permittivity of the studied ceramic reveals a temperature-independent region with elevated values of ε_r' and excellent temperature stability. In addition to the high relative permittivity value (ε_r') with good temperature stability (**Figure 18(a)**) and a low dielectric loss factor (tanδ), the temperature coefficient of the dielectric constant ($\Delta\varepsilon'$) is a crucial factor to consider when analyzing the dielectric performance of commercial electronic devices. The relationship below was used to calculate the variation of $\Delta\varepsilon'(T)$ for the $GdCa_2Cu_3O_\delta$ ceramic at select frequency values.

$$\Delta\varepsilon' = \frac{\varepsilon'_T - \varepsilon'_{30}}{\varepsilon'_{30}} \times 100 \tag{20}$$

where ε_T' and ε_{30}' are ε' at T °C and 30°C, respectively.

Over a wide temperature range (−30–125°C), the $\Delta\varepsilon'(T)$ value of the $GdCa_2Cu_3O_\delta$ sample is less than ±15% (**Figure 18(b)**). Therefore, it is recommended that the $GdCa_2Cu_3O_\delta$ ceramic has potential use for the EIA code Y5R and Y6R capacitors. Y5R and Y6R capacitor codes have been designated by the American Electronic Industries Alliance (EIA) Standards (**Table 3**). 'Y' specifies a minimum operating temperature of – 30°C, whereas the '5' and '6' specify maximum temperatures of 105°C and 125°C, respectively [55].

6. Conclusion

To conclude, studying the correlation between polaron conduction and colossal permittivity in $GdCa_2Cu_3O_\delta$ ($\varepsilon' = 10^4$) has provided valuable insights into the electrical properties of the material. We have established a clear link between these two phenomena using a modified equivalent circuit based on the internal barrier layer (IBL) model. This correlation highlights the complex interaction between the structure and the dielectric properties.

To comprehend the dielectric polarization of $GdCa_2Cu_3O_\delta$, it is necessary to examine the various mechanisms that contribute to its high dielectric constant and its behavior at various temperatures. The defect chemistry of the compound can explain electronic heterogeneity in $GdCa_2Cu_3O_8$ ceramics, which consist of insulating GB and semiconducting grains regions. Two distinct defect mechanisms may explain the giant dielectric response. According to the Maxwell-Wagner behavior observed, the high dielectric constant is caused by enhanced interfacial polarization rather than relaxation polarization.

By using the distribution function, a hopping conduction model of localized electrons can successfully explain DC conductivity (by variations in Cu content). It has been proven that dielectric relaxation is strongly linked to the hopping electrons in the localized states. The small polaron hopping processes (SPH) at high temperatures and variable range hopping (VRH) processes at low temperatures were responsible for the electrical transport. The double Jonscher power law has been utilized to analyze the conductivity spectrum. The frequency exponent's variances are responsible for describing the AC electrical characteristics of the studied system by activating correlated barrier hopping (CBH) and quantum mechanical tunneling (QMT) conduction processes.

The findings contribute to both fundamental understanding and practical applications of $GdCa_2Cu_3O_8$ in advanced electronic devices, sensors, capacitors, and related technologies. Future research should continue to explore how these insights can be harnessed to optimize material design and enhance performance in various electrical and energy storage applications. By further refining our understanding of polaron dynamics and permittivity mechanisms in $GdCa_2Cu_3O_8$, we can pave the way for innovative solutions that leverage its unique electrical properties. Astonishingly, $GdCa_2Cu_3O_8$ ceramic specimens sintered at 950°C exhibited high ε_r' with temperature stability of $<\pm15\%$ over a wide range of −30–105°C. Therefore, this sample is proposed to be a promising material for the EIA code Y5R, and Y6R capacitors.

Acknowledgements

This work was supported by Faculty of Science of Bizerte, University of Carthage, Tunisia. Laboratory of Physics of Materials: Structure and Property.

Conflict of interest

The authors whose names are listed immediately below certify that they have no affiliations with or involvement in any organization or entity with any financial interest in the subject matter or materials discussed in this manuscript.

Author details

Khouloud Moualhi* and Mouldi Zouaoui*
Faculty of Science of Bizerte, Laboratory of Physics of Materials, Structure and Property LR01ES15, University of Carthage, Tunisia

*Address all correspondence to: khouloud.moualhi@fsb.u-carthage.tn and zouaoui.mouldi@fsb.u-carthage.tn

References

[1] Tabares FL, Junkar I. Cold plasma systems and their application in surface treatments for medicine. Molecules. 2021;**26**:1903. DOI: 10.3390/molecules26071903

[2] Autret-Lambert C, Merad S, De Almeida-Didry S, Roger S, Gervais F. Effect of MnO on the dielectric properties of Nb-$SrTiO_3$ ceramics for capacitor applications. ACS Applied Engineering Materials. 2023;**1**:68-79. DOI: 10.1021/acsaenm.2c00013

[3] Hao H, Liu M, Liu H, Zhang S, Shu X, Wang T, et al. Design, fabrication and dielectric properties in core-double shell $BaTiO_3$-based ceramics for MLCC application. RSC Advances. 2015;**5**: 8868-8876. DOI: 10.1039/C4RA13367F

[4] Pecharroman C, Esteban-Betegón F, BartolomØ J, López-Esteban S, Moya J. New percolative $BaTiO_3$-Ni composites with a high and frequency-independent dielectric constant ($\epsilon r \approx 80000$). Advanced Materials. 2001;**13**:1541. DOI: 10.1002/1521-4095

[5] Shirakawa H. Discovery of conductive polymers and conductive mechanism. Journal of Chemistry and Education. The Chemical Society of Japan. 2019;**67**:82-85. DOI: 10.20665/kakyoshi.67.2_82

[6] Almeida-Didry SD, Merad S, Autret-Lambert C, Nomel MM, Lucas A, Gervais F. A core-shell synthesis of $CaCu_3Ti_4O_{12}$ (CCTO) ceramics showing colossal permittivity and low electric losses for application in capacitors. Solid State Sciences. 2020;**109**:106431. DOI: 10.1016/j.solidstatesciences.2020.106431

[7] Feng L, Denghui J, Yinghao Z, Yumei T, Hongyu Y, Zhimin L. Achieving large strain and low hysteresis in $BiFeO_3$-$BaTiO_3$-based piezoelectric ceramics. Ceramics International. 2024;**50**:27846-27855. DOI: 10.1016/j.ceramint.2024.05.081

[8] Nickolay G, Studenikin SA, Michael C. Properties of dielectric $BaTiO_3$ thin films prepared by spray pyrolysis. Chemistry of Materials. 1998;**10**:2000-2005. DOI: 10.1021/cm980153

[9] Shi JB, Hsu Y, Lin CT. Dielectric properties of Gd_2CuO_4. Physica C: Superconductivity. 1998;**299**:272-278. DOI: 10.1016/S0921-4534(98)00038-0

[10] Ilyas J, Fatima C, Wissal B, El Hassan Y, Amine B, Soufian EL B, et al. Synthesis and characterization of new doped dielectric materials based on $CaCu_3Ti_4O_{12}$ (CCTO) applied at high temperature. Journal of Solid State Chemistry. 2024;**331**:124506. DOI: 10.1016/j.jssc.2023.124506

[11] Moualhi K, Chouk W, Moualhi Y, Othmani A, Zouaoui M. Novel high intrinsic permittivity in new perovskite ceramic $GdCa_2Cu_3O_8$. Materials Chemistry and Physics. 2023;**312**:128663. DOI: 10.1016/j.matchemphys.2023.128663

[12] Moualhi K, Moualhi Y, Zouaoui M. Investigation of conduction mechanisms and permittivity–conductivity correlation in a Gd-based perovskite structure. RSC Advances. 2024;**14**: 4142-4152. DOI: 10.1039/D3RA08703D

[13] Mazlan MR, Jamadon NH, Rajabi A, Sulong AB, Intan FM, Farazila Y, et al. Necking mechanism under various sintering process parameters–a review. Journal of Materials Research and Technology. 2023;**23**:2189-2201. DOI: 10.1016/j.jmrt.2023.01.013

[14] Elbashir S, Broström M, Skoglund N. Thermodynamic modelling assisted three-stage solid state synthesis of high purity β-$Ca_3(PO_4)_2$. Materials and Design. 2024;**238**:112679. DOI: 10.1016/j.matdes.2024.112679

[15] Stefano G, Thompson VP, Ferencz JL, Nelson RFA, Estevam BA. A new classification system for all-ceramic and ceramic-like restorative materials. International Journal of Prosthodontics. 2015;**28**:227. DOI: 10.11607/ijp.4244

[16] Ballou G. Resistors, capacitors, and inductors. In: Ballou GM, editor. Handbook for Sound Engineers. Routledge; 2013. pp. 253-284. DOI: 9780080927619

[17] Cyriac J, Kalarikkal N, Mathew S, Augustine S. Rare earth doped $BiFeO_3$ multiferroic system: Optical, dielectric, and magnetoelectric coupling properties and applications. In: Thomas S, Kalarikkal N, Abraham AR, editors. Applications of Multifunctional Nanomaterials. Elsevier; 2023. pp. 43-55. DOI: 10.1016/B978-0-12-820557-0.00001-1

[18] Kim MP, Um DS, Shin YE, Ko H. High-performance triboelectric devices via dielectric polarization: A review. Nanoscale Research Letters. 2021;**16**:35. DOI: 10.1186/s11671-021-03492-4

[19] Du G, Wang J, Liu Y, Yuan J, Liu J, Cai C, et al. Fabrication of advanced cellulosic triboelectric materials via dielectric modulation. Advanced Science. 2023;**10**:2206243. DOI: 10.1002/advs.202206243

[20] Wu M, Bichler KJ, Jakobi B, Schneider GJ. On the Determination of the Transition to Pure Reptation by Dielectric Spectroscopy. arXiv preprint arXiv:2104.12841. 2021. DOI: 10.48550/arXiv.2104.12841

[21] Liu S, Shen B, Zhai HHJ. Glass-ceramic dielectric materials with high energy density and ultra-fast discharge speed for high power energy storage applications. Journal of Materials Chemistry C. 2019;**7**:15118-15135. DOI: 10.1039/C9TC05253D

[22] Gong G, Ding J, Wang C, Zhang Y, Guo Y, Song K, et al. Defect-induced dipole polarization engineering of electromagnetic wave absorbers: Insights and perspectives. Composites Part B: Engineering. 2023;**252**:110479. DOI: 10.1016/j.compositesb.2022.110479

[23] Liu J, Smith RW, Mei WN. Synthesis of the giant dielectric constant material $CaCu_3Ti_4O_{12}$ by wet-chemistry methods. Chemistry of Materials. 2007;**19**:6020-6024. DOI: 10.1021/cm0716553

[24] Schmidt R, Sinclair DC. Anomalous increase of dielectric permittivity in Sr-doped CCTO ceramics $Ca_{1-x}Sr_xCu_3Ti_4O_{12}$. Chemistry of Materials. 2010;**22**:6-8. DOI: 10.1021/cm903220z

[25] Mao P, Wang J, Xiao P, Zhang L, Kang F, Gong H. Colossal dielectric response and relaxation behavior in novel system of Zr^{4+} and Nb^{5+} co-substituted $CaCu_3Ti_4O_{12}$ ceramics. Ceramics International. 2021;**47**:111-120. DOI: 10.1016/j.ceramint.2020.08.113

[26] Palneedi H, Peddigari M, Hwang GT, Jeong DT, Ryu J. High-performance dielectric ceramic films for energy storage capacitors: Progress and outlook. Advanced Functional Materials. 2018;**28**:1803665. DOI: 10.1002/adfm.201803665

[27] Elliott SR, A. c. Conduction in amorphous chalcogenide and pnictide semiconductors. Advances in Physics. 1987;**36**:135-217. DOI: 10.1080/00018738700101971

[28] Ghosh A, Sural M. Conductivity spectra of sodium fluorozirconate glasses. The Journal of Chemical Physics. 2001;**114**:3243-3247. DOI: 10.1063/1.1343073

[29] Singh L, Rai US, Mandal KD, Singh NB. Progress in the growth of $CaCu_3Ti_4O_{12}$ and related functional dielectric perovskites. Progress in Crystal Growth and Characterization of Materials. 2014;**60**:15-62. DOI: 10.1016/j.pcrysgrow.2014.04.001

[30] Liu L, Fan H, Fang P, Jin. Electrical heterogeneity in $CaCu_3Ti_4O_{12}$ ceramics fabricated by sol–gel method. Solid State Communications. 2007;**142**:573-576. DOI: 10.1016/j.ssc.2007.04.005

[31] Orazem ME, Frateur I, Tribollet B, Vivier V, Marcelin S, Pébère N, et al. Dielectric properties of materials showing constant-phase element (CPE) impedance response. Journal of the Electrochemical Society. 2013;**160**:215. DOI: 10.1149/2.033306jes

[32] Abram EJ, Sinclair DC, West AR. A strategy for analysis and modelling of impedance spectroscopy data of Electroceramics: Doped lanthanum gallate. Journal of Electroceramics. 2003;**10**:165. DOI: 10.1023/B:JECR.0000011215.56084.87

[33] Balaguera EH, Romero B, Najafi M, Galagan Y. Analysis of light-enhanced capacitance dispersion in perovskite solar cells. Advanced Materials Interfaces. 2022;**9**:2102275. DOI: 10.1002/admi.202102275

[34] Wang Y, Jie W, Yang C, Wei X, Hao J. Colossal permittivity materials as superior dielectrics for diverse applications. Advanced Functional Materials. 2019;**29**:1808118. DOI: 10.1002/adfm.201808118

[35] Didry SDA, Autret C, Lucas A, Pacreau F, Gervais F. Comparison of colossal permittivity of $CaCu_3Ti_4O_{12}$ with commercial grain boundary barrier layer capacitor. Solid State Sciences. 2019;**96**:105943. DOI: 10.1016/j.solidstatesciences.2019.105943

[36] Luo XJ, Liu YS, Yang CP, Chen SS, Tang SL, Bärner K. Oxygen vacancy related defect dipoles in $CaCu_3Ti_4O_{12}$: Detected by electron paramagnetic resonance spectroscopy. Journal of the European Ceramic Society. 2015;**32**:2073-2081. DOI: 10.1016/j.jeurceramsoc.2015.01.024

[37] Li M, Feteira A, Sinclair DC, West AR. Influence of Mn doping on the semiconducting properties of $CaCu_3Ti_4O_{12}$ ceramics. Applied Physics Letters. 2006;**88**:23. DOI: 10.1063/1.2200732

[38] Li J, Sleight AW, Subramanian MA. Evidence for internal resistive barriers in a crystal of the giant dielectric constant material: $CaCu_3Ti_4O_{12}$. Solid State Communications. 2005;**135**:260-262. DOI: 10.1016/j.ssc.2005.04.028

[39] Feng L, Tang X, Yan Y, Chen X, Jiao Z, Cao G. Decrease of dielectric loss in $CaCu_3Ti_4O_{12}$ ceramics by La doping. Physica Status Solidi (a). 2006;**203**:R22-R24. DOI: 10.1002/pssa.200622038

[40] Cole KS, Cole RH. Dispersion and absorption in dielectrics I. Alternating current characteristics. The Journal of Chemical Physics. 1941;**9**:341-351. DOI: 10.1063/1.1750906

[41] Li W, Auciello O, Premnath RN, Kabius B. Giant dielectric constant dominated by Maxwell–Wagner relaxation in Al_2O_3/TiO_2 nanolaminates synthesized by atomic layer deposition. Applied Physics Letters. 2019;**96**:16. DOI: 10.1063/1.3413961

[42] Feldman Y, Puzenko A, Ryabov Y. Non-Debye dielectric relaxation in complex materials. Chemical Physics. 2002;**284**:139-168. DOI: 10.1016/S0301-0104(02)00545-1

[43] Woodward WHH. Broadband dielectric spectroscopy—A practical guide. In: Woodward WHH, editor. Broadband Dielectric Spectroscopy: A Modern Analytical Technique. American Chemical Society; 2021. pp. 3-59. DOI: 10.1021/bk-2021-1375.ch001

[44] Davidson DW, Cole RH. Dielectric relaxation in glycerol, propylene glycol, and n-propanol. The Journal of Chemical Physics. 1951;**19**:1484-1490. DOI: 10.1063/1.1748105

[45] Holt DB. The remote electron beam-induced current analysis of grain boundaries in semiconducting and semi-insulating materials. Scanning. 2000;**22**: 28-25. DOI: 10.1002/sca.4950220106

[46] Bruce P. High and low frequency Jonscher behavior of an ionically conducting glass. Solid State Ionics. 1985; **15**:247-251. DOI: 10.1016/0167-2738(85)90010-4

[47] Mott NF. Polarons. Materials Research Bulletin. 1978;**13**:1389-1394. DOI: 10.1016/0025-5408(78)90130-7

[48] Jonscher AK. Dielectric relaxation in solids. Journal of Physics D: Applied Physics. 1999;**32**:57. DOI: 10.1088/0022-3727/32/14/201

[49] Mott NF. Conduction in glasses containing transition metal ions. Journal of Non-Crystalline Solids. 1968;**1**:1-17. DOI: 10.1016/0022-3093(68)90002-1

[50] Bhunia AK, Pradhan SS, Bhunia K, Pradhan AK, Saha S. Study of the optical properties and frequency-dependent electrical modulus spectrum to the analysis of electric relaxation and conductivity effect in zinc oxide nanoparticles. Journal of Materials Science: Materials in Electronics. 2021; **32**:22561-22578. DOI: 10.1007/s10854-021-06742-4

[51] Elliott SR. A theory of ac conduction in chalcogenide glasses. Philosophical Magazine. 1977;**3**:1291-1304. DOI: 10.1080/14786437708238517

[52] Qasrawi A, Taleb M. Enhancement of electrical performance of ZnSe thin films via Au nanosandwiching. Materials Science-Poland. 2020;**38**:174-180. DOI: 10.2478/msp-2020-0009

[53] Belmonte M. Advanced ceramic materials for high temperature applications. Advanced Engineering Materials. 2006;**8**:693-703. DOI: 10.1002/adem.200500269

[54] Engel GF. Material requirements for power and high temperature multilayer ceramic capacitors (MLCC). Additional Papers and Presentations. 2015;**2015**: 000021-000028. DOI: 10.4071/CICMT-TA12

[55] Wang C, Cao W, Liang C, Zhao H, Cheng C, Huang S, et al. Ultrahigh energy-storage density of $BaTiO_3$-based ceramics via the interfacial polarization strategy. ACS Applied Materials and Interfaces. 2023;**15**:42774-42783. DOI: 10.1021/acsami.3c08168

Chapter 5

Color Effects in Dental Ceramic Laminate Veneers

Carlos Eduardo Francci, Samara Silva and Mylena Régis

Abstract

We live in the esthetic age, and dentistry is no exception. The pursuit of the perfect smile is a constant in today's society, and ceramic veneers are among the most commonly performed treatments. With the trend toward minimally invasive dentistry, these veneers are becoming more conservative, with thicknesses less than 0.5 mm. To perform such treatments, the ceramic restorative material must adhere to a dental substrate using a bonding agent. Each of these individual factors, which together form a restorative system, has different optical characteristics that will influence the final color of the restoration. For ceramics, the type, microstructure, thickness, translucency, and color need to be analyzed beforehand. For the dental substrate, saturation and tissue type (dentin or enamel) will have an impact; and for the bonding agent, its color, translucency/opacity, brand, thickness, internal composition, and color changes with aging will be important considerations. This chapter is aimed at clinicians who seek to make esthetic restorative planning for ceramic veneers more precise, meet patient expectations, and successfully navigate the challenges of complex cases. Here, we present the variables involving the optical and color effects that influence conservative ceramic restorations and their behavior in terms of longevity and esthetic maintenance. Enjoy!

Keywords: ceramic, color, resin cement, dental esthetics, dental bonding

1. Introduction

Esthetic dentistry is increasingly gaining ground in society. This is linked to its relationship with patients' quality of life, since it has implications for psychological factors, such as greater self-esteem, security in interpersonal relationships and satisfaction with each individual's self-image [1–5]. More recently, the "minimally invasive" approach has gained ground in dentistry, which seeks to refund esthetics and function, along with restoring the patient's self-esteem through less wear of the dental hard tissues, mimicking the dental structure through restorations adhered preferably to enamel, with tiny volumes/thickness [6–9].

Ceramics are the materials of choice for esthetic restorations, as they can mimic the natural tooth and promote great longevity, even in thin layers, rehabilitating the smile [7]. Among ceramics, lithium disilicate-based glass ceramics are an excellent

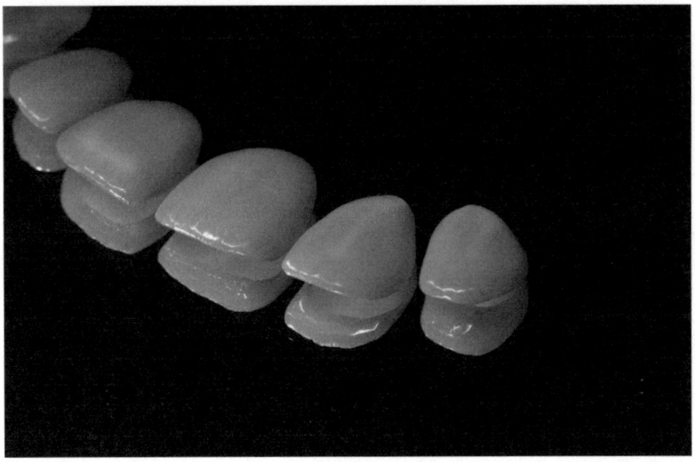

Figure 1.
Conservative lithium disilicate ceramic laminate veneers.

option for creating conservative laminates on anterior teeth, as in addition to favorable esthetics, they have good mechanical properties (**Figure 1**) [10, 11].

In restoration with ceramic laminates, due to their thinness and greater translucency, the effect of the luting material and substrate on the final color will be more impactful, which will require care and planning from the clinician to perform the case accurately [12–15]. Since resin cement is more influential in these cases, it can be used to help mask the saturated substrate or neutralize the color difference between substrates, when in high opacity, as it will serve as a background blocker, similar to opacifiers in composite resin restorations [13, 14, 16–19]. In this way, the masking effect associated with opaque material is achieved, but the natural esthetics are maintained, considering the use of a ceramic that is not completely opaque, but rather allows translucency and a depth effect of light, such as dental enamel. This will reduce the effect of the ceramic on the final color of the restoration, which will reduce the thickness of the preparation and, consequently, of the ceramic piece, promoting the preservation of healthy tooth structure, long-lasting and stable adhesion because it is made of enamel, and not compromising pulp and gingival health [20].

In this sense, it is important to evaluate the change in optical properties to which the cementation line is subject, since it is the least mechanically resistant material among the restorative set [21–25]. Because it is formed by a low molecular weight polymer network, resin cements are susceptible to hydrolytic degradation, which will lead to opacification and yellowing of the material, especially noticeable in high-value chromatic resin cements and when associated with translucent ceramics [26, 27]. This fact may compromise the esthetic result in the long term, leading to the replacement of the restoration [28].

2. Esthetic dentistry: Impact on quality of life

Esthetics is deeply rooted in our society, often dictating our behavior, ways of thinking and how we relate to other individuals. Since ancient times, it has been sought after as a way of acquiring advantages for those considered beautiful. However, the esthetic standard is not unique, and has changed throughout history and in different societies and respective cultures. Present in all social segments, dentistry is no different. In recent years, esthetics has been gaining more space and notoriety within dentistry, due to the great search by patients for harmonious smiles that meet their esthetic expectations. There is no doubt that the smile plays a central role in facial esthetics, and it is a challenge for the clinician to deliver highly esthetic results, but that value correct function and that harmoniously relate the dental elements to each other and to the facial elements [1–3, 29].

In this sense, dentistry has a strong impact on people's quality of life, interfering with their self-esteem, the way they see themselves and their security in relationships with other individuals [1, 4, 5]. Therefore, a patient who is dissatisfied with the appearance of their smile and, therefore, stops smiling, is not only repressing a feeling, but is also impacting their health, since this is a combination of physical, mental and social factors and not just the absence of disease. This observation is supported by the definition of health by the World Health Organization (WHO), which, in 1946, defined health as a state of complete physical, mental and social well-being, not just as the absence of disease or illness. The search for a perfect smile and the solution to personal dissatisfaction makes many patients willing to undergo invasive, irreversible and painful procedures in the name of beauty. This is because solving a problem, changing the smile, considered by the patient can restore their self-esteem and personal confidence again, and the appreciation of their own smile [30].

3. Minimally invasive or sufficiently invasive esthetic dentistry: Advantages and challenges

No restorative material can surpass natural dental tissues. In this regard, a conservative approach has gained traction in dentistry, advocating for the maximum preservation of hard dental tissues, which cannot be regenerated once lost [6, 9, 31]. Viewing the dental element as an organ with a unique and complex internal system supports this minimally invasive ideology, where restorative dental procedures aim to remove only the necessary amount of mineralized tissue, ensuring that preparations remain in enamel. This approach promotes the conservation of dental tissue and preserves the vitality and health of the dentin/pulp complex [9]. Another advantage is improved adhesion and longevity of the restoration, which is a growing concern for dentists [7]. Adhesion to dentin, being a more heterogeneous tissue constantly moist due to the presence of dentinal tubules and associated tubular fluid, poses challenges for adhesion, making it more critical and less durable than adhesion to enamel [8].

However, this makes the planning and execution of esthetic restorative work more challenging, as the substrate and the cementing material in indirect restorations have a greater influence on the final color result, given the thinner ceramic layers. This thickness makes it more difficult to mask unfavorable, saturated, and darkened substrates. The concern for conserving as much healthy dental structure as possible does not preclude the need for prosthetic preparations when indicated, such as in cases of masking substrates and achieving the necessary thickness to accommodate the material, restoring both function and esthetics. Failure to perform preparations in cases like ceramic veneers, for example, can lead to overcontours, cervical steps, and periodontal issues that compromise patient health and case esthetics due to an inadvertent interpretation of this approach [32]. A study [33] analyzed conventional preparations and no-prep/minimally invasive veneers over a 9-year period and found higher survival rates for no-prep/minimally invasive veneers.

Another major concern for clinicians is achieving a natural appearance in their restorations, which should mimic the healthy dental element in terms of anatomical shape, color, and translucency. The choice of material is crucial in restorative esthetics to restore function and appearance in harmony with the biological aspects of teeth. Therefore, the selection of restorative material must be based on its optical and mechanical properties [31]. One such material for esthetic restorations with excellent mechanical properties is dental ceramics.

4. Evolution of esthetic indirect restorations

Fully ceramic restorations first appeared in the dental field around 1960. However, it was not until the 1980s, with the advent of acid-etchable porcelains, that they could be bonded to hard dental tissues, particularly enamel, due to the improved longevity of adhesion [11].

Initially, fully ceramic crowns were made with significant thickness to compensate for the absence of a metal coping, which enhanced the mechanical properties of the restoration. By increasing the thickness of the piece, the intrinsic strength of the ceramic material was improved. It was only with advances in adhesion knowledge and the development of acid-etchable ceramics, combined with high-strength ceramics such as lithium disilicate and zirconia, that it became possible to create more conservative ceramic restorations with thinner layers [34]. This reduces the likelihood of endodontic treatments and hypersensitivity in extensively prepared teeth compared to minimally invasive preparations, while preserving healthy dental structure [9].

The emergence of mechanically more resistant and adhesive all-ceramic restorations has reduced the use of metal-ceramic restorations. This is due to the advantage of using more translucent ceramics that better mimic natural teeth by allowing more light to pass through. Unlike metal materials, which require an opaque ceramic layer to mask their presence and necessitate greater preparation thickness to accommodate different layers of feldspathic ceramics (translucent, opaque, and effect) for esthetic purposes, modern ceramics provide a more effective solution [11, 12, 35]. Furthermore, with the development of ceramic restoration techniques, it is now possible to combine different types of ceramics in multi-layered pieces. This approach leverages the virtues of each ceramic material, typically with better mechanical properties in the infrastructure and superior esthetics in the final layer, unlike monolithic pieces [36].

Currently, due to the advent of ceramics suitable for conditioning, it is possible to create restorations that rely exclusively on adhesion without the need for geometric preparation and mechanical retention. This makes the fabrication of ceramic restorations, including crowns or onlays, more conservative compared to metallic and metal-ceramic restorations. This is particularly true when using resin cement for the cementation of ceramic pieces, which contributes to increased mechanical strength of the restoration, making the bonded piece more durable than the pure ceramic before cementation [37]. This perspective encourages clinicians to pursue increasingly minimally invasive preparations, minimizing damage to the tooth.

Lithium disilicate-based glass ceramics have intermediate properties between feldspathic porcelains and zirconias, offering excellent mechanical and esthetic properties, making them suitable for both posterior and anterior restorations. They are composed of an amorphous phase of crystalline quartz and lithium disilicate crystals in varying proportions, which determines different levels of translucency, ranging from high translucency to high opacity. The higher the proportion of crystals, the opaquer the material becomes [38]. This allows lithium disilicate to be used in highly esthetic conservative restorations, as it maintains good mechanical and optical properties even in thin layers [10]. Furthermore, its suitability for anterior esthetics is largely based on its various translucency levels, which can ensure natural-looking restorations that mimic healthy teeth. Literature indicates that one of the key aspects of naturalness in restoration is the translucency of the restorative material [39, 40]. Since dental enamel is highly translucent, the opaquer the dental material, the more noticeable the restoration will be. Conversely, materials with high translucency can more effectively mimic natural dental tissue and achieve an "invisible restoration" appearance. It is important to note that translucency is affected by the thickness and type of ceramic material, with greater thickness resulting in more light scattering, thus creating an opaque effect [41–43]. The reduction in translucency of lithium disilicate veneers is a crucial factor in masking saturated substrates and a key element in the final color of the restoration [16, 17].

5. Color effects in conservative indirect restorations

The optical properties of materials directly influence their esthetic indication and, consequently, the success of a harmonious and natural smile rehabilitation, particularly when biological limits are respected and natural tooth characteristics are used as guiding parameters. Mimicking the natural characteristics of a tooth to achieve esthetically satisfactory and imperceptible restorations, while also being functionally adequate, is not a simple task for the dentist. This is increasingly demanded by patients and requires not only technical skill from the professional but also a good understanding of the optical characteristics of teeth, such as morphology, texture, and color, as well as the properties of restorative materials suitable for each case and their ability to mimic natural dental structures [29, 31, 44–46].

These properties result from the interaction between incident light and the physical-chemical structure of the material. One such property is translucency, which reveals the relative amount of light that is transmitted and how much is diffusely reflected through a turbid medium [47, 48]. This optical phenomenon of light transmission and reflection is considered an intermediate state between transparent (total transmission) and opaque (total reflection), described as partial opacity [49, 50].

Reproducing the natural characteristics of dental elements, such as shape, color, and texture, in restorative work to achieve satisfactory esthetic results and mimic adjacent natural teeth is the clinician's goal and undoubtedly one of their greatest challenges [44–46]. This is primarily due to the complex color of a dental element, which is not monochromatic, having a more translucent enamel and more saturated dentin, variable according to the location and histo-anatomical arrangement of these tissues. This variability causes differences in saturation and luminosity in different regions of the same dental element [44]. Therefore, understanding the factors involved in an esthetic restoration is essential for better predictability of restorative work.

Many variables affect the final color of an indirect restoration, such as the color and translucency of the ceramic, substrate saturation, and the color and translucency of the resin cement [13, 14, 17, 19, 44, 51, 52]. Among these variables, the ceramic has the greatest influence on the restoration color, with thickness and translucency being key factors. Greater thickness and lower translucency reduce light transmission and reflection of the substrate and resin cement color, highlighting the characteristics of the ceramic material [12–15].

The choice of ceramic material type and translucency remains the most influential factor for thinner ceramic pieces, such as veneers. However, the color of the resin cement also significantly impacts these cases, especially as the ceramic veneer becomes thinner and more translucent [13–15, 53]. Therefore, careful attention is needed when choosing the resin cement for bonding conservative ceramic veneers, focusing on translucency and value, which affect the restoration more than chroma and hue [53]. In this regard, resin cement can aid in masking unfavorable substrates, as shown in studies [16, 17, 19], which observed that combining less translucent veneers with more opaque resin cements positively contributes to this process and maintains the restoration's esthetics.

However, maintaining the color obtained immediately after restoration over time is a challenging factor for long-term esthetic success [27, 54]. Additionally, literature indicates that restorations with ceramic veneers cemented with resin cement show some pigmentation with aging, which is important for the long-term esthetics of the restoration [55–59].

5.1 Ceramics

Understanding the optical characteristics of indirect restorative materials used in clinical practice is essential for the dentist. This knowledge enables the precise selection of the most suitable material for each clinical situation, leading to more accurate color outcomes [13–15, 53].

Glass ceramics are highly recommended for anterior esthetic work, such as veneers, due to their adequate translucency, though they have a lower masking ability [48]. The translucency of an object directly influences the material's masking capability, as it affects the amount of light absorption, reflection, and transmission, and consequently, how much of the underlying structure is visible [47, 48]. Small variations in thickness in more translucent ceramics are more noticeable compared to opaque ceramics, which also makes the substrate more easily perceived and influences the color of the restoration [48, 53, 60].

Translucency in dental ceramics is significantly dependent on the ceramic composition, including the size and number of crystals, pigments, and the size, quantity, and distribution of internal defects such as porosities and microcracks [60, 61]. The greater the difference in refractive indices between the phases of a ceramic, the lighter scattering occurs between particles, resulting in lower light transmission and higher opacity [47, 60–62]. When the crystals are smaller than the visible light spectrum (400 to 740 nm), the material is transparent. However, larger crystal diameters increase light scattering, making the restoration opaquer. For example, the translucency difference between the high translucency (HT) and Low Translucency (LT) ceramics from the IPS e.max CAD line is due to HT having fewer large-diameter crystals, while LT has more of these larger crystals, resulting in HT having a more apparent glassy phase and less light scattering, making it more translucent than LT [47, 61, 62].

In addition to composition, translucency in ceramics is inversely affected by the thickness of the piece, for the same type and brand of ceramic [62, 63]. The type of ceramic and its thickness (intrinsic and extrinsic translucencies) affect the amount of light absorption, reflection, and transmission, and consequently, how much of the underlying structure is perceived [13, 47, 48, 60]. Furthermore, surface alterations caused by abrasive procedures can change the surface texture of the ceramic material, affecting its translucency [47, 61, 62].

The stability of these optical characteristics in restorations is crucial for maintaining the esthetics of the smile. Ceramics are more resistant to the chemical challenges of the oral environment compared to resin composites, suffering less from color and translucency changes [21, 28, 64, 65]. According to the study by Subasi et al. [64], which evaluated different thicknesses of ceramic veneers (0.5, 0.7, 1 mm) and types of ceramics subjected to aging, the type and thickness of veneers were more significant in altering color and translucency than aging itself, remaining stable over time. This contrasts with resin materials and even the dental substrate, which undergoes mineral content changes, increasing translucency [47, 57, 58, 66–68]. The composition of the ceramic material is more significant in determining its optical properties than thickness and the technique used to fabricate the ceramic piece (milled, pressed, and layered) [65].

Thus, a significant portion of color alteration in a ceramic restoration set comes from the dental substrate and the underlying resin cement, which undergoes water absorption and pigmentation during the restoration's aging process, potentially affecting the esthetic outcome [58, 64, 69].

5.2 Substrate

The substrate will also significantly influence the final color of a conservative ceramic veneer restoration and will be one of the most challenging factors for the clinician, as it is intrinsic to the case and cannot be completely altered. For cases involving preparations for ceramic veneers on discolored teeth from trauma, endodontically treated teeth, or teeth with intrinsically altered formation processes, masking these substrates to achieve satisfactory esthetics can be challenging. Additionally, when covering multiple adjacent teeth with varying saturation levels with ceramic veneers, the dentist faces a major challenge in neutralizing the effect of these substrates conservatively while maintaining the requested esthetics with thin veneers and without using highly opaque materials. In this regard, esthetic restorative planning is crucial, and understanding the masking potential of the chosen ceramic and the influence of the resin cement is also fundamental.

Pre-restorative whitening in such cases can be advantageous as it reduces the color difference between elements and the chosen shade, minimizing the necessary preparation thickness [20]. This is particularly important when using translucent pieces. Moreover, the chromatic effect of the substrate will vary depending on the dental tissue where the veneer is bonded. More invasive preparations, such as those for full crowns, will highlight the saturation and hue of the tooth, also increasing the restoration's opacity due to the greater visible dentin content. Therefore, the color difference between the tooth before and after veneer cementation will be more noticeable [52]. Conversely, if the veneer is bonded to enamel, there will be less impact from substrate saturation and hue, although there will be a greater effect of intrinsic translucency to natural dental enamel, which can lead to greater reduction in the restoration's value, requiring careful attention to avoid graying. When analyzing these hard tissues separately, it is evident that enamel has greater translucency than dentin of the same thickness, with dentin being the primary contributor to the visible color of the tooth [63]. The translucency of enamel and dentin is directly related to the orientation of enamel prisms and dentinal tubules, impacting how light spreads through these structures [47, 70]. This contributes to the notion that the final color of a ceramic restoration will be affected not only by the substrate color/saturation but also by the opacity/translucency of the tissue to which the piece is bonded. Errors in selecting material translucency/opacity and, consequently, the reflection/value of a restoration are considered the most critical and noticeable since human eyes are more sensitive to brightness differences, making accurate selection in restorative work crucial for the restoration's esthetics, more so than color and hue [61].

The substrate will also undergo optical changes with the aging of the dental structure, with a tendency for increased translucency. Aging will increase the translucency of enamel due to significant loss of surface structure over the dental lifespan, making this structure thinner, as well as greater alignment and organization of enamel prisms transversely [70]. Meanwhile, in dentin, there will be a reduction in the pulp chamber volume and decreased blood supply, with dentin deposition in dentinal tubules (reducing their diameter), increasing its chroma and translucency due to increased mineral content [47, 71].

5.3 Cement line

In conservative veneer restorations, the cement line will affect the final color of the restoration and should be considered in esthetic restorative planning, as it can negatively impact the color result if its color/opacity is not planned. On the other hand, it can help neutralize the effect of unfavorable substrates (**Figure 2**), as the ceramic material's masking ability and its translucency play a significant role, but the resin cement also affects the final color of the restoration, contributing to masking discolored substrates when used with high opacity [13, 16–18].

In a resin cement, its color, opacity, composition, and brand will directly affect its optical properties and influence the final color result of a restoration with ceramic

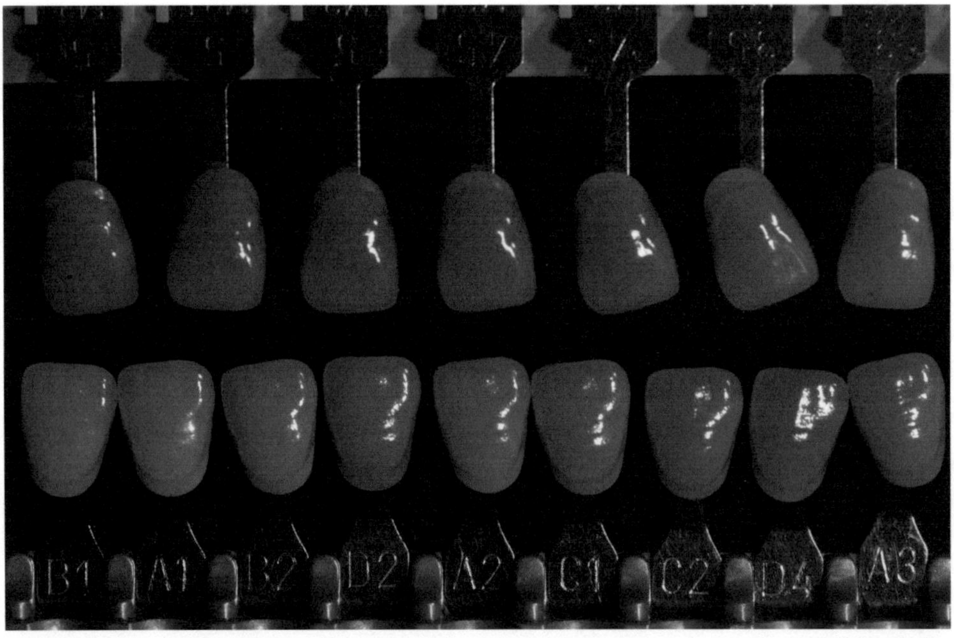

Figure 2.
Contrast between favorable substrates (bottom) and unfavorable substrates (top).

laminates. The study by Silva et al., which evaluated the influence of different translucency levels of lithium disilicate (IPS e.max CAD – HT, MT, and LT) and the color/opacity of resin cements from the Variolink Esthetic light-cure (LC) line (light+, high opacity, and Neutral, high translucency) (**Figure 3**), aimed at masking the difference between substrate shades A1 and A4, showed that all groups with light+ resin cement (opaque) had smaller color differences compared to the groups with Neutral cement, regardless of the translucency level of the disilicate [16].

This demonstrates not only that the opaque resin cement contributed to substrate masking but also that it reduced the influence of the ceramic on the final color of the restoration. This supports the use of this cement as an adjunct in neutralizing unfavorable substrates in a more conservative manner while maintaining natural esthetics. The opaque resin cement acts as an opacifier, blocking the substrate and allowing it to be paired with a less opaque ceramic laminate, resulting in partial light transmission through the restoration (**Table 1**). Additionally, when comparing color differences between opaque and translucent cements, all groups showed significant and clinically impactful differences, according to the parameters developed by Paravina et al., reinforcing the statement that resin cement affects the esthetic color result [72].

The brand of resin cement will also affect the chromatic characteristics of a restoration, such as its ability to mask substrates [73]. This is because the intrinsic composition of each manufacturer will vary, leading to different optical properties, with different refractive indices/opacities between brands, which was significant in the study.

Moreover, [13] a study demonstrated that the thickness of the cement line will also affect the final color of the restoration, but in a less significant and clinically visible manner than the variation in ceramic thickness. For ceramic pieces thicker than 2 mm, such as traditional crowns, the resin cement's influence is not clinically visible, indicating that, although opaque resin cement positively affects masking, the main factor determining the final color result of the restoration is the ceramic material itself. However, for ceramic pieces less than 1.5 mm thick, the color of the cement has a greater influence, necessitating careful selection to avoid errors in the final

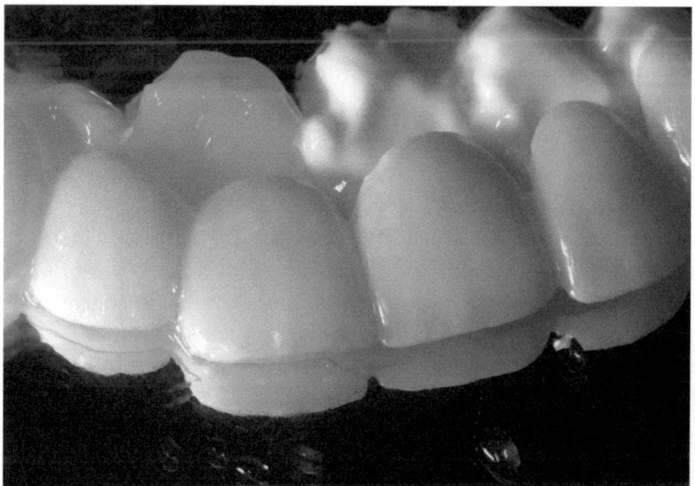

Figure 3.
Lithium disilicate ceramic laminates with Neutral resin cement (Ivoclar) on the left and Light+ resin cement (Ivoclar) on the right.

Laminate translucency	Resin cement shade	Substrate saturation	Mean ± SD
			ΔE00
HT	Light+	A1 vs. A4	$(3.6 \pm 0.7)_B$
	Neutral	A1 vs. A4	$(5.7 \pm 0.5)_A$
MT	Light+	A1 vs. A4	$(4.1 \pm 0.4)_B$
	Neutral	A1 vs. A4	$(5.2 \pm 0.2)_A$
LT	Light+	A1 vs. A4	$(2.7 \pm 0.2)_C$
	Neutral	A1 vs. A4	$(4.3 \pm 0.4)_B$

Source: da Silva et al. [16].

Table 1.
Mean and Standard Deviation (SD) of color differences (ΔE00) between A1 and A4 substrates (Influence of substrate on restoration final color). Subscript equal letters indicate lack of statistical difference among groups evaluated from Tukey test with 95% confidence intervals ($\rho < 0.0001$).

restoration color. This can be mitigated by using try-in pastes. Similar results were obtained by Cubas et al. [14], who evaluated the masking ability of ceramic systems in groups with 1, 1.5, and 2 mm thicknesses cemented with opaque, intermediate, and no resin cements. The conclusions were consistent: groups cemented with opaque resin cement showed a significant improvement in masking ability, especially in thinner ceramics, with the 2 mm groups achieving the best masking results due to the considerable ceramic thickness.

This effect is intensified with thinner thicknesses, such as in ceramic laminates. According to Begum et al., in their study comparing ceramic pieces with 0.5, 1, and 1.5 mm thicknesses cemented with opaque and translucent resin cements over a darkened substrate (C3), it was shown that the color difference between the 1 mm lithium disilicate group with translucent cement and the 0.5 mm group with opaque cement is small [19]. This indicates that the masking ability of a 0.5 mm laminate with opaque resin cement is equivalent to that of a 1 mm thick piece with translucent cement, but with the advantage of greater preservation of healthy dental structure and a more natural final appearance. However, the color of the substrate will be decisive for achieving optimal esthetic results. The study by Kandil et al. not only compared the influence of resin cement opacity on the masking ability of a darkened substrate but also evaluated the role of the translucency of the restorative ceramic material in this process [17]. They cemented 0.5 mm thick ceramic pieces with opaque and translucent cements in two translucency levels and concluded that the group cemented with opaque cement was more effective in masking the darkened substrate compared to the translucent cement, although all groups showed detectable color differences. This demonstrates the advantages of opaque cement and its contribution to substrate masking compared to translucent cement. However, the degree of translucency of the ceramic had a greater impact on this process, with opaque ceramics providing the best masking of the substrate.

Thus, using thin lithium disilicate-based ceramic veneers with high-opacity resin cement, as opposed to high-translucency cements which only promote cementation, ensures various benefits such as masking saturated substrates, achieving a natural restoration appearance, better mechanical properties, and longevity. This approach promotes a conservative esthetic, providing good esthetic outcomes and improved quality of life for the patient [14, 16, 17, 19]. However, even though this material can positively influence masking, it is crucial to properly choose the translucency level and thickness of the ceramic material, which is primarily responsible for the final color of the restoration, to effectively mask the substrate and perform a more conservative preparation on the remaining dental structure.

However, even if this material plays a beneficial chromatic role, the thickness of the cement line should be as thin as possible, as it is the weakest link in the restoration system. There is still a gap regarding the cement line in ceramic laminates, as ISO 4049:200 [74] standards require a resin cement thickness not exceeding 50 μm, but the majority of studies use 100 μm as a standard thickness [75–77]. It is indicated [78] that the ideal thickness for cement in laminates is around 90 μm. At this thickness, the maximum stress transferred to the ceramic is minimized compared to other thicknesses, reducing stress levels in the restoration, which aligns with [79], who found 100 μm as an acceptable cement thickness.

However, regarding current popular manufacturing processes, according to Aboushelib et al. [80], CAD/CAM-milled ceramics result in an average cement thickness of 340.35 and 106.74 μm for the injected technique, and this thick cement under the cemented laminate can interfere with the mechanical integrity of the restoration, increase polymerization stress, or influence the color and translucency of the restoration [35, 81]. Additionally, there is an increased chance of iatrogenic gingivitis caused by the increased porosity in the thick resin cement layer, as the cementing agent layer is the most critical and prone to degradation when dealing with enamel-cement-laminate systems.

5.3.1 Alteration of optical properties of the cement line due to aging

Thin, translucent ceramic veneers will have their color influenced by the color of the cement. Therefore, the stability of the optical properties of the cement line is crucial for the long-term esthetic success of restorations, as it is affected by the aging of the restoration [54]. Changes in color and translucency that ceramic restorations undergo over time, along with challenges in the oral environment, can significantly impact the esthetic outcome, especially in the anterior region where there is a high demand for color and shine maintenance, potentially leading to restoration replacements [28].

During the aging process, the cement line will undergo more significant alterations in physical and chemical characteristics than the ceramic, which has stable color and shine and does not degrade in the oral environment like resin composites do. These composites represent instability in their optical properties, with resin cement being the main contributor to changes in color and translucency in the indirect restoration system [21, 56, 67, 68]. Over time and with exposure to oral challenges, there is a tendency for the cement line to become more opaque, which clinicians must consider, particularly for already opaque resin cements, as it can reduce the natural effect of the restoration [23, 67, 82, 83].

The optical properties of resin cements, such as translucency and color, are influenced by the type and brand of the material. This can be explained by differences in composition, the amount of polymer matrix, the quantity, type, and distribution of filler particles, and the degree of monomer conversion due to differences in refractive index between the filler and matrix of these cements [21, 54, 56, 84, 85]. Additionally, the addition of pigments also impacts translucency, with more chromatic composites having greater opacity than less saturated and translucent ones, where generally fewer pigments are added. There is a significant correlation between translucency and chroma [26, 49].

The average particle size of the resin cement will directly affect the material's degree of translucency, as it relates to the amount of light transmission/scattering. Smaller particles, around 0.1 μm, allow greater light transmission and higher translucency, while particles of 10 μm in diameter cause greater light scattering, making the material more opaque [22, 65]. Thus, the quantity and size of filler particles and the type of monomers present will also affect the translucency changes a material undergoes with aging in the mouth [22, 56, 65–67, 84, 86, 87]. Therefore, the cement line will be more resistant to degradation and staining and will have more stable color and translucency if it has a higher proportion of filler and a lower organic portion in its composition, as the organic portion is more susceptible to hydrolysis in the oral environment [54, 82, 88].

However, it will also be affected by extrinsic factors such as polymerization, food, beverages, and medications [85, 89–91]. In this sense, a parallel can be drawn between the color stability of composite resins and resin cements, as both are materials with a polymeric network subject to degradation in a moist environment and have characteristics similar to flowable resins, with lower filler particle proportions to achieve lower viscosity and higher flow, making them more prone to degradation in the oral environment [52, 55, 67].

Hydrolytic degradation, which alters the color of resin composites, may be related to the distribution of filler particles, with composites having smaller particles and better distribution being more resistant to hydrolysis, and to the composition of the polymer matrix network, related to the hydrophilicity of the monomers present, such as TEGDMA and UDMA, with polar functional groups more susceptible to water absorption [56, 67, 86, 87]. Additionally, as studied by [82], resin composites with different average particle sizes, with higher packing of particles, and with a higher volume percentage of filler showed greater color stability when exposed to pigmented beverages. This study also demonstrates that this degradation leads to changes in opacity/translucency of the composites after aging, with increased reflectance for most groups.

With water absorption and the breakdown of the polymer network of resin cement, there is a rearrangement of the non-organic particles in the material, causing a change in the refractive index of the cement line after aging compared to the immediate moment. This is because changes in the physical and chemical characteristics of the polymer network will affect the refractive index, altering the amount of light reflected/transmitted by the restoration, which will affect its translucency, with lower translucency as the difference in refractive index increases [22–24]. The greater the difference between the refractive indices of the filler particles and the matrix of a resin material, the greater its opacity, according to [23]. Additionally, the breakdown of monomers can lead to changes in the pigment distribution of resin cements, with lighter cements being more prone to greater color changes after aging, as suggested by [26]. After aging, there is a change in the color of the cement line, with a tendency for yellowing of resin cements (increased values on the b^* coordinate, blue-yellow axis), according to studies [26, 27]. Besides the breakdown of resin monomer bonds, water absorption leads to the expansion and plasticization of the polymer network, producing microcracks that allow greater penetration of pigmented substances into the cement line [83].

Chemical composition, and consequently the type of cement (light-cured, dual-cured, self-adhesive), will imply different responses to aging, with a tendency for the cement line to suffer from both opacification and marginal pigmentation. Cements with chemical activation are more prone to changes in optical properties due to lower chemical stability and higher susceptibility to water absorption [22, 54, 67, 83, 84, 88, 92]. Resin composites, such as resin cements, with higher proportions of TEGDMA compared to UDMA, for example, have a greater tendency to suffer from water absorption, as TEGDMA is a low-viscosity monomer with high hydrophilicity, while a cement with higher proportions of UDMA will be more stable, as the polymer network of the cement line will be less degraded and less prone to pigmentation [56, 67, 84, 86, 87]. For ceramic piece cementation, light-cure (LC) or dual-cure (DC) resin cements are the first choice. Light-activated cements tend to be more stable against oral challenges due to their composition of aliphatic amines, which are more resistant to oxidation than tertiary amines present in dual-cure and self-adhesive (chemically activated) resin cements. This results in lower resistance to water absorption and greater marginal pigmentation of restorations, as the oxidation of residual monomers forms yellowish compounds, which is more likely to occur in dual-cure and self-adhesive cements [24, 57, 58, 66–68, 83, 88, 92]. This process can be intensified in cases of thicker and more opaque ceramic pieces, due to the tendency for subpolymerization of the cement line, impacting the degree of monomer conversion in both light-cured and dual-cured cements [21, 56, 68, 93, 94].

According to a systematic review [21], light-cure cement showed a higher degree of monomer conversion compared to dual-cure cement when used for cementing ceramic pieces up to 2 mm thick, providing greater color stability to the restoration. Additionally, due to higher degrees of conversion, light-cure cements have improved mechanical properties, making them less prone to water absorption in areas such as the gingival sulcus where the cement line is more exposed. However, even dual-cure cements are affected by reduced light exposure, as chemical polymerization alone is not sufficient to compensate for inadequate light polymerization. Both physical and chemical stimuli are necessary for proper polymerization of dual-cure cements, as noted [21]. The less effective the light polymerization of the cements, with lower irradiance received, the lower the degree of monomer conversion, resulting in a mechanically weakened cement line with poorer long-term esthetic properties [93]. According to Janda et al. [89], the type, curing time, and photopolymerizer influence the photopolymerization and, consequently, the color stability of a resin composite over time.

Dual-cure resins generally show less color stability due to the possibility of degradation of the tertiary amines used for the chemical polymerization reaction of monomers and oxidation of residual monomers. This modification in the chemical structure of the cement line contributes to the formation of yellowish compounds and changes the color of the cement [24, 25, 27, 57, 58, 67, 68, 86]. Unlike dual-cure cements, light-cure cements are more resistant to staining, as they contain aliphatic amines, which are more resistant to oxidation than tertiary amines, co-initiators less resistant to degradation in the oral environment [83, 86].

6. Conclusion

Knowledge of the color, translucency/opacity of both the natural tooth and the restorative materials is directly related to the success of esthetic dental treatments. Mastery of each of these factors is essential. In a conservative indirect restoration using ceramic veneers, many factors will influence the chromatic outcome. The ceramic will have a significant impact, but so will the resin cement and the substrate, considering the thin thickness of the piece. Therefore, understanding these factors not only at the time of cementation but also regarding how the optical behavior of the restoration will evolve over time is important.

Acknowledgements

This research was funded by the Institutional Scientific Initiation Scholarship Program of the National Council for Scientific and Technological Development (PIBIC – CNPq), number: 143597/2021-3, and the Coordination of superior Level Staff Improvement (CAPES), number: 88882.376604/2019-2101.

Author details

Carlos Eduardo Francci*, Samara Silva and Mylena Régis
University of São Paulo, Sao Paulo, Brazil

*Address all correspondence to: francci@usp.br

References

[1] Klages U, Bruckner A, Zentner A. Dental aesthetics, self-awareness, and oral health-related quality of life in young adults. European Journal of Orthodontics. 2004;**26**(5):507-514. DOI: 10.1093/ejo/26.5.507

[2] Bersezio C, Estay J, Jorquera G, et al. Effectiveness of dental bleaching with 37.5% and 6% hydrogen peroxide and its effect on quality of life. Operative Dentistry. 2019;**44**(2):146-155. DOI: 10.2341/17-229-C

[3] Martin J, Rivas V, Vildósola P, et al. Personality style in patients looking for tooth bleaching and its correlation with treatment satisfaction. Brazilian Dental Journal. 2016;**27**(1):60-65. DOI: 10.1590/0103-6440201600127

[4] Campos LA, Costa MA, Bonafé FSS, et al. Psychosocial impact of dental aesthetics on dental patients. International Dental Journal. 2020;**70**(5):321-327. DOI: 10.1111/idj.12574

[5] Santa-Rosa TTDA, Ferreira RC, Drummond AMA, et al. Impact of aesthetic restorative treatment on anterior teeth with fluorosis among residents of an endemic area in Brazil: Intervention study. BMC Oral Health. 2014;**14**(1):52-59. DOI: 10.1186/1472-6831-14-52

[6] Ricci WA, Fahl N. Nature-mimicking layering with composite resins through a bio-inspired analysis: 25 years of the polychromatic technique. Journal of Esthetic and Restorative Dentistry. 2023;**35**(1):7-18. DOI: 10.1111/jerd.13021

[7] Imburgia M, Cortellini D, Valenti M. Minimally invasive vertical preparation design for ceramic veneers: A multicenter retrospective follow-up clinical study of 265 lithium disilicate veneers. The International Journal of Esthetic Dentistry. 2019;**14**(3):286-298. DOI: 10.1016/j.cden.2017.06.001

[8] Bedran-Russo A, Leme-Kraus AA, Vidal CMP, Teixeira EC. An overview of dental adhesive systems and the dynamic tooth–adhesive Interface. Dental Clinics of North America. 2017;**61**(4):713-731. DOI: 10.1016/j.cden.2017.06.001

[9] Edelhoff D, Sorensen JA. Tooth structure removal associated with various preparation designs for anterior teeth. The Journal of Prosthetic Dentistry. 2002;**87**(5):503-509. DOI: 10.1067/mpr.2002.124094

[10] Soares PV, Spini PHR, Carvalho VF, et al. Esthetic rehabilitation with laminated ceramic veneers reinforced by lithium disilicate. Quintessence International. 2014;**45**(2):129-133. DOI: 10.3290/j.qi.a31009

[11] Spear F, Holloway J. Which all-ceramic system is optimal for anterior esthetics? Journal of the American Dental Association (Chicago, IL). 2008;**139**(4):S19-S24. DOI: 10.14219/jada.archive.2008.0358

[12] Boscato N, Hauschild FG, Kaizer MR, Moraes RR. Effectiveness of combination of dentin and enamel layers on the masking ability of porcelain. Brazilian Dental Journal. 2015;**26**(6):654-659. DOI: 10.1590/0103-6440201300463

[13] Vichi A, Ferrari M, Davidson CL. Influence of ceramic and cement thickness on the masking of various types of opaque posts. The Journal of Prosthetic Dentistry. 2000;**83**(4):412-417. DOI: 10.1016/s0022-3913(00)70035-7

[14] Cubas GBA, Camacho GB, Demarco FF, Pereira-Cenci T. The effect of luting agents and ceramic thickness on the color variation of different ceramics against a chromatic background. Eur. Journal of Dentistry. 2011;**5**(3):245-252

[15] Liebermann A, Mandl A, Eichberger M, Stawarczyk B. Impact of resin composite cement on color of computer-aided design/computer aided manufacturing ceramics. Journal of Esthetic and Restorative Dentistry. 2021;**33**(5):786-794. DOI: 10.1111/jerd.12738

[16] da Silva SF, de Araújo RM, Francci CE. The capacity of conservative preparations for lithium disilicate glass–ceramic laminates luted with different resin cements to mask different substrate shades: A minimally invasive esthetic approach. Journal of Esthetic and Restorative Dentistry. 2024;**36**(5):761-769. DOI: 10.1111/jerd.13176

[17] Kandil BSM, Hamdy AM, Aboelfadl AK, El-Anwar MI. Effect of ceramic translucency and luting cement shade on the color masking ability of laminate veneers. Dental Research Journal (Isfahan). 2019;**16**(3):193-199

[18] Basegio MM, Pecho OE, Ghinea R, et al. Masking ability of indirect restorative systems on tooth-colored resin substrate. Dental Materials. 2019;**35**(6):e122-e130. DOI: 10.1016/j.dental.2019.03.001

[19] Begum Z, Pratik C, Shruthi CS, Sonika R. Effect of ceramic thickness and luting agent shade on the color masking ability of laminate veneers. Journal of Indian Prosthodontic Society. 2014;**14**(1):S46-S50. DOI: 10.1007/s13191-014-0362-2

[20] Coachman C, Gurel G, Calamita M, et al. The influence of tooth color on preparation Design for Laminate Veneers from a minimally invasive perspective: Case report. The International Journal of Periodontics & Restorative Dentistry. 2014;**34**(4):453-459. DOI: 10.11607/prd.1900

[21] David-Pérez M, Ramírez-Suárez JP, Latorre-Correa F, Agudelo-Suárez AA. Degree of conversion of resin-cements (light-cured/dual-cured) under different thicknesses of vitreous ceramics: Systematic review. Journal of Prosthodontic Research. 2022;**66**(3):385-394. DOI: 10.2186/jpr.JPR_D_20_00090

[22] Fidan M. Accelerated aging effects on color change, translucency parameter, and surface hardness of resin composites. BioMed Research International. 2022;**2022**:6468281

[23] Lee YK, Powers JM. Color changes of resin composites in the reflectance and transmittance modes. Dental Materials. 2007;**23**(3):259-264. DOI: 10.1016/j.dental.2006.01.019

[24] Pissaia JF, Correr GM, Gonzaga CC, Cunha LF. Influence of shade, curing mode, and aging on the color stability of resin cements. Brazilian Journal of Oral Sciences. 2015;**14**(4):272-275. DOI: 10.1590/1677-3225v14n4a04

[25] Ramos NC, Luz JN, Valera MC, et al. Color stability of resin cements exposed to aging. Journal of Operative Dentistry. 2019;**44**(6):609-614. DOI: 10.2341/18-064-L

[26] Uchida H, Vaidyanathan J, Viswanadhan T, Vaidyanathan TK. Color stability of dental composites as a function of shade. The Journal of Prosthetic Dentistry. 1998;**79**(4):372-377. DOI: 10.1016/s0022-3913(98)70147-7

[27] Gürdal I, Atay A, Eichberger M, et al. Color change of CAD-CAM materials

and composite resin cements after thermocycling. Journal of Prosthodontic Research. 2018;**120**(4):546-552. DOI: 10.1016/j.prosdent.2017.12.003

[28] Mazzetti T, Collares K, Rodolfo B, et al. 10-year practice-based evaluation of ceramic and direct composite veneers. Dental Materials. 2022;**38**(5):898-906. DOI: 10.1016/j.dental.2022.03.007

[29] Zagar M, Zlatarić DK. Influence of esthetic dental and facial measurements on the Caucasian patients' satisfaction. Journal of Esthetic and Restorative Dentistry. 2011;**23**(1):12-20. DOI: 10.1111/j.1708-8240.2010.00381.x

[30] Gatto RCJ, Garbin AJI, Corrente JE, Garbin CAS. The relationship between oral health-related quality of life, the need for orthodontic treatment and bullying, among Brazilian teenagers. Dental Press Journal of Orthodontics. 2019;**24**(2):73-80. DOI: 10.1590/2177-6709.24.2.073-080.oar

[31] Reis GR, Vilela ALR, Silva FP, et al. Minimally invasive approach in esthetic dentistry: Composite resin versus ceramics veneers. Bioscience Journal. 2017;**33**(1):238-246. DOI: 10.14393/BJ-v33n1a2017-34617

[32] Arif R, Dennison JB, Garcia D, Yaman P. Gingival of porcelain laminate veneered teeth: A retrospective assessment. Operative Dentistry. 2019;**44**(5):452-458. DOI: 10.2341/18-088-C

[33] Smielak B, Armata O, Bojar W. A prospective comparative analysis of the survival rates of conventional vs no-prep/minimally invasive veneers over a mean period of 9 years. Clinical Oral Investigations. 2022;**26**(3):3049-3059. DOI: 10.1007/s00784-021-04289-6

[34] Cho YE, Lim YJ, Han JS, et al. Effect of Yttria content on the translucency and masking ability of Yttria-stabilized tetragonal zirconia polycrystal. Materials. 2020;**13**(21):1-10. DOI: 10.3390/ma13214726

[35] Li Q, Yu H, Wang N. Spectrophotometric evaluation of the optical influence of core build-up composites on all-ceramic materials. Dental Materials. 2009;**25**(2):158-165. DOI: 10.1016/j.dental.2008.05.008

[36] Bacchi A, Boccardi S, Alessandretti R, Pereira GK. Substrate masking ability of bilayer and monolithic ceramics used for complete crowns and the effect of association with an opaque resin-based luting agent. Journal of Prosthodontic Research. 2019;**63**(3):321-326. DOI: 10.1016/j.jpor.2019.01.005

[37] Heboyan A, Vardanyan A, Karobari MI, Marya A, Avagyan T, Tebyaniyan H, et al. Dental luting cements: An updated comprehensive review. Molecules. 2023;**28**(4):1619-1634. DOI: 10.3390/molecules28041619

[38] Anusavice KJ, Rawls HR, Shen C. Phillips Materiais Dentários. 12th ed. Rio de Janeiro: Elsevier; 2013

[39] Heffernan MJ, Aquilino SA, Diaz-Arnold AM, et al. Relative translucency of six all-ceramic systems. Part II: Core and veneer materials. The Journal of Prosthetic Dentistry. 2002;**88**(1):10-15. DOI: 10.1067/mpr.2002.126795

[40] Sravanthi Y, Ramani Y, Rathod AM, et al. The comparative evaluation of the translucency of crowns fabricated with three different all-ceramic materials: An in vitro study. Journal of Clinical and Diagnostic Research. 2015;**9**(2):ZC30-ZC34. DOI: 10.7860/JCDR/2015/12069.5559

[41] Alayad AS, Alqhatani A, Alkatheeri MS, et al. Effects of CAD/

CAM ceramics and thicknesses on translucency and color masking of substrates. The Saudi Dental Journal. 2021;**33**(7):761-768. DOI: 10.1016/j.sdentj.2020.03.011

[42] Alfouzan AF, Al-Otaibi HN, Labban N, et al. Influence of thickness and background on the color changes of CAD/CAM dental ceramic restorative materials. Materials Research Express. 2020;**7**(5):1-9. DOI: 10.1088/2053-1591/ab9348

[43] Shono NN, Nahedh HNAA. Contrast ratio and masking ability of three ceramic veneering materials. Operative Dentistry. 2012;**37**(4):406-416. DOI: 10.2341/10-237-L

[44] Bazos P, Magne P. Bio-emulation: Biomimetically emulating nature utilizing a histo-anatomic approach; structural analysis. The European Journal of Esthetic Dentistry. 2011;**6**(1):8-19

[45] Kim-Pusateri S, Brewer JD, Davis EL, Wee AG. Reliability and accuracy of four dental shade-matching devices. The Journal of Prosthetic Dentistry. 2009;**101**(3):193-199. DOI: 10.1016/S0022-3913(09)60028-7

[46] Vichi A, Louca C, Corciolani G, Ferrari M. Color related to ceramic and zirconia restorations: A review. Dental Materials. 2011;**27**(1):97-108. DOI: 10.1016/j.dental.2010.10.018

[47] Lee YK. Translucency of human teeth and dental restorative materials and its clinical relevance. Journal of Biomedical Optics. 2015;**20**(4):045002. DOI: 10.1117/1.JBO.20.4.045002

[48] Wang J, Yang J, Lv K, Zhang H, Huang H, Jiang X. Can we use the translucency parameter to predict the CAD/CAM ceramic restoration aesthetic? Dental Materials. 2023;**39**(3):e1-e10. DOI: 10.1016/j.dental.2023.01.002

[49] Yu B, Lee YK. Influence of color parameters of resin composites on their translucency. Dental Materials. 2008;**24**(9):1236-1242. DOI: 10.1016/j.dental.2008.01.016

[50] Johnston WM. Review of translucency determinations and applications to dental materials. Journal of Esthetic and Restorative Dentistry. 2014;**26**(4):217-223. DOI: 10.1111/jerd.12112

[51] Czigola A, Abram E, Kovacs ZI, et al. Effects of substrate, ceramic thickness, translucency, and cement shade on the color of CAD/CAM lithium-disilicate crowns. Journal of Esthetic and Restorative Dentistry. 2019;**31**(5):457-464. DOI: 10.1111/jerd.12470

[52] Perroni AP, Bergoli CD, Santos MBF, et al. Spectrophotometric analysis of clinical factors related to the color of ceramic restorations: A pilot study. The Journal of Prosthetic Dentistry. 2017;**118**(5):611-616. DOI: 10.1016/j.prosdent.2016.12.010

[53] Perroni AP, Kaizer MR, Della Bona A, et al. Influence of light-cured luting agents and associated factors on the color of ceramic laminate veneers: A systematic review of in vitro studies. Dental Materials. 2018;**34**(11):1610-1624. DOI: 10.1016/j.dental.2018.08.298

[54] Vichi A, Ferrari M, Davidson CL. Color and opacity variations in three different resin-based composite products after water aging. Dental Materials. 2004;**20**(6):530-534. DOI: 10.1016/j.dental.2002.11.001

[55] Almeida JR, Schmitt GU, Kaizer MR, et al. Resin-based luting agents and color stability of bonded ceramic veneers.

The Journal of Prosthetic Dentistry. 2015;**114**(2):272-277. DOI: 10.1016/j.prosdent.2015.01.008

[56] Marchionatti AME, Wandscher VF, May MM, et al. Color stability of ceramic laminate veneers cemented with light-polymerizing and dual-polymerizing luting agent: A split-mouth randomized clinical trial. The Journal of Prosthetic Dentistry. 2017;**118**(5):604-610. DOI: 10.1016/j.prosdent.2016.11.013

[57] Mina NR, Baba NZ, Al-Harbi FA, et al. The influence of simulated aging on the color stability of composite resin cements. The Journal of Prosthetic Dentistry. 2019;**121**(2):306-310. DOI: 10.1016/j.prosdent.2018.03.014

[58] Lee SM, Choi YS. Effect of ceramic material and resin cement systems on the color stability of laminate veneers after accelerated aging. The Journal of Prosthetic Dentistry. 2018;**120**(1):99-106. DOI: 10.1016/j.prosdent.2017.09.014

[59] Rodrigues RB, Lima E, Roscoe MG, et al. Influence of resin cements on color stability of different ceramic systems. Brazilian Dental Journal. 2017;**28**(2):191-195. DOI: 10.1590/0103-6440201700170

[60] Pop-Ciutrila IS, Ghinea R, Dudea D, et al. The effects of thickness and shade on translucency parameters of contemporary, esthetic dental ceramics. Journal of Esthetic and Restorative Dentistry. 2021;**33**(5):795-806. DOI: 10.1111/jerd.12733

[61] Awad D, Stawarczyk B, Liebermann A, Ilie N. Translucency of esthetic dental restorative CAD/CAM materials and composite resins with respect to thickness and surface roughness. The Journal of Prosthetic Dentistry. 2015;**113**(6):534-540. DOI: 10.1016/j.prosdent.2014.12.003

[62] Sulaiman TA, Abdulmajeed AA, Donovan TE, et al. Optical properties and light irradiance of monolithic zirconia at variable thicknesses. Dental Materials. 2015;**31**(10):1180-1187. DOI: 10.1016/j.dental.2015.06.016

[63] Yu B, Ahn JS, Lee YK. Measurement of translucency of tooth enamel and dentin. Acta Odontologica Scandinavica. 2009;**67**(1):57-64. DOI: 10.1080/00016350802577818

[64] Subaşı MG, Alp G, Johnston WM, Yilmaz B. Effect of thickness on optical properties of monolithic CAD-CAM ceramics. Journal of Dentistry. 2018;**71**(1):38-42. DOI: 10.1016/j.jdent.2018.01.010

[65] Bagis B, Turgut S. Optical properties of current ceramics systems for laminate veneers. Journal of Dentistry. 2013;**41**(3):e24-e30. DOI: 10.1016/j.jdent.2012.11.013

[66] Karaokutan I, Aykent F, Özdoğan MS. Comparison of the color change of porcelain laminate veneers produced by different materials after luting with three resin cements. Operative Dentistry. 2023;**48**(2):166-175. DOI: 10.2341/21-099-L

[67] Archegas LRP, Freire A, Vieira S, et al. Colour stability and opacity of resin cements and flowable composites for ceramic veneer luting after accelerated ageing. Journal of Dentistry. 2011;**39**(11):804-810. DOI: 10.1016/j.jdent.2011.08.013

[68] Kilinc E, Antonson SA, Hardigan PC, Kesercioglu A. Resin cement color stability and its influence on the final shade of all-ceramics. Journal of Dentistry. 2011;**39**(1):e30-e36. DOI: 10.1016/j.jdent.2011.01.005

[69] Silami FDJ, Tonani R, Alandia-Román C, Pires-de-Souza FCP.

Influence of different types of resin luting agents on color stability of ceramic laminate veneers subjected to accelerated artificial aging. Brazilian Dental Journal. 2016;**27**(1):95-100. DOI: 10.1590/0103-6440201600348

[70] Hariri I, Sadr A, Shimada Y, et al. Effects of structural orientation of enamel and dentine on light attenuation and local refractive index: An optical coherence tomography study. Journal of Dentistry. 2012;**40**(5):387-396. DOI: 10.1016/j.jdent.2012.01.017

[71] Dietschi D, Ardu S, Krejci I. A new shading concept based on natural tooth color applied to direct composite restorations. Quintessence International. 2006;**37**(2):91-102

[72] Paravina RD, Ghinea R, Herrera LJ, et al. Color difference thresholds in dentistry. Journal of Esthetic and Restorative Dentistry. 2015;**27**(1):S1-S9. DOI: 10.1111/jerd.12149

[73] Dede D, Ceylan G, Yilmaz B. Effect of brand and shade of resin cements on the final color of lithium disilicate ceramic. The Journal of Prosthetic Dentistry. 2016;**117**(4):539-544. DOI: 10.1016/j.prosdent.2016.07.014

[74] ISO 4049:2000. Dentistry: Polymer-Based Filing, Restorative and Luting Materials. International Organization for Standardization; 2000

[75] Pires A, Novais R, Araújo D, Pegoraro F. Effects of the type and thickness of ceramic, substrate, and cement on the optical color of a lithium disilicate ceramic. The Journal of Prosthetic Dentistry. 2017;**117**(1):144-149. DOI: 10.1016/j.prosdent.2016.04.003

[76] Chen XD, Hong G, Xing WZ, Wang YN. The influence of resin cements on the final color of ceramic veneers. Journal of Prosthodontic Research. 2015;**59**(3):172-177. DOI: 10.1016/j.jpor.2015.03.001

[77] Igiel C, Weyhrauch M, Mayer B, et al. Effects of ceramic layer thickness, cement color, and abutment tooth color on color reproduction of feldspathic veneers. The International Journal of Esthetic Dentistry. 2018;**13**(1):110-119

[78] Liu B, Lu C, Wu Y, Zhang X, Arola D, Zhang D. The effects of adhesive type and thickness on stress distribution in molars restored with all-ceramic crowns. Journal of Prosthodontics. 2011;**20**(1):35-44. DOI: 10.1111/j.1532-849X.2010.00650.x

[79] Uzgur R, Ercan E, Uzgur Z, et al. Cement thickness of inlay restorations made of lithium dissilicate, polymer-infiltrated ceramic and nano-ceramic CAD/CAM materials evaluated using 3D X-ray micro-computed tomography. Journal of Prosthodontics. 2018;**27**(5):456-460. DOI: 10.1111/jopr.12521

[80] Aboushelib MN, Elmahy WA, Ghazy MH. Internal adaptation, marginal accuracy and microleakage of a pressable versus a machinable ceramic laminate veneers. Journal of Dentistry. 2012;**40**(8):670-677. DOI: 10.1016/j.jdent.2012.04.019

[81] Lasserre JF, Pop-Ciutrila IS, Colosi HA. A comparison between a new visual method of colour matching by intraoral camera and conventional visual and spectrometric methods. Journal of Dentistry. 2011;**39**(3):e29-e36. DOI: 10.1016/j.jdent.2011.11.002

[82] Lopes-Rocha L, Mendes JM, Garcez J, et al. The effect of different dietary and therapeutic solutions on the color stability of resin-matrix composites used in dentistry: An In vitro study.

Materials. 2021;**14**(1):1-10. DOI: 10.3390/ma14216267

[83] Yu H, Cheng SL, Jiang NW, Cheng H. Effects of cyclic staining on the color, translucency, surface roughness, and substance loss of contemporary adhesive resin cements. The Journal of Prosthetic Dentistry. 2018;**120**(3):462-469. DOI: 10.1016/j.prosdent.2017.10.009

[84] Turgut S, Bagis B, Turkaslan SS, Bagis YH. Effect of ultraviolet aging on translucency of resin-cemented ceramic veneers: An in vitro study. Journal of Prosthodontics. 2014;**23**(1):39-44. DOI: 10.1111/jopr.12061

[85] Marufu C, Kisumbi BK, Osiro OA, Otieno FO. Effect of finishing protocols and staining solutions on color stability of dental resin composites. Clinical and Experimental Dental Research. 2022;**8**(1):561-570. DOI: 10.1002/cre2.555

[86] Haralur SB, Alfaifi M, Almuaddi A, et al. The effect of accelerated aging on the colour stability of composite resin luting cements using different bonding techniques. Journal of Clinical and Diagnostic Research. 2017;**11**(4):57-60. DOI: 10.7860/JCDR/2017/25491.9681

[87] Ahmed MA, Jouhar R, Vohra F. Effect of different pH beverages on the color stability of smart monochromatic composite. Applied Sciences. 2022;**12**(9):1-11. DOI: 0.3390/app12094163

[88] Ghavam M, Amani-Tehran M, Saffarpour M. Effect of accelerated aging on the color and opacity of resin cements. Operative Dentistry. 2010;**35**(6):605-609. DOI: 10.2341/09-161-L

[89] Janda R, Roulet JF, Latta M, et al. Effect of exponential polymerization on color stability of resin-based filling materials. Dental Materials. 2007;**23**(6):696-704. DOI: 10.1016/j.dental.2006.06.009

[90] Shetty P, Purayil TP, Ginjupalli K, Pentapati KC. Effect of polishing technique and immersion in beverages on color stability of nanoceramic composites. Journal of Oral Biology and Craniofacial Research. 2021;**11**(1):53-56. DOI: 10.1016/j.jobcr.2020.11.011

[91] Valizadeh S, Asiaie Z, Kiomarsi N, Kharazifard MJ. Color stability of self-adhering composite resins in different solutions. Dental and Medical Problems. 2020;**57**(1):31-38. DOI: 10.17219/dmp/114099

[92] Zhou W, Liao ZX, Chen JH, et al. Color change of glass ceramic restorations cemented by four types of dual-cured resin luting agents with different initiator systems. Dental Materials Journal. 2022;**41**(6):833-842. DOI: 10.4012/dmj.2022-044

[93] Strazzi Sahyon HB, Chimanski A, Yoshimura HN, Santos PH. Effect of previous photoactivation of the adhesive system on the color stability and mechanical properties of resin components in ceramic laminate veneer luting. The Journal of Prosthetic Dentistry. 2018;**120**(4):631.e1-631.e6. DOI: 10.1016/j.prosdent.2018.06.014

[94] Oh S, Shin SM, Kim HJ, et al. Influence of glass-based dental ceramic type and thickness with identical shade on the light transmittance and the degree of conversion of resin cement. International Journal of Oral Science. 2018;**10**(1):1-6. DOI: 10.1038/s41368-017-0005-7

Chapter 6

Perovskite Ceramics: Promising Materials for Solar Cells (Photovoltaics)

Shah Aarif Ul Islam and Edson Leroy Meyer

Abstract

This chapter discusses the future of perovskite solar cells (PSCs) as a new generation of photovoltaic technologies to replace traditional silicon-based solar cells. PSCs have properties such as high efficiency, low processing cost, and flexibility in form, and, therefore, can be implemented in various applications such as building-integrated photovoltaics (BIPV), flexible electronics, and wearable electronics. Nevertheless, some issues still need to be solved in commercialising PSCs, such as stability issues, scaling-up issues, and policy barriers. However, the prospects for market development are vast, and PSCs can revolutionise the solar industry on the planet. In this chapter, the most recent methods for the synthesis of small- and large-scale perovskite-based solar cells are described. This chapter also explores some of the new research areas of interest, including tandem solar cells, perovskite-based multi-junction solar cells, and perovskite quantum dots, all expected to advance the photovoltaic efficiency and versatility further. Further, the evolution of perovskite-silicon heterojunctions, all perovskite tandem cells, and indoor photovoltaics show the growing area of perovskite utilisation. If PSCs are to overcome certain challenges and further the research, it can change the face of solar energy as a clean, efficient, and diverse option for the future.

Keywords: perovskite solar cells (PSCs), photovoltaics, high efficiency, tandem solar cells, building-integrated photovoltaics (BIPV), flexible electronics

1. Introduction

With the world's energy consumption increasing and the onset of climate change, the role of renewable power sources can hardly be overestimated. The oil, gas, and coal that have been the foundation of the global economy for more than a century are now recognised as a significant source of environmental issues, including pollution and climate change. This has shifted the world energy mix where renewable energy is essential, particularly solar energy, in addressing climate change. Solar energy, as energy from the sun, is one of the abundant sources of renewable energy [1, 2]. It is a clean energy source that can meet the world's energy demands without the social and environmental vices associated with fossil energy resources. Hence, solar energy technologies, particularly solar cells, have been more focused on research and development in the quest for green power technologies [3–6].

Solar or photovoltaic cells are electrical appliances that generate electric power through the photovoltaic process. These are the basic building blocks of solar panels widely applied in residential, commercial, and industrial applications. Solar cells are particularly attractive for their capacity to generate power without generating any greenhouse gases [7, 8]. This capacity makes solar cells a vital weapon in the war against global warming [9]. Over the past few decades, there has been a huge advancement in the solar cell and, therefore, the efficiency, cost, and overall scalability of solar cells [10, 11]. The first generation of solar cells is the crystalline silicon solar cells that have gained the market due to the efficiency and the durability of the solar cells [12]. But in the need to push these machines further and bring down the cost of production even more, the search for other materials and technologies has been called for. Out of all the new materials for solar cells, perovskite and double perovskite have been researched more because they possess certain features and benefits over silicon solar cells [13, 14]. Perovskites are a class of materials with the general formula ABX3, where 'A' and 'B' are cations of different sizes, and 'X' is an anion that coordinates with both [15]. These materials have displayed good optoelectronic properties such as high absorbance, tunable band gap, and long carrier diffusion that make them suitable for photovoltaic applications [16]. Other structures have also been investigated in photovoltaics, including double perovskites, which are obtained when two different cations occupy the 'B' site cation. These materials can offer increased stability, reduced toxicity, and electronic character adjustability by selection of cations and anions. For instance, the double perovskite La_2NiMnO_6 with the designed bandgap is the great advancement in achieving high efficiency and stability of the solar cell materials [17]. The fact that perovskite and double perovskite materials have some advantages over silicon-based solar cells has led to global interest in the study of their properties and the development of ways to improve their efficiency and address issues related to stability and production. In this chapter, the reader will be acquainted with the fundamentals of perovskite solar cells and the materials used in the formation of perovskite solar cells, the manufacturing process of perovskite solar cells, and the efficiency and stability of the perovskite solar cells, and the prospects of perovskite solar cells for the future market of solar cells.

2. Fundamentals of perovskite solar cells

2.1 Crystal structure and chemical composition

The perovskite materials are defined by their specific crystal structure and are generally represented by the formula ABX_3. In this structure:

'A' cation: Usually a big organic or inorganic cation (e.g. methylammonium (MA^+), formamidinium (FA^+), or caesium (Cs^+)).

'B' cation: A smaller metal ion, usually lead (Pb^{2+}) or tin (Sn^{2+}), situated in the centre of the coordination sphere.

'X' anion: A halide ion meaning an ion of chlorine, bromine, or iodine [18].

This results in a three-dimensional structure in which the 'B' cation is surrounded by six 'X' anions to form an octahedral structure, and the 'A' cation is located in the space in between the octahedra. This is because the 'A', 'B', and 'X' positions can be occupied by many elements, and these can be adjusted to provide the most suitable physical or chemical characteristic for a particular use (**Figure 1**) [20].

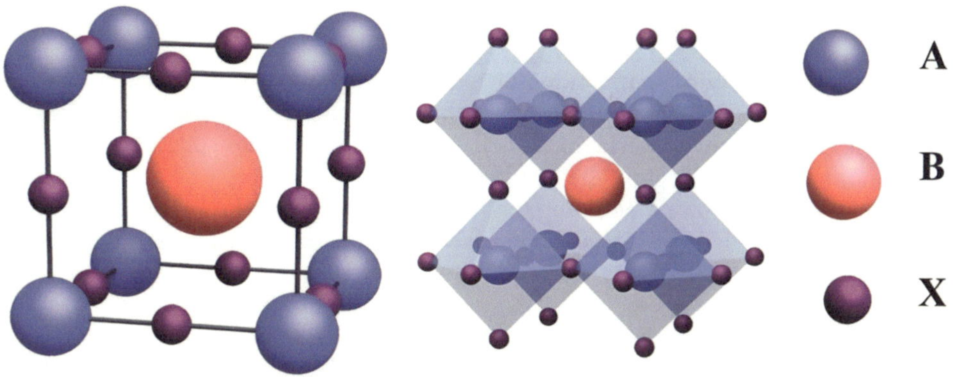

Figure 1.
A typical cubic unit cell (left) and the crystal (right) of perovskite ABX_3. Reproduced from Ref. [19] under CC BY 4.0.

The double perovskites are slightly more complex than the traditional perovskite structure; here, two different cations are present in the B′ site and are denoted by B′ and B″. In this manner, one can get the formula $A_2B'B''X_6$ (**Figure 2**).

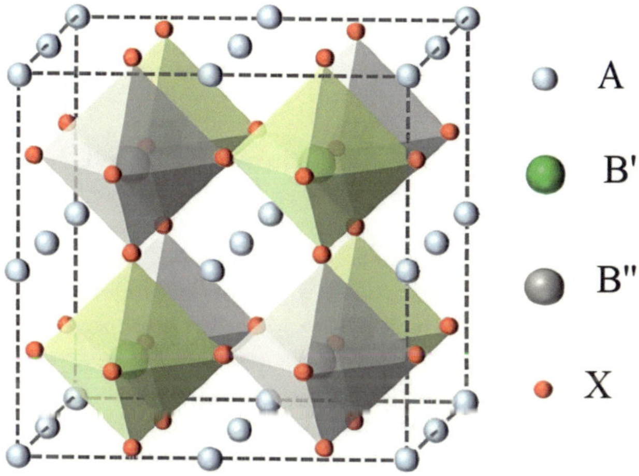

Figure 2.
Double perovskite crystal structure, with B′ and B″ lying in alternate octahedra $B'X_6$ and $B''X_6$. Adapted from Ref. [21], under CC BY 4.0.

The two different cations can change the electronic, stability, and band gap of the material making the double perovskites of particular interest in photovoltaic applications [22]. For example, La_2NiMnO_6 is a double perovskite in which Ni^{2+} and Mn^{4+} occupy the B′ and B″ sites, respectively; they exhibit extraordinary magnetic and electrical properties that can be applied in solar cell [23]. Another important feature of perovskite materials which opens the possibility to tune them for the best performance in solar cells is the possibility to control the chemical composition and structure of the material [24].

2.2 Bandgap engineering

A crucial parameter of semiconductor material used in photovoltaic is its bandgap, which is the energy gap between the valence band and the conduction band. This defines the ability of the material to absorb and convert the solar spectrum to electricity depending on the bandgap. The bandgap for the best efficiency of the solar cell should be around 1.3–1.5 eV, enabling the material to capture a large part of the solar spectrum while providing a high voltage in the open circuit [25]. The bandgap can also be engineered in perovskite materials by altering the composition of the 'A', 'B', and 'X' sites to fit the range required [26]. For example, in the most basic perovskite-type material, methylammonium lead iodide ($MAPbI_3$), the bandgap is about 1.55 eV, which is very close to the maximum achievable for single-junction solar cells. When one or several halides at the 'X' site are replaced, for example, iodine at the 'X' place can be replaced by bromine to get $MAPbBr_3$, the bandgap increases, and the material can absorb higher-energy lower wavelength light. Likewise, the substitution of lead with tin (in $MASnI_3$) results in a decrease in bandgap to allow the material to absorb lower energy photons [27]. In double perovskites, the cations B′ and B″ also contribute to the bandgap engineering. For instance, in the perovskite La_2NiMnO_6, the bandgap can be controlled by varying the concentrations of Ni^{2+} and Mn^{4+} or by introducing impurities in the structure [26]. That is, the ability to fine-tune the bandgap through careful control of composition can be a powerful asset in the development of perovskite solar cells by enhancing light absorption and thus increasing the device's efficiency.

2.3 Charge carrier dynamics

Apart from light absorption, the performance of a solar cell is equally measured by the ability to create and collect the photogenerated charge carriers, electrons, and holes. In perovskite solar cells, the processes of generation, separation, and transport of charge carriers depend on the perovskite material's electronic structure and defect states [28]. In the absorption process, when a photon is incident on the perovskite material, it provides energy to the valence band electron to move to the conduction band, creating a hole in the valence band. This process creates an electron-hole pair, which has to be transported separately and turned into a current at the electrodes. The efficiency of this process depends on several factors, including:

- *Carrier mobility*: High mobility so that the electrons and holes in the material can freely shuttle between them and reduce the chances of recombination.

- *Diffusion length*: The distance that, on average, a carrier can travel before recombining is usually the definition of the letter. Due to the long diffusion lengths of more than a 100 nanometres, perovskites are applicable for charge collection.

- *Recombination rates*: It is also essential to have a low recombination frequency to facilitate the flow of a high number of carriers to the electrode.

Some perovskites have exhibited dynamic features of charge carriers, low density of defects, and good crystallinity, which is beneficial for efficient charge production and transport. Nevertheless, vacancies, interstitials, and grain boundaries are always available and can pose as recombination sites, reducing the solar cell's efficiency [29].

Some of these defects originate from material chemistry, while others come from the fabrication processes, and these are areas of active research in perovskite photovoltaics. Thus, the crystal structure, band gap control, and charge carrier behaviour are vital factors of perovskite solar cells. For these reasons and the fact that perovskites are characterised by adjustable band gaps and outstanding charge transport properties, these materials are considered the future for the development of photovoltaic technologies. Still, stability, toxicity, and scalability issues arose, which is why the researchers must address them in the future.

3. Perovskite materials for solar cells

3.1 Single perovskite materials-methylammonium lead iodide ($MAPbI_3$)

One of the best-investigated perovskite materials for solar cells is methylammonium lead iodide ($MAPbI_3$). $MAPbI_3$ has attracted interest because of its properties, such as a suitable bandgap of about 1.55 eV, high absorption coefficient, and long carrier diffusion lengths, which makes it suitable for high photovoltaic performance [30]. $MAPbI_3$ has the simplest structure of the perovskite family, which is a three-dimensional structure with an ABX_3 formula, where A is the methylammonium (MA^+), B is lead (Pb^{2+}), and X is iodine (I^-). This has made it to record high PCE efficiency over the last decade, with recorded efficiencies standing at over 25% for the most efficient materials in the system [31]. These results give $MAPbI_3$ and other lead-based perovskites the promise of being new-generation solar cells that can replace traditional silicon-based solar cells. However, the utilisation of $MAPbI_3$ in commercial products is limited by several factors as follows. The most critical issue of the material is that it is not as stable as it should be, but that $MAPbI_3$ is quite sensitive to environmental conditions such as moisture, oxygen, and heat. The sensitivity towards these degradation conditions decreases the devices' effective usage rates over time and, thus, reduces their actual service life [32]. Methods used to enhance the stability of $MAPbI_3$ are encapsulation, compositional engineering, and synthesising of new stable forms of $MAPbI_3$ such as $FAPbI_3$ and mixed cation perovskites. Another critical concern is lead, which is a toxic heavy metal that has adverse effects on the environment and human health. Lead integrated into perovskite solar cells is an issue in mass production and recycling. Current research is being conducted to come up with lead-free perovskites with high efficiency and low toxicity as lead-based perovskites [33].

3.2 Lead-free perovskites

Due to potential lead toxicity issues, there have been significant attempts to synthesise lead-free perovskite materials. Among the most studied alternatives is a group of Tin (Sn)-based perovskites, for example methylammonium tin iodide ($MASnI_3$) [34]. Tin is also a member of group 14 and can also be incorporated into perovskite structures with similar electronic characteristics to lead. Nevertheless, tin-based perovskites have the issue of the fast oxidation of Sn^{2+} to Sn^{4+}, which causes instability and performance degradation of the devices. To overcome these challenges, researchers have tried out different approaches, one of them being the partial substitution of lead by tin to decrease the toxicity level and increase the efficiency of the perovskite solar cells [35]. The second approach is doping Sn-based perovskites with other elements to improve stability and electronic characteristics. Bismuth (Bi) and antimony (Sb) have also been proposed to replace lead in perovskite solar cells. The organic-inorganic hybrid bismuth perovskites, including $(CH_3NH_3)_3Bi_2I_9$, possess stability and non-toxicity issues. However, the efficiency of these perovskites in photovoltaic applications is slightly lower than that of lead perovskites at present [36]. More investigations are directed towards enhancing these materials using compositional design and optimising the processing methods. In general, lead-free perovskites are a good approach to addressing the problem of the environmental impact of lead; however, matching their efficiency to that of lead-based perovskites is a major concern. One of the main current research topics is the improvement of perovskite materials, which should have high efficiency and stability and should not harm the environment.

3.3 Double perovskite materials

Double perovskites, which have two different cations at the B site, are promising materials for developing new photovoltaic materials. Of these, La_2NiMnO_6 has received considerable attention because of its magnetic and electronic characteristics. In La_2NiMnO_6, the B sites are occupied by Ni^{2+} and Mn^{4+} ions, which make them ordered in such a way that may affect the electronic band structure and magnetic coupling in the material. Since doping and variations in the composition of La_2NiMnO_6 can be used to control the properties of this material, it is appropriate for photovoltaic uses. For instance, doping of La_2NiMnO_6 by Sr. may lead to an increase of the density of states, the change of the band structure, and the enhancement of the electronic properties of the material [26]. Sr-doped La_2NiMnO_6 has been found to improve the efficiency of solar cells since the ratio of light absorption to charge carrier kinetics is improved. In this regard, it has been discovered that the bandgap of La_2NiMnO_6 can be adjusted to the range of values that is suitable for the conversion of solar photons by the incorporation of dopants and the control of cationic ordering. This flexibility can be employed to fine-tune the material for particular application, which can be a step towards the development of efficient solar cells with longer durability.

However, synthesising high-quality double perovskite thin films has the following challenges. Among the most important factors that determine the device performance and must therefore be controlled during the synthesis process are the cation ordering and the extent of defects that are introduced. More advanced techniques such as PLD and MBE are employed to have a better control over the film forming material and the structure. However, they are costly and difficult to apply on a commercial scale and hence may not be suitable for large-scale production of commercial solar cells. Since it is possible to tune the properties of double perovskites in such a manner, these materials may be crucial in the next generation of photovoltaic uses [37].

4. Fabrication techniques for perovskite solar cells

4.1 Solution-based methods

4.1.1 Spin coating

Spin coating is one of the most popular perovskite solar cell deposition methods because of its simplicity, low cost, and capability to deposit a uniform thin layer [38]. The technique entails the preparation of a perovskite precursor solution and then placing the substrate on which the solution is to be coated and spinning at high speeds. The

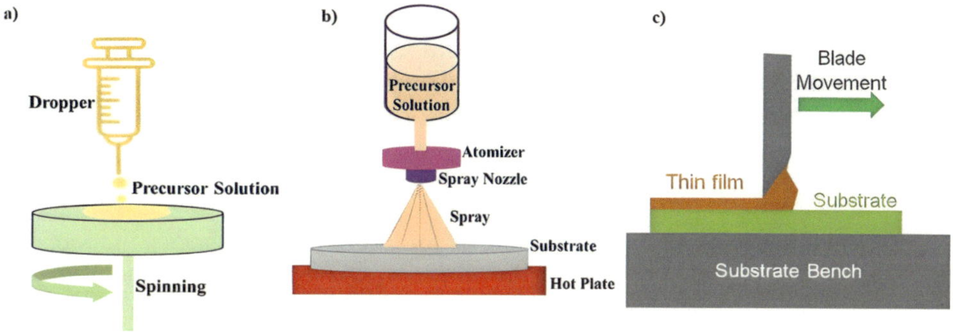

Figure 3.
Schematics of various solution-based fabrication techniques: (a) spin coating, (b) spray coating, and (c) doctor blade coating. Adapted from Ref. [39], under CC BY-NC 3.0.

centrifugal force then disperses the solution on the substrate in a thin and uniform layer (**Figure 3a**). This technique is useful for small-scale laboratory applications and has been instrumental in the initial stages of perovskite solar cell development. The spin coating process typically consists of two stages: deposition and drying. The perovskite precursor solution comprising 'A', 'B', and 'X' in the deposition stage is spread on the substrate. The substrate is then spun at high speeds (between 1000 and 5000 rpm) to ensure an even distribution of the solution. The spin speed and concentration of the solution can be used to regulate the thickness of the film. During the drying stage, the solvent is removed while the film is still spinning, resulting in the formation of the perovskite material in the form of a crystalline structure. The drying process should be well controlled to produce high-quality films, uniform grain size, and almost no defects. This can be achieved by solvent engineering, whereby an anti-solvent is added dropwise to the spinning substrate to cause rapid crystallisation or by regulating the temperature and relative humidity during the process. The spin coating method is very efficient for perovskite film deposition in laboratory conditions but unsuitable for large-scale production. The technique is rather complicated to apply over large areas, and the thickness of the film may vary towards the edges. Nonetheless, spin coating continues to be one of the most commonly used techniques for the deposition of perovskite films, especially for materials research and improvement [38].

4.1.2 Spray coating

Spray coating is another large-area solution-processed deposition technique for perovskite solar cells. Perovskite precursor solution is atomized into fine particles and then coated onto a substrate. The process can be taken outside or inside a controlled setting in order to enhance the quality of the film (**Figure 3b**). Compared to other deposition techniques, spray coating is used mostly for the deposition of perovskite films on flexible or complex-shaped substrates, which may improve the flexibility of the device. The main advantage of spray coating is its ability to be used for large-area coatings, and thus, it can be used in mass production [40]. The thickness and uniformity of the film that is being formed can be controlled with the help of the characteristics of spraying, for example, size of droplets, distance from the substrate, and the concentration of the solution. Also, the ability to use several spray heads at once allows for the deposition of the layers consecutively—something that is critical when forming multi-layered structures. But spray coating does come with some drawbacks;

the thickness of the film is not uniform; the droplets tend to coalesce and there is poor coverage. These problems can be solved by optimising the spray process as well as the post-deposition treatment like thermal anneal or solvent anneal, which will help in getting better quality of the films and higher degree of crystallinity [41].

4.1.3 Doctor blading

Doctor blading or blade coating is another solution-based method even more scalable than spin coating [42]. In this technique, a perovskite precursor solution is applied on a substrate, and then, a blade or knife is used to bar the solution into a layer (**Figure 3c**). The thickness of the film can be regulated through the distance between the blade and the substrate and the speed of the blade. Doctor blading is advantageous in the sense that it is relatively easy to implement and may be more suited for large-area substrates; therefore, it is more likely to be used in large-scale production. It can be readily incorporated into roll-coating systems in which a thin and continuous material layer is applied to a flexible support medium, such as plastic or metal foil. This makes it possible to produce perovskite solar cells on a large scale, which may lead to a reduction in the cost of production. A major difficulty of the doctor's blading process is to produce a homogeneous film thickness over a wide area. Some of these factors include the viscosity of the precursor solution, the surface energy of the substrate as well as the drying conditions, and these factors must be well controlled in order to prevent some of the defects that are commonly associated with the process, such as pinholes, non-uniform thickness, or incomplete coverage. Nonetheless, progress in process control and the development of new materials have rendered doctors blading a viable method for the up-scaling of perovskite solar cells [42].

4.2 Vacuum-based methods

4.2.1 Thermal evaporation

Thermal evaporation is one method that works under vacuum, and it is used to deposit perovskite on the substrate. In this process, the perovskite precursor materials are heated at such a level that the material evaporates in a vacuum chamber. The vaporised species then condense to form a thin film on a more relaxed substrate because heat transfers from the substrate to the vaporised species. Thermal evaporation has some merits, such as depositing high-purity films, and the thickness and composition of the films can be well controlled [43]. The other advantage of thermal evaporation is that the deposition of perovskite layers can be done in a clean environment, enabling the control of the film's composition. This technique is very efficient for depositing multilayer structures where each layer can be deposited rapidly to the next layer. Thermal evaporation is often applied when depositing perovskite onto other materials, such as silicon, to develop tandem solar cells, which are more effective. The high ability of perovskite films to be deposited on different substrates, including flexible and transparent ones, makes thermal evaporation one of the most appropriate techniques for future sun power cell designs [44]. Thermal evaporation, on the other hand, is expensive because it requires vacuum equipment and is unsuitable for large-area deposition, as is the case with solution-based techniques. The process also has energy costs since the precursor materials require heating to high temperatures to evaporate; this may be a limitation to mass production (**Figure 4**).

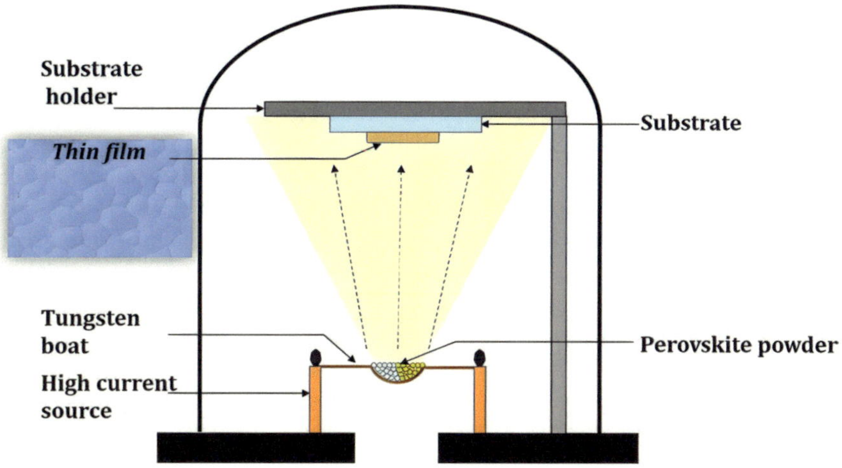

Figure 4.
Schematic diagram depicting the fabrication process using thermal evaporation technique. Adapted from Ref. [45] under CC BY licence.

4.2.2 Pulsed laser deposition (PLD)

Another vacuum-based technique employed for the synthesis of perovskite solar cells is called pulsed laser deposition (PLD) which is also used for the deposition of high-quality perovskite thin film with precise control over composition and thickness [44]. In PLD, a high-power laser pulse is incident on a target material and due to this, target material is vaporised and converted into a plasma plume. The material then vaporises and condenses on a substrate in the form of a thin layer of a film. Some of

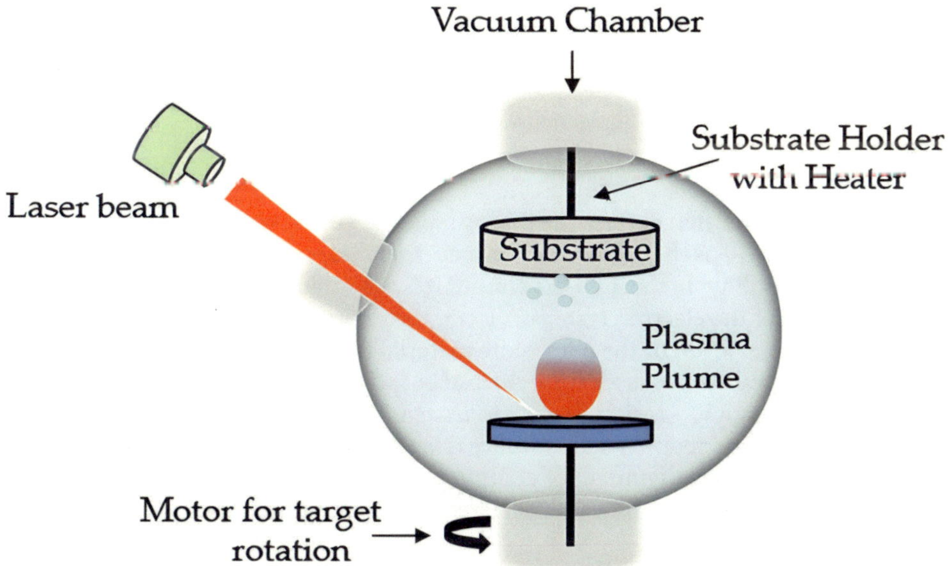

Figure 5.
Depiction of a pulsed laser deposition mechanism for fabrication of perovskite layers. Adapted from Ref. [47] under CC BY licence.

the advantages of PLD are as follows: It is easy to deposit multi-component materials and a stoichiometric transfer of the target material to the film. This is especially helpful in the formation of double perovskite materials because the cation content defines the electronic properties of the compound. The high quality of the films that are produced with smooth surface and clear interfaces is one of the major benefits of the PLD process and it makes it very suitable for creating complex structures of the solar cells like the heterojunction or the tandem cells [46]. It is also versatile in terms of the substrate it can be used on and can be used on flexible materials leading to new device structures. However, PLD is slow, it has low deposition rates, and it involves complicated vacuum systems; therefore, it is not suitable for large-scale production. Another disadvantage of using high energy is that it results to target degradation and non-uniform film in laser ablation. However, as is mentioned above, PLD is still an effective method for R&D of perovskite PVs, especially when the new materials and structures are to be found (**Figure 5**).

5. Performance and stability metrics of perovskite solar cells

5.1 Power conversion efficiency (PCE)

The most significant characteristic of solar cells is the power conversion efficiency or PCE, which defines the capability of the solar cell to convert light into electricity [48]. Lately, perovskite solar cells have had a very high PCE improvement in the past 10 years; laboratory perovskite-sized devices have PCEs above 25%. This development is because of the properties of the perovskite materials such as high absorption coefficients, tunable band gap, and long diffusion lengths of the carriers. The PCE of a perovskite solar cell consequently depends on the quality of the perovskite layer, the efficiency of the charge transport layers, and the structure of the solar cell. The films should not have any defects, and the grain size should be small and homogenised to enhance the probability of light absorption and reduce the recombination losses. Another aspect that is important for the efficiency is the choice of charge transport materials that assist in the extraction of electrons and holes from the perovskite layer.

5.2 Open-circuit voltage (V_{oc})

Another parameter that defines the performance of perovskite solar cells is open-circuit voltage (V_{OC}). V_{OC} is the maximum open circuit voltage obtained from the solar cell when no current is flowing through the circuit. It is determined by the difference of the conduction band of the electron transport layer, the valence band of the hole transport layer, and the energy levels of the perovskite material [49]. Perovskite solar cells generally have high V_{OC} values because of the steep absorption edge and low non-radiative recombination. A high V_{OC} is significant in enhancing the PCE since it contributes to the device's conversion efficiency. Scientists have tried to improve V_{OC} by several approaches, which include the proper alignment of energy in the perovskite and charge transport layers and the minimisation of the defect states in the perovskite film.

5.3 Short-circuit current density (J_{SC})

J_{SC} stands for short-circuit current density, which is the current density of the solar cell when the terminals of the cell are shorted and is the maximum current that the cell can deliver under standard test conditions. J_{SC} is dependent on the absorption spectrum of the perovskite material, thickness of the perovskite layer, and the quality of the interfaces between the perovskite layer and the charge transport layers [50]. The perovskite material incorporated should be able to capture sunlight especially in the visible light range for the achievement of high J_{SC} values in the solar cells. The thickness of perovskite layer must be tuned to ensure that it can capture the light and at the same time, to reduce the recombination of the carriers. Moreover, the contact between the perovskite layer with the charge transport layers should be designed in a manner that will enhance the extraction of charges while on the same reducing the recombination of charges. Innovations in the structure of perovskite material such as the use of mixed halide perovskites and the deposition of passivation layers have boosted J_{SC} in the recent past. All these advancements have helped to improve the overall performance of PCE in perovskite solar cells, making them potential candidates for commercialisation.

5.4 Fill factor (FF)

The fill factor (FF) measures the quality of the solar cell output characteristics and is given by the ratio of the maximum power output Pm to the product of V_{OC} and J_{SC} [51]. FF depends on the series resistance and the shunt resistance of the solar cell, as well as the recombination activities in the device. A high FF implies that charges are well extracted from the active layer, and there are low resistive losses, which favour high PCE. Measures to enhance FF in perovskite solar cells are the perovskite film surface quality, CTLs' conductivity, and the interface defect.

6. Potential applications and future prospects of perovskites

6.1 Applications

6.1.1 Building-integrated photovoltaics (BIPV)

The perovskite solar cells (PSCs) can be integrated into building structural components due to their versatility in terms of shape and, form and their aesthetics [52]. Perovskites, unlike silicon-based panels, are flexible in colour and transparency, making it possible to integrate them into windows, façades, and roofing systems. For instance, the semitransparent perovskite films allow architects to develop unique designs of buildings and make them energy-efficient and visually appealing.

6.1.2 Flexible electronics

Perovskite films can be easily integrated into flexible electronics. They are also low-density and in thin film form, so they can be applied to flexible systems such as fabrics and curved surfaces to fabricate solar panels [53]. It also opens up new product opportunities in portable and wearable electronics like Foldable solar panels for charging or energy fabrics.

6.1.3 Wearable devices

The application of perovskite material has been considered in wearables because this material is light and has high energy conversion efficiency. Integrating perovskite solar cells with wearable devices will produce a power supply for gadgets like smartwatches, fitness trackers, and other products [54]. Because they can easily be made in different sizes and shapes, they will be suitable for incorporating into the fashions that are worn daily.

6.1.4 Solar-powered vehicles

Perovskite solar cells will probably revolutionise the automotive industry, whereby the cells are installed in cars to provide extra or even primary power. Due to their low weight, flexibility, and high efficiency of operation, they can be mounted on roofs, windows, and other parts of the body of a car. This can help charge electric vehicles (EVs), avoid reliance on external charging, and even increase the driving range by charging all the time. In solar-powered vehicles, perovskites could also be helpful in energy control systems to power the vehicle electronics and reduce the effect on the environment [55].

6.1.5 Portable solar chargers

Perovskite solar cells are light and highly efficient; therefore, they are suitable for portable solar chargers. These chargers can be used to charge small portable electronics like smartphones, tablets, cameras, and the like in a convenient and environment-friendly way. As perovskite-based chargers are flexible, they can be rolled up, folded, or placed in a backpack and other outdoor equipment, which can be very useful when camping, hiking, and other activities without direct access to the main power supply [56].

6.1.6 Space applications

Due to the low weight and high efficiency of perovskite solar cells, they are suitable for use in space, including satellite and space station operation, as well as in space exploration. In space, weight is a significant concern, and since perovskite solar panels are relatively lighter than traditional silicon panels, the cost of launching and operating them is considerably lower [57]. In addition, they can be easily arranged in a compact, foldable format that can be carried around and then unfolded in space. Efforts are still being made to make perovskite more radiation tolerant, thus making them suitable for long-term space expeditions.

6.2 Unique properties of perovskites

Perovskite materials exhibit several unique properties that contribute to their suitability for these applications. The following are some of the properties of perovskite materials that make them suitable for these applications:

6.2.1 High efficiency

Perovskite solar cells have been researched for the last few years due to their high PCE, which is at par with silicon-based solar cells. Perovskites have other favourable optoelectronic characteristics, including a high absorption coefficient, long carrier diffusion length, and low recombination rate, hence high efficiency. Such characteristics enable perovskite solar cells to absorb and convert more sunlight into electricity. The perovskite cells' laboratory efficiencies are now over 25%, thus at par with the best silicon technologies. Likewise, the versatility of incorporating perovskites with other materials in tandem structures is beneficial for further efficiency improvement, making the perovskite solar cells a premier contender for the next generation of photovoltaic systems.

6.2.2 Low processing costs

This is one of the greatest strengths of perovskite solar cells, mainly because it has been predicted that they will not be costly to produce. Silicon solar cells, on the other hand, involve high-temperature and energy-consuming processes such as crystal growth and wafer fabrication. At the same time, perovskite can be synthesised and deposited at low temperatures through a solution process. Such deposition methods as spin coating, inkjet depositing, and spray coating allow manufacturing at a large scale and a low cost. Moreover, the roll-to-roll printing technique, which has been used for flexible electronics, can also be used to produce perovskite solar cells on flexible substrates continuously. This low-temperature processing, combined with the possibility of scaling up the production, significantly reduces the cost of production of perovskite solar cells and thus can be suitable for large-scale production.

6.2.3 Tunable bandgap

One of the most outstanding characteristics of perovskite materials is the ability to control the bandgap's value by changing the material's chemical composition, for example, replacing one or more of the halogen atoms or the cation. This tunability makes it possible for perovskites to be tailored to fit in the various segments of the solar spectrum and is hence suitable for photovoltaic applications. For instance, by choosing particular compositions, it is possible to enhance absorption in specific spectrum regions, thus enhancing the performance of tandem solar cells or multi-junction devices. This flexibility in adjusting the bandgap will allow for creating new perovskite-based photovoltaic applications that are more specialised and will add to the versatility of perovskites in innovative solar systems.

6.2.4 Lightweight structure

Perovskite solar cells are much lighter than conventional silicon solar cells, which is a significant advantage in several ways. Firstly, perovskite solar cells are lightweight and, as such, can be applied in aerospace applications, portable electronics, and wearable technology. The low density of perovskites also makes them portable and easy to transport and install, thus cutting the costs that accompany the transport and installation of large solar plants. This property is precious for such applications where conventional rigid and massive silicon panels are not suitable, thus opening new opportunities for the development of solar technology.

6.2.5 Wide absorption spectrum

Perovskites are also characterised by a high absorption coefficient value that is over the spectrum range of UV and NIR. This broad absorption characteristic makes perovskite solar cells able to capture more energy from the sun—another reason for high PCE. Perovskites also have a feature of trapping low-energy photons that other materials may not trap, and this makes them effective even in polluted or low-light conditions, such as those that prevail in cloudy or indoor environments. This broad-spectrum absorption makes perovskites suitable for any lighting conditions and, therefore, suitable for any application in solar energy conversion.

6.2.6 Defect tolerance

Perovskite materials are very immune to defects in their structures, which is a plus for their use. However, they can have high performance, so the quality of the material can be a little low, making producing these cells less sensitive and cheaper. This defect tolerance also enables perovskite solar cells to have better stability and reliability in real life since there will always be some defects in the solar cells. For this reason, the ability to operate at the highest efficiency with imperfections in the material structure makes perovskites stable and reliable for large-scale applications in various configurations and applications.

All of these properties contribute to the conclusion that perovskite materials can change the field of solar energy and further extend their application to other new and advanced technologies.

6.3 Emerging research directions

Because of the fast development of perovskite materials, much attention has been paid to it in both academic research and the industry. Indeed, it is evident that as the field advances new ideas and potential research areas that may help to enhance the performance, stability, and scalability of the perovskite-based technologies are emerging. These include following areas depicting the versatility of perovskite materials in future generation energy technologies:

6.3.1 Tandem solar cells

Tandem solar cells are the most promising development in the photovoltaic field, where perovskite materials are incorporated into silicon or other photovoltaic materials to increase efficiency. These tandem structures can be created by stacking perovskites on top of the conventional silicon cells, and this is because different materials have different absorption characteristics. The first perovskite layer is designed to capture high-energy photons, and the second silicon layer is responsible for capturing low-energy photons for optimum energy conversion. It has the possibility of attaining efficiencies that are much higher than those of single-junction cells that are restricted by the Shockley-Queisser limit. With research advancing further, the concept behind the tandem cells is expected to gain higher levels of commercial appeal, which will lead to better solar solutions that can revolutionise the energy market [58].

6.3.2 Perovskite-based multi-junction solar cells

Multi-junction solar cells using perovskite materials are another way to develop solar cells. These cells accumulate one layer of various photovoltaic materials, each intended to capture specific parts of the solar spectrum [59]. Thus, by controlling the composition and thickness of these layers, scientists strive to go beyond the efficiency of p-n cells. In theory, the various junctions in multi-junction cells can provide efficiencies of over 40% and are thus some of the most effective solar cells. The problem here is how to achieve stable, high-quality interfaces between the layers and the fact that each layer must function as well as possible. Future developments in material technology and in manufacturing perovskite-based multi-junction photovoltaic cells may create a new benchmark for the entire solar power system.

6.3.3 Next-generation photovoltaic technologies

Combining perovskites with future-oriented PV technologies such as OPVs and QDSCs is a current research field. These so-called hybrid technologies are designed to optimise the properties of each technology, thus creating new generations of photovoltaic devices with higher efficiency, flexibility, and adaptability [60]. For instance, perovskite-organic hybrid cells could combine the advantages of being flexible and cheap to produce organic photovoltaics and the high efficiency of perovskites. Likewise, incorporating perovskites with quantum dots—nanoscale semiconductor particles with adjustable photovoltaic characteristics—can boost light absorption and carrier dynamics in solar cells. These next-generation photovoltaic technologies are seen to offer the potential for delivering more flexible and high-performance solar systems for various uses, including portable electronic devices and large power stations.

6.3.4 Perovskite-silicon heterojunctions

Perovskite-silicon heterojunctions significantly improve solar cell design as they take the best of the perovskite and silicon properties. In this case, researchers can stack a perovskite on top of a silicon solar cell to form a heterojunction that offers both the efficiency of perovskite and the stability of silicon [61]. This combination not only improves the total conversion efficiency of power but also leverages the mature silicon manufacturing system and demand. The invention of high-performance and stable perovskite/silicon heterojunctions can change the solar industry and provide a cost-effective route towards exceeding the efficiency limits of conventional silicon-based photovoltaics. These hybrid cells could become one of the most popular technologies in the solar industry as research continues, thus bringing down costs and increasing efficiency.

6.3.5 Perovskite quantum dots for photovoltaics

Perovskite QDs are a relatively new research area with many advantages over bulk perovskite materials. Quantum dots are semiconductor nanocrystals that have their dimensions in the nanometre size and are capable of quantum confinement effects for their band gap and optical characteristics. In the context of photovoltaics, perovskite QDs can be employed to form high efficiency and stability of the solar cells and enhance light trap and charge carrier dynamics. The size of QDs is small, allowing for the fine-tuning of the electronic characteristics of the material, and the ability to decide what these characteristics are is a significant advantage that the technology has over other approaches. Also, perovskite QDs can be solution-processed, ensuring they can be used in large-scale, low-cost production. Perovskite QDs are still under investigation, and future research may lead to higher efficiency and new applications such as solar windows and portable, flexible solar cells [61].

6.3.6 All-perovskite tandem solar cells

Tandem solar cells made only of perovskite materials incorporate two or more layers of perovskite materials with varying bandgaps arranged in a tandem manner.

While conventional tandem cells use perovskites with other materials, such as silicon, all-perovskite tandem cells utilise various perovskite layers designed for different solar spectrum parts [62]. It is possible to achieve very high utilisation of sunlight, and thus, the power conversion efficiencies are higher than those of single-junction perovskite cells. Current work in this field is dedicated to modifying perovskite to achieve high stability and efficiency, achieving the desired bandgap, and optimising the fabrication of the structure to form perfect interfaces between the layers. All-perovskite tandems are considered one of the most promising approaches to creating the next-generation ultra-efficient, lightweight, flexible, and low-cost solar cells.

6.3.7 Perovskite solar cells for indoor photovoltaics

Perovskite solar cells for indoor photovoltaics (IPV) are a fast-emerging area that aims at harvesting energy from artificial lighting [63]. Indoor conditions are characterised by lower light intensity and different spectrums compared to sunlight, and perovskites are well suited to indoor light conditions because of their ability to tune the absorption and their efficiency at low light. Indoor perovskite materials may pave the way to highly efficient IPV systems that can power small devices in smart homes, as well as IoT devices such as sensors, small electronics, wearables, and more, without the need for batteries. This research direction is most suitable for developing indoor energy self-sufficient spaces, minimum battery replacement or recharging requirements, and the emergence of a new generation of energy-efficient indoor technologies.

These new directions demonstrate the richness of perovskite materials and promise to go beyond the state of the art in photovoltaic applications. Therefore, the potential application of perovskite solar cells is limited to conventional solar cells and extended to many other fields. Nevertheless, commercialisation has problems, but constant research and development offer new possibilities for advancing photovoltaic technologies.

7. Conclusion

This chapter seeks to present the subject of this study, that is, perovskite solar cells (PSCs), as the object that holds the potential to transform the solar energy sector. PSCs are morphologically distinct from conventional photovoltaics based on silicon due to their high efficiency, low processing costs, and flexibility, which can be adapted to BIPV, flexible electronics, and wearable devices. Thus, it is creative in addressing energy issues and ensures that solar technology is integrated into people's lifestyles. However, there are some issues related to the development of PSCs to reach full-scale commercialisation. There are stability issues as PSCs are equally affected by environmental factors that could affect their performance in the long run. Moreover, the transition of the process from a pilot scale to an industrial scale has its own problems in terms of technology and the economy. However, the future of PSCs is not so gloomy, as the development in this field is still ongoing. It is for this reason that tandem solar cells that are combined with other materials, such as PSCs, to offer higher efficiency and perovskite-based multi-junction cells that are developed to capture more of the solar spectrum are believed to be among the most promising novelties of the photovoltaic industry. Perovskite QDs have great potential for fabricating efficient and stable solar cells with excellent optical properties. Some improvements have been reported in perovskite-silicon heterojunction and all perovskite tandem solar cells, which combine the best of both worlds and expand the applicability of the technology. Apart from improving PSC's performance and flexibility, these advances generate new applications such as high-efficiency solar panels, energy-harvesting devices for smart homes and wearable technologies. If the current challenges are met and the research continues, PSCs can significantly revolutionise the solar industry with efficient, clean, and flexible energy solutions. Hence, as renewable energy is adopted more often, PSCs can assist in providing the world with energy without being detrimental to the environment.

Acknowledgements

The authors are thankful to PV Spoke National Science and Innovation Energy Research Programme and National Research Foundation (GUN:137944 and 118947) and Govan Mbeki Research and Development Centre (GMRDC), South Africa, University of Fort Hare.

Author details

Shah Aarif Ul Islam* and Edson Leroy Meyer
Fort Hare Institute of Technology, University of Fort Hare, Alice, Eastern Cape, South Africa

*Address all correspondence to: aarifulislam111@gmail.com; ashah@ufh.ac.za

References

[1] Hosseini SE, Wahid MA. Hydrogen from solar energy, a clean energy carrier from a sustainable source of energy. International Journal of Energy Research. 2020;**44**(6):4110-4131

[2] Dambhare MV, Butey B, Moharil S. Solar photovoltaic technology: A review of different types of solar cells and its future trends. In: International Conference on Research Frontiers in Sciences (ICRFS 2021) 5th-6th February 2021, Nagpur, India. Journal of Physics: Conference Series. 2021;**1913**:012053. DOI: 10.1088/1742-6596/1913/1/012053

[3] Hayat MB et al. Solar energy—A look into power generation, challenges, and a solar-powered future. International Journal of Energy Research. 2019;**43**(3):1049-1067

[4] Kabir E et al. Solar energy: Potential and future prospects. Renewable and Sustainable Energy Reviews. 2018;**82**:894-900

[5] Maka AO, Alabid JM. Solar energy technology and its roles in sustainable development. Clean Energy. 2022;**6**(3):476-483

[6] Ashraf M et al. Recent trends in sustainable solar energy conversion technologies: Mechanisms, prospects, and challenges. Energy & Fuels. 2023;**37**(9):6283-6301

[7] Ravishankar E et al. Achieving net zero energy greenhouses by integrating semitransparent organic solar cells. Joule. 2020;**4**(2):490-506

[8] Victoria M et al. Solar photovoltaics is ready to power a sustainable future. Joule. 2021;**5**(5):1041-1056

[9] Jacobson MZ. Review of solutions to global warming, air pollution, and energy security. Energy & Environmental Science. 2009;**2**(2):148-173

[10] Lee SW et al. Historical analysis of high-efficiency, large-area solar cells: Toward upscaling of perovskite solar cells. Advanced Materials. 2020;**32**(51):2002202

[11] Li Z et al. Scalable fabrication of perovskite solar cells. Nature Reviews Materials. 2018;**3**(4):1-20

[12] Glunz SW, Preu R, Biro D. Crystalline silicon solar cells: State-of-the-art and future developments. Comprehensive Renewable Energy. 2012;**1**:353-387

[13] Olaleru S et al. Perovskite solar cells: The new epoch in photovoltaics. Solar Energy. 2020;**196**:295-309

[14] Petrović M, Chellappan V, Ramakrishna S. Perovskites: Solar cells & engineering applications–materials and device developments. Solar Energy. 2015;**122**:678-699

[15] Kumar D et al. Synthesis techniques and applications of perovskite materials. In: Tian He, editor. Perovskite Materials, Devices and Integration. UK: IntechOpen; 2020

[16] Adinolfi V et al. The electrical and optical properties of organometal halide perovskites relevant to optoelectronic performance. Advanced Materials. 2018;**30**(1):1700764

[17] Sangavi T et al. Synergizing experimental and theoretical insights: Unveiling the solar potential of La_2NiMnO_6 double perovskite for

enhanced efficiency and sustainability in photovoltaics. Chemical Engineering Journal. 2024;**486**:150216

[18] Brittman S, Adhyaksa GWP, Garnett EC. The expanding world of hybrid perovskites: Materials properties and emerging applications. MRS Communications. 2015;**5**(1):7-26

[19] Zhou J, Huang J. Photodetectors based on organic–inorganic hybrid lead halide perovskites. Advanced Science. 2018;**5**(1):1700256

[20] King G, Woodward PM. Cation ordering in perovskites. Journal of Materials Chemistry. 2010;**20**(28):5785-5796

[21] Akinpelu A, Bhullar M, Yao Y. Discovery of novel materials through machine learning. Journal of Physics: Condensed Matter. 2024;**36**(45):453001

[22] Igbari F, Wang ZK, Liao LS. Progress of lead-free halide double perovskites. Advanced Energy Materials. 2019;**9**(12):1803150

[23] Connor BA et al. Understanding the evolution of double perovskite band structure upon dimensional reduction. Chemical Science. 2023;**14**(42):11858-11871

[24] Jeon NJ et al. Compositional engineering of perovskite materials for high-performance solar cells. Nature. 2015;**517**(7535):476-480

[25] Polman A et al. Photovoltaic materials: Present efficiencies and future challenges. Science. 2016;**352**(6283):aad4424

[26] Aarif Ul Islam S, Ikram M. Structural stability improvement, Williamson Hall analysis and band-gap tailoring through A-site Sr doping in rare earth based double perovskite La_2NiMnO_6. Rare Metals. 2019;**38**:805-813

[27] Ahmad K, Kumar P, Kim H. Recent progress in lead free tin-halide perovskite materials based solar cells via SCAPS based numerical simulation. ChemistrySelect. 2024;**9**(31):e202402044

[28] Wang B et al. The charge carrier dynamics, efficiency and stability of two-dimensional material-based perovskite solar cells. Chemical Society Reviews. 2019;**48**(18):4854-4891

[29] Maiti A et al. Defects and their passivation in hybrid halide perovskites toward solar cell applications. Solar RRL. 2020;**4**(12):2000505

[30] Sojati R. Solvent Annealing as a Post-Treatment: Towards Single Crystalline Epitaxial $MAPbI_3$. Enschede, Netherlands: University of Twente; 2023. Available from: https://purl.utwente.nl/essays/97337

[31] Dahal B, Li W. Configuration of methylammonium lead iodide perovskite solar cell and its effect on the device's performance: A review. Advanced Materials Interfaces. 2022;**9**(19):2200042

[32] Zhang Y. Metal Halide Perovskite Nanocrystals: Synthesis, Stability, and Lead-Free Alternatives. Austin: University of Texas; 2021. DOI: 10.26153/tsw/52014

[33] Schileo G, Grancini G. Lead or no lead? Availability, toxicity, sustainability and environmental impact of lead-free perovskite solar cells. Journal of Materials Chemistry C. 2021;**9**(1):67-76

[34] Martin ES. The Photophysics of Metal Halide Perovskites for Next-Generation Solar Cells. UK: University of Oxford; 2019

[35] Kour R et al. Potential substitutes for replacement of lead in perovskite solar cells: A review. Global Challenges. 2019;**3**(11):1900050

[36] Tasleem S, Tahir M. Current trends in strategies to improve photocatalytic performance of perovskites materials for solar to hydrogen production. Renewable and Sustainable Energy Reviews. 2020;**132**:110073

[37] Prakash J et al. Progress in tailoring perovskite based solar cells through compositional engineering: Materials properties, photovoltaic performance and critical issues. Materials Today Energy. 2018;**9**:440-486

[38] Saki Z et al. Solution-processed perovskite thin-films: The journey from lab- to large-scale solar cells. Energy & Environmental Science. 2021;**14**(11):5690-5722

[39] Miah MH et al. Perovskite materials in X-ray detection and imaging: Recent progress, challenges, and future prospects. RSC Advances. 2024;**14**(10):6656-6698

[40] Chou L-H et al. Scalable ultrasonic spray-processing technique for manufacturing large-area $CH_3NH_3PbI_3$ perovskite solar cells. ACS Applied Materials & Interfaces. 2018;**10**(44):38042-38050

[41] Shamardin A et al. Quality improvement of CZTS thin films deposited by spray pyrolysis method using pulsed Nd: YAG laser irradiation. Applied Surface Science. 2019;**488**:827-835

[42] Deng Y et al. Scalable fabrication of efficient organolead trihalide perovskite solar cells with doctor-bladed active layers. Energy & Environmental Science. 2015;**8**(5):1544-1550

[43] Vaynzof Y. The future of perovskite photovoltaics—Thermal evaporation or solution processing? Advanced Energy Materials. 2020;**10**(48):2003073

[44] Zhang J et al. Critical review of recent progress of flexible perovskite solar cells. Materials Today. 2020;**39**:66-88

[45] Qaid SMH et al. Single-source thermal evaporation growth and the tuning surface passivation layer thickness effect in enhanced amplified spontaneous emission properties of $CsPb(Br_{0.5}Cl_{0.5})_3$ perovskite films. Polymers. 2020;**12**(12):2953

[46] Fara L et al. Review: Heterojunction tandem solar cells on Si-based metal oxides. Energies. 2023;**16**(7):3033

[47] Lu X et al. Review on preparation of perovskite solar cells by pulsed laser deposition. Inorganics. 2024;**12**(5):128

[48] Kim JY et al. High-efficiency perovskite solar cells. Chemical Reviews. 2020;**120**(15):7867-7918

[49] Elumalai NK, Uddin A. Open circuit voltage of organic solar cells: An in-depth review. Energy & Environmental Science. 2016;**9**(2):391-410

[50] Maka AOM. Performance Analysis and Characterisation of a High Concentrating Solar Photovoltaic Receiver. Edinburgh, Scotland, UK: Heriot-Watt University; 2020. Available from: http://hdl.handle.net/10399/4304

[51] Emery K. Measurement and characterization of solar cells and modules. In: Handbook of Photovoltaic Science and Engineering. Chichester, West Sussex, United Kingdom: John Wiley & Sons Ltd. The Atrium Southern Gate; 2011. pp. 797-840

[52] Koh TM et al. Halide perovskite solar cells for building integrated photovoltaics: Transforming building façades into power generators. Advanced Materials. 2022;**34**(25):2104661

[53] Corzo D, Tostado-Blázquez G, Baran D. Flexible electronics: Status, challenges and opportunities. Frontiers in Electronics. 2020;**1**:594003

[54] Balilonda A et al. Perovskite fiber-shaped optoelectronic devices for wearable applications. Journal of Materials Chemistry C. 2022;**10**(18):6957-6991

[55] Yamaguchi M et al. Development of high-efficiency and low-cost solar cells for PV-powered vehicles application. Progress in Photovoltaics: Research and Applications. 2021;**29**(7):684-693

[56] Yang Y et al. Perovskite solar cells based self-charging power packs: Fundamentals, applications and challenges. Nano Energy. 2022;**94**:106910

[57] Tu Y et al. Perovskite solar cells for space applications: Progress and challenges. Advanced Materials. 2021;**33**(21):2006545

[58] Lal NN et al. Perovskite tandem solar cells. Advanced Energy Materials. 2017;**7**(18):1602761

[59] Yamaguchi M et al. Multi-junction solar cells paving the way for super high-efficiency. Journal of Applied Physics. 2021;**129**(24):240901, 1-15

[60] Parisi ML et al. Prospective life cycle assessment of third-generation photovoltaics at the pre-industrial scale: A long-term scenario approach. Renewable and Sustainable Energy Reviews. 2020;**121**:109703

[61] Albrecht S et al. Monolithic perovskite/silicon-heterojunction tandem solar cells processed at low temperature. Energy & Environmental Science. 2016;**9**(1):81-88

[62] Wen J, Tan H. Present status and future prospects for monolithic all-perovskite tandem solar cells. Science China Materials. 2022;**65**(12):3353-3360

[63] Muhammad BT et al. Halide perovskite-based indoor photovoltaics: Recent development and challenges. Materials Today Energy. 2022;**23**:100907

Chapter 7

Hot Isostatic Pressing (HIP) in Advanced Ceramics Production

Pouria Dehghani, Seyed Salman Seyed Afghahi and Farshad Soleimani

Abstract

Hot isostatic pressing (HIP) is pivotal in advancing ceramic materials by consolidating and densifying them through high temperature and pressure. This technique significantly improves mechanical, thermal, and electrical properties, resulting in ceramics with enhanced structural integrity and reduced porosity. HIP addresses challenges of traditional fabrication methods by achieving near-theoretical density while minimizing residual pores and defects. The outcome is ceramics with superior mechanical strength, wear resistance, and thermal stability, making them ideal for demanding applications such as cutting tools, biomedical implants, and high-performance engine parts. Additionally, HIP facilitates the production of complex-shaped ceramic components, enabling intricate designs and the integration of multiple functionalities within a single part. It can also consolidate ceramic-based composites with reinforcing fibers or particles, providing tailored properties for specific applications. Overall, HIP is essential in the development of high-performance, reliable, and cost-effective ceramic components necessary for various industries, including aerospace, energy, and healthcare.

Keywords: hot isostatic press, HIP, capsulation advanced ceramics, transparent ceramics, ceramic manufacturing techniques

1. Introduction

Hot isostatic pressing (HIP) is a manufacturing process designed to decrease the porosity and increase the density of various ceramic materials. This technique improves the mechanical properties and performance of these materials by subjecting components to high temperatures and isostatic gas pressure in a high-pressure chamber. Argon is commonly used as a pressurizing gas due to its inert nature, which prevents chemical reactions with the materials. As the chamber is heated, the internal pressure rises, often aided by continuous gas pumping to maintain the desired pressure. The uniform application of pressure from all directions is why this process is called "isostatic." Currently, HIP is extensively used in the production of ceramic components. The advantages of the HIP process have led to its widespread adoption in shaping structural ceramics. Non-oxide ceramics can be converted into high-density components using this method. Additionally, components produced through HIP usually have small grain sizes and often do not require additives. The combination of high-density and fine grain structure results in components with outstanding properties. In the HIP forming process, temperatures typically range from 500°C to 1900°C, and pressures range from 500 to 2000 kgf/cm^2. Key applications of these high temperatures and pressures include: (1) producing components capable of withstanding high operational pressures, (2) enhancing physical properties through phase transformations under pressure, (3) accelerating solid-gas chemical reactions, (4) significantly reducing the need for sintering additives in ceramics, (5) enabling the manufacture of parts that closely match their final shapes, minimizing the need for complex machining, and (6) preventing excessive grain growth There are generally two main methods for shaping parts using hot isostatic pressing: (a) pre-shaping through techniques such as slip casting, cold isostatic pressing (CIP), additive manufacturing (e.g., 3D printing), and plastic forming methods (e.g., injection molding). After pre-shaping, the products are sealed in chambers or glass capsules to undergo the HIP process. (b) The pre-shaped components initially undergo pressureless sintering, achieving a density of 90–95% with minimal porosity. To reach near-theoretical maximum density, the components are then subjected to a post-HIP process [1–3].

2. Encapsulation method

The encapsulation method is a widely utilized technique for high-density sintering of powders. This process involves placing the powder into a mold, sealing the inlet, and applying isostatic pressing to the capsule. By using hot gas to exert pressure, this technique enables the production of parts with enhanced density [4].

The applications of encapsulation are diverse, encompassing the sintering of powders that are difficult to process by conventional methods, the creation of composites, and the impregnation of porous materials with a molten phase. A crucial element of the encapsulation process is the effective sealing of the capsule, which ensures it remains gas-tight. The capsule's function is to uniformly deliver gas pressure to the powder, facilitating compaction and the subsequent formation of the component [5].

Various materials and techniques can be used for encapsulation, including metals, glasses, and coatings. Metal encapsulation typically employs materials with high softening points and adequate plasticity. Glass encapsulation is particularly beneficial for producing ceramic parts, as it minimizes surface contamination and allows for easier removal. Coated capsules may utilize a chemical vapor deposition (CVD) process to apply a micrometer-thick coating on the component's surface [4].

Distinct encapsulation methods each offer unique advantages:

a. They enable high-density fabrication of materials that are challenging to sinter with traditional techniques, such as spherical powders, superalloys, and nitride ceramics like boron nitride and silicon nitride,

b. They provide some control over the capsule atmosphere, allowing for processes such as vacuum conditions when purity is essential,

c. They incorporate preheating systems tailored for specific heat treatment needs,

d. They facilitate the creation of hollow cylinders using the hard-core technique, employing a material with superior temperature resistance and mechanical strength compared to the original powder.

Nevertheless, the encapsulation method has its limitations:

a. It requires a complex preprocessing phase, including the shaping and sealing of capsules, to successfully perform the hot isostatic pressing (HIP) process.

b. There are restrictions in choosing encapsulation materials due to differences in the thermal expansion coefficients between the encapsulating material and the part.

2.1 Metal encapsulation

The powder or component formed from the powder is sealed in a metal capsule and subjected to the hot isostatic pressing (HIP) process. The choice of material for encapsulation depends on factors such as the heat treatment temperature, the reactivity of the powder with other materials (type of capsule), and cost considerations. Typically, the maximum recrystallization temperature for metals and superalloys is below 1500°C. In heat treatment processes involving alloys and metals, stainless steel capsules are commonly employed due to their ease of fabrication and welding. A primary reason for encapsulation is to achieve precise geometrical shapes; thus, metals with good malleability facilitate the shaping of the metal powder for sintering. However, encapsulation is generally limited to simple geometries, such as cylindrical shapes. The composition, impurity levels, and average particle size of the powder significantly influence the mechanical properties of the final sintered component. When employing the HIP process for encapsulation to produce high-density parts, the bulk density of the components is crucial. Consequently, the use of spherical particles with high compressibility is a key aspect of this process. For powders that cannot be compressed to 60% density, preforming techniques—such as cold isostatic pressing (CIP)—are recommended. The dimensions of the capsule required for the encapsulation process are determined by the relative density of the molded or sintered powder or component [6].

As illustrated in **Figure 1**, metal capsules typically consist of a thin welded metal body, with the top and bottom sealed using tubes that serve as powder and vacuum

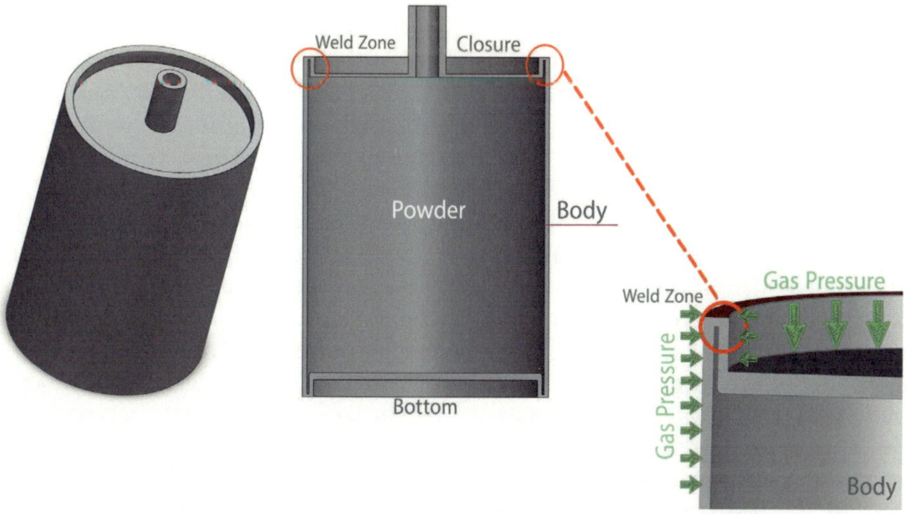

Figure 1.
Typical mild steel capsule.

inlets. Once the powder is packed into the capsule, air, moisture, and gases are evacuated *via* a vacuum process. The heat treatment process, initiated by HIP, begins at lower temperatures. Due to the softening of the steel at elevated temperatures, heating continues until the ductility temperature of the steel is reached. Subsequently, gas pressure is incrementally increased until both pressure and temperature reach their maximum values. During the HIP process, the capsule experiences uniform pressure from all directions, allowing the materials undergoing heat treatment to approach their final properties [6]. After the hot isostatic pressing process, the capsule must be removed. Its shape and structure significantly differ from those of the initial capsule and powder, and in many cases, the capsule may have reacted with the formed component, leading to penetration. Consequently, the removal of the capsule can be a time-consuming and costly endeavor. Mechanical methods, such as turning, and chemical dissolution are potential techniques for this removal process [4].

2.2 Glass encapsulation method

Another encapsulation method involves the use of glass capsules, which are then sealed under vacuum. This process is akin to metal encapsulation; however, it becomes challenging to apply when the components exhibit irregular shapes or high porosity. Glass capsule technology is particularly effective for fabricating engineering ceramic parts that require high sintering temperatures, such as silicon nitride and silicon carbide. The mechanisms and loading conditions for the powder within the sample are largely consistent with those used in metal capsules. The advantages of utilizing glass capsules include their availability, cost-effectiveness, and reduced preparation time [6].

2.2.1 Glass encapsulation material

To address the challenges associated with the metal encapsulation process—such as weld points and the brittleness of steel—while accommodating the higher temperatures required for sintering high-temperature ceramics like silicon nitride, a glass encapsulation process has been developed. This method offers several advantages, including the production of components that are closer to their final shape and improved cost-effectiveness due to the easier removal of glass material. It is important to note that the glass encapsulation technique is still in the developmental stages and has shown promising results at the laboratory scale, though further advancements will require time. The glasses used in encapsulation must exhibit high-temperature tolerance, typically ranging from 1600°C to 2000°C, making them unsuitable for metal capsules. Additionally, the softening point and viscosity of the glass are critical for effectively sintering engineering ceramic components, as these properties ensure efficient performance and optimal pressure transfer. During the initial heating stage, glass capsules surrounding the raw material powders act as a barrier to neutral gas flow, and given the potential reactions between the encapsulating material and the glass, it is essential for the glass to have a softening point lower than the sintering temperature of the components. As the temperature increases, the layers of material gradually migrate toward the component, and the lack of gas penetration within these layers leads to pressure buildup and compaction of the powder. A table detailing various types of glasses and their characteristics is provided below [7, 8]. The properties needed for encapsulation glasses are highly dependent on the glass composition, and all the composition constraints mentioned earlier must be followed to ensure that all the necessary properties are integrated into a single glass.

The properties required for encapsulation glasses depend critically upon the composition of the glass, and all of the composition limitations hereinabove set forth must be observed if all of the required properties are to be combined in a single glass. The coefficient of thermal expansion (CTE) of glass significantly influences the glass encapsulation process. A CTE value that closely matches that of the material being processed under hot isostatic pressing (HIP) helps mitigate the formation of micro-cracks. Furthermore, the glass utilized as a capsule must be precisely bonded to the bulk material. The strain point refers to the temperature at which internal stresses in the glass are relaxed over a few hours. In contrast, the annealing point indicates the temperature where these internal stresses are relieved within a few minutes [9].

Key performance indicators (KPIs) in the glass encapsulation method include strain points, annealing points, softening points, and working temperatures, as outlined in **Table 1**. Additionally, **Figure 2** illustrates the relationship between temperature and viscosity in graphical form. Once the temperature reaches the annealing point, deformation of the glass occurs, driven by applied forces. It is crucial to achieve the glass's annealing temperature; if stress is applied before this temperature is reached, small cracks and breakage may develop in the glass capsule [10].

In a research study, an intermediate layer is used, which can be created by different methods, such as spraying or deep coating, and has a thickness ranging from 0.5 to 1 mm. After coating, the ceramic piece is degassed at room temperature and then placed into the hot isostatic pressing (HIP) system, where the temperature is raised until the outer layer is completely melted. The powder between the part and the glass layer prevents the penetration of the glass into the part. Applying pressure after changing the initial state of the outer layer of glass takes place in the temperature range of 600–1100°C, while the inner layer of glass powder also changes state at 1300–1600°C. Then, the hot isostatic pressing process is carried out in the temperature range of 1600°C with a pressure of 100–200 MPa. After the glass processing is finished, the glass is removed from the piece using the sandblasting process. Additionally, BN powder is used as a spacer layer between the part and the

Composition		**PYREX ©**		**VYCOR ©**	**QUARTZ**
		SiO_2: 80.9%	**Al_2O_3: 2.3%**	**SiO_2: 96%**	**SiO_2: 100%**
		Fe_2O_3: 0.03%	**B_2O_3: 12.7%**	**B_2O_3: 2.6%**	
		Na_2O: 4.0%	**K_2O: 0/04%**	**Others**	
Thermal expansion coefficients	0–300°C	32.5		7.5	5.5
	25°C (strain point)	35		5.5	3.5
Characteristic temperature (°C)	Strain point	510		890	956
	Slow-cooling point	560		1020	1084
	Softening point	821		1530	1580
	Operation point	1252		—	—
Density	(g/cm^3)	2.23		2.18	2.20
Youngs modulus	(×10 kg/mm^2)	6.4		6.9	7.4

Table 1.
Properties of glasses employed in the glass encapsulation process.

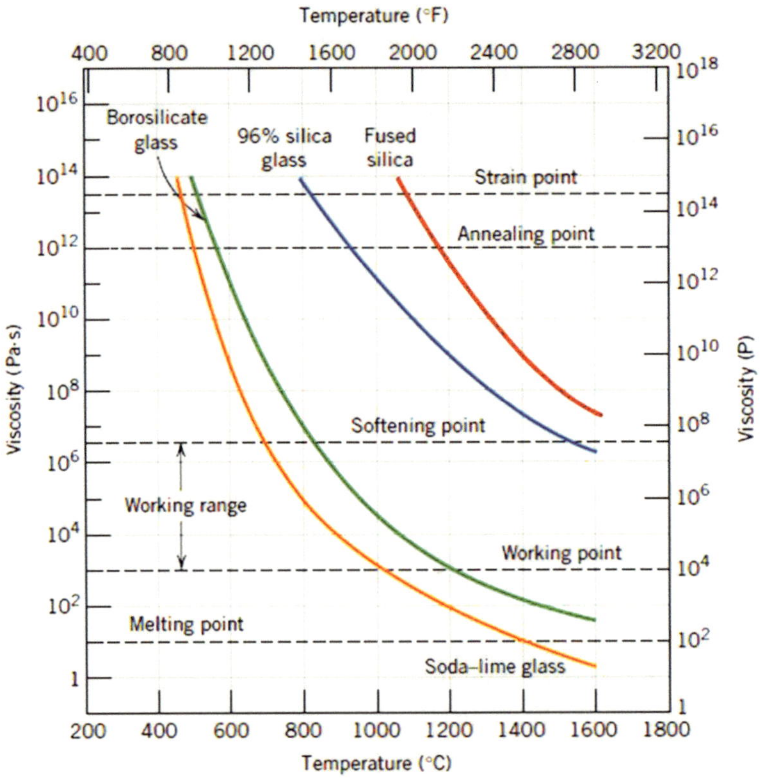

Figure 2.
The correlation between glass temperature and viscosity [10].

glass to facilitate the separation of the capsule from the part, and B_2O_3 can also be used as a protective layer by mixing it with Si_3N_4 at a ratio of 50:50 by weight, where the B_2O_3 reacts with the surface layer of Si_3N_4 at a temperature of approximately 1000°C, forming a boron nitride layer that prevents the penetration of glass into the silicon nitride ceramic. The pre-sintered body of silicon nitride is placed inside a capsule made of SiO_2 or borosilicate glass, which is then kept in an oven at 100 with a vacuum of 0.1 Pa for 8 hours before being sealed and heated to a temperature of 1250°C, where the softening point of the glass is reached, and the glass changes its shape. The capsule is then placed in the HIP chamber, and a pressure between 200 and 300 MPa and a temperature between 1700°C and 1800°C are applied to the parts for at least 2 hours. Using this method, more complex shapes can be produced, and compression of Si_3N_4 using HIP in SiO_2 powder is another method, where silicon nitride ceramic is placed in a bed of SiO_2 powder inside a capsule of borosilicate glass, usually Pyrex [11, 12].

2.2.2 Sealing of glass capsules

Solder glasses can be classified as either vitreous or devitrifying. Vitreous glasses are thermoplastic materials that melt and flow at consistent temperatures each time they are heated. On the other hand, devitrifying solder glasses, often referred to as frits, are surface-nucleating thermosetting materials. When heated, they crystallize, resulting in their unique expansion properties [13]. The proximity of the glass to the main body raises the potential for chemical interaction due to temperature and pressure variations during the hot isostatic pressing (HIP) process. Therefore, it is advisable to select materials that exhibit minimal reactivity. Additionally, it is essential to ensure that the sintering temperature remains below the glass's softening point. Several techniques can be employed for sealing components intended for encapsulation with glass during the HIP process, including Ref. [2]:

- Sealing using the glass ampoule method
- Sealing under applied pressure
- Sealing by immersion in powder
- Sealing via the sintered glass method
- Sealing through a coating method

2.2.2.1 Glass ampoule method

In this method, the article is placed inside a glass ampoule and surrounded by glass powder, which is sealed using oxyacetylene flames. During sealing, a vacuum is simultaneously applied to ensure the interior of the parts is free of gases. Subsequently, the ampoule is heated in a hot isostatic pressing (HIP) unit to reach the glass softening temperature under reduced pressure before applying gas pressure. However, this technique has several disadvantages. First, the vacuum sealing process is sensitive; any leaks can cause the glass capsule to rupture when gas pressure is increased, potentially damaging heating elements. Second, the method is limited to samples with simple geometries. Lastly, the choice of glass composition is critical, as the capsule must soften at a temperature slightly below the sintering temperature while maintaining high viscosity throughout the densification process. A schematic representation of this process is shown in **Figure 3** [14].

2.2.2.2 Pressurize method

In this method, the component is initially shaped using powder metallurgy techniques and then surrounded by glass powder. Following this, heat treatment is performed until the glass powder softens. A uniaxial press is then employed to apply pressure, facilitating the welding of the powders and ensuring they bond together [14, 15].

2.2.2.3 Sintered glass method

In this process, glass powder is initially mixed with an appropriate solution to create a slurry, which is then applied to the surface of the preformed component. Following the application, the slurry is dried and sintered, resulting in a sealing coating that prevents gas penetration. If this coating process is performed in a single step, there is a risk of cracking and delamination. To mitigate this, the coating procedure is repeated multiple times until the desired thickness is achieved [14–16].

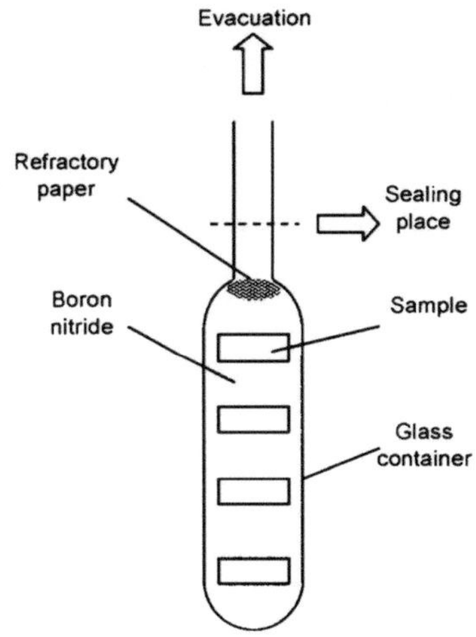

Figure 3.
Sealing process using the glass ampoule method [14].

2.2.2.4 The glass bath method

The method described in Ref. [6] involves processing a porous body by immersing it in powdered glass with a low softening temperature. This approach is similar to the encapsulation method but without the need for the complex encapsulation technique. However, this method has some drawbacks—the low-density green body tends to float on the molten glass, requiring a means to keep the workpiece submerged below the glass surface. Additionally, the low-softening-temperature glass often penetrates the pores of the workpiece, necessitating a specialized coating to be applied over the entire surface of the workpiece. The shaped components are initially encased in glass powder and then subjected to heat treatment, followed by the hot isostatic pressing (HIP) process. It is crucial that the sintering temperature exceeds the softening temperature of the glass, with the process beginning at the glass's softening point, at which point argon gas is injected to apply pressure, and the temperature is further elevated to reach the sintering point of the components during the immersion in glass powder [15, 17].

2.2.2.5 Coating method

The rationale for this method dates back to the proposal of coating the pores of a preformed raw material to prevent argon gas penetration. Instead of employing various types of capsules and sealing methods, the idea was to utilize a thin coating to obstruct argon gas during the hot isostatic pressing (HIP) process. Coating techniques considered for this purpose include plasma spraying, which creates a protective layer, as well as Physical Vapor Deposition (PVD) and Chemical Vapor Deposition (CVD). However, a major drawback of this approach is its inability to effectively fill all pores and create a uniform barrier against argon gas penetration. Conversely, one of the significant advantages of this method lies in its simplicity and cost-effectiveness, as well as the rapid removal of the coated layer using acids. **Table 2** presents the advantages and disadvantages of each method for sealing glass capsules [15].

Method	Advantages	Disadvantages
Glass ampoule	Simple method for forming cylindrical components	Needs a conduit to create a vacuum
	Exact regulation of the piece's environment	Shaping intricate components using this method is challenging
Sealing through the application of pressure	The sealing process can occur at a reduced temperature relative to alternative methods	The risk of microcracks and deformation in the component is markedly elevated
	It is a straightforward method for sealing rectangular and square components	It requires special sealing devices
	Precise control of the atmosphere of the piece	
Glass immersion method	No specialized equipment is necessary	It necessitates a significant quantity of glass powder
	Minimizing the steps involved in the sealing and encapsulation process	It requires a higher temperature for sealing
	Capability to utilize various types of glass, including powder, flat, and cylindrical forms	
Sintered glass method	It needs only a small quantity of glass	The likelihood of microcracks and deformation in the component is notably increased
	The ability to form parts with complex shapes	

Table 2.
Analysis of glass capsule sealing techniques [15].

During the design, simulation, and manufacturing of capsules for the hot isostatic pressing process, several critical factors warrant careful consideration [18]:

a. Assessment of variables: It is essential to analyze various parameters influencing the rheological behavior of capsule materials and granules, including geometric configurations and the preparation conditions employed during compression.

b. Material research: Conducting thorough research to select the appropriate materials for the capsules, as well as determining the optimal wall thickness, is crucial for ensuring effectiveness and durability.

c. Thermal and pressure optimization: Investigating and optimizing the thermal and pressure profiles that affect both the capsule and the granulated materials is necessary for successful outcomes.

d. Softening point analysis: Evaluating the softening points of the capsule materials under pressure is important, particularly in the formation of powder granules in geometrically complex regions.

e. Advancements in technology: The development and implementation of innovative technologies in capsule design and fabrication can significantly enhance their physical, mechanical, and weldability characteristics.

3. Advanced HIPed ceramics

Oxide and non-oxide ceramics represent two distinct categories of advanced ceramics with numerous industrial and technological applications. Within the oxide ceramics, notable examples include alumina and zirconia which are stabilized with magnesia, yttria, or ceria. In contrast, advanced non-oxide ceramics encompass silicon-based materials, such as silicon nitride and silicon carbide. Hot isostatic pressing (HIP) is a widely utilized technique for the fabrication of both ceramics and metals, including components featuring intricate geometries, high-density ceramics, and ceramic matrix composites (CMCs). This method's primary advantage lies in its ability to produce and densify advanced ceramic components characterized by complex shapes. The sintering process employed for the fabrication of ceramics and ceramic matrix composites can focus on either shaping and densification or solely on densification. Noteworthy benefits of HIP include a substantial enhancement of the mechanical properties of the resultant materials [4, 19].

3.1 Silicon nitride (Si_3N_4)

There are various processes for shaping silicon nitride ceramics, with one of the most prevalent and established methods being cold isostatic pressing (CIP) combined with sintering. In one study, silicon alpha-nitride powder was utilized alongside alumina and yttria as sintering aids, applying weight percentages of 3% and 5%, respectively. The reaction between the SiO_2 formed on the surface of the silicon nitride powder and the additives enhanced the mechanical strength of the silicon nitride components [19].

Typically, sintering processes differ from hot isostatic pressing (HIP). Commonly employed nonpressure sintering techniques include vacuum sintering and the use of inert gases like argon. Additionally, gas pressure sintering of silicon nitride (GPSSN) is an effective method for the initial production of high-density silicon nitride parts, achieving significant density improvements. Silicon nitride is recognized as a distinctive material suitable for high-temperature applications and the ability to withstand substantial mechanical loads. Dense silicon nitride components are generally sintered utilizing oxide sintering and liquid phase formation. However, the presence of an intergranular glass phase in sintered parts can notably reduce their mechanical strength. To address this, research into simultaneous temperature and pressure sintering technologies, such as HP-GPS-HIP, has been undertaken. Hot pressing, another technique in this field, has notable limitations, particularly regarding the complexity of the parts that can be produced; it is often impractical to fabricate complex shapes using this method. The GPS process, known as GPSSN, employs simultaneous gas pressure and temperature during the sintering process. This approach facilitates the sintering of complex geometries due to the isostatic pressure from the gas. However, a significant drawback of this method is the relatively low mechanical pressure used, which may prevent the final components from achieving theoretical density [2].

When considering non-oxide ceramics, it is crucial to employ processes that minimize the formation of oxide phases while remaining cost-effective. Recently, the generation of oxide phases has garnered significant attention. In research conducted by Zehui Du, silicon nitride ceramics were produced using a 3D printing process to create cellular structures referred to as "honeycomb." This innovative approach

Materials	Sintering condition	Relative green density	Relative density	Open porosity	Thermal diffusivity	Thermal conductivity
		%	%	%	$mm^2\ s^{-1}$	$W\ m^{-1}\ K^{-1}$
Si_3N_4	1800°C HIP 1800°C	60.5	98.5	1.6	14.0	11.4
Si_3N_4 + 1 wt.% nSiC	1800°C HIP 1800°C	62.3	98.2	0.3	13.9	15.6
Si_3N_4 + 5 wt.% nSiC	1800°C HIP 1800°C	61.3	94.7	3.7	13.4	13.2
Si_3N_4 + 10 wt.% nSiC	1800°C HIP 1800°C	62.1	94.6	17.5	13.2	11.7

Table 3.
Impact of sintering conditions and the concentration of nanosilicon carbide particles on the thermal and mechanical properties of silicon nitride [21].

replaces traditional shaping methods, such as extrusion. To achieve densification, the samples underwent hot isostatic pressing at 1700°C for 2 hours [20].

In a 2023 study by Gizowska and colleagues, silicon nitride ceramics and silicon nitride reinforced with silicon carbide nanoparticles (Si_3N_4-nSiC) were examined, with weight percentages ranging from 1% to 10%. The samples were produced using two sintering methods: under ambient conditions and high isostatic pressure. The researchers investigated the influence of sintering conditions and the concentration of nano silicon carbide particles on the thermal and mechanical properties, as summarized in **Table 3**. The inclusion of highly conductive silicon carbide particles enhanced thermal conductivity, particularly in composites containing 1% by weight of the carbide phase (115.6 $W\ m^{-1}\ K^{-1}$) compared to the thermal conductivity of silicon nitride ceramics (111.4 $W\ m^{-1}\ K^{-1}$) under identical conditions. However, an increase in the carbide phase led to a reduction in densification efficiency during sintering, which negatively affected both thermal and mechanical performance. Sintering performed via hot isostatic pressing demonstrated advantages in terms of mechanical properties.

3.2 SiAlON

Sialon ceramics are among the most widely utilized ceramics. Their densification occurs through a liquid-phase sintering process, where a liquid phase is initially created from molten SiO_2 and Al_2O_3 and subsequently removed in the final stage. Given the constraints of the Sialon sintering process, employing a high-speed sintering method can significantly aid in densification and achieving the desired properties of this ceramic. To this end, hot isostatic pressing and encapsulation techniques are utilized to achieve near-theoretical density in Sialon ceramics. A Vycor glass capsule is used in this process. During heat treatment, the temperature is initially raised to 1300°C (the glass softening temperature), followed by an increase in pressure to 200 MPa. The sintering occurs at a maximum temperature of 1780°C, with a dwell time of 2 hours at this temperature. The results indicate that the combination of high temperature and short sintering times facilitates the complete formation of Sialon and reduces the particle size to 500 nm. Additionally, employing hot isostatic pressing as a supplemental process can further enhance the density of the sintered components [22–24].

3.3 Alumina

Alumina is one of the most widely utilized materials in the densification of ceramics, recognized for its exceptional properties as a ceramic oxide, particularly when subjected to the hot isostatic pressing (HIP) process. Its superior mechanical characteristics make it a cost-effective choice for applications such as foot joints. The hot isostatic pressing process enhances the mechanical properties of alumina by uniformly applying temperature and pressure, which effectively reduces the open pores on the surface of the closed alumina ceramic and increases the bulk density of the components [25].

Various methods exist for the performing of alumina ceramics, with the most significant techniques including cold isostatic pressing, slurry casting, and plastic forming methods such as injection molding and extrusion. For instance, in a research study aimed at shaping alumina components, submicron powder was subjected to a pre-sintering process at 1240°C for 2 hours. This was followed by hot isostatic pressing at a temperature of 1400°C and a pressure of 150 MPa, conducted in an argon atmosphere [26].

Typically, preformed alumina parts are coated with boron nitride powder to minimize reaction between the glass and the bulk material during the sintering process. To encapsulate alumina parts at a sintering temperature of 1400°C, Vycor glass capsules are employed [27].

3.4 Transparent ceramics

Transparent ceramics have numerous potential applications, including armor, windows, and various optical elements. Key challenges in the development of transparent ceramics include microporosity, the presence of disruptive secondary phases, and wide grain boundaries. These factors adversely affect the refractive index between air and the material, leading to significant light scattering and a reduction in the transparency of the final products. Extensive research has been conducted to achieve transparent ceramics, focusing on the synthesis of high-purity powders, the selection of appropriate additives and sintering aids, and the enhancement of shaping and sintering techniques. Generally, raw ceramic components do not attain densities close to theoretical values. Consequently, employing methods that apply pressure and temperature simultaneously is crucial for achieving high optical transparency, specifically targeting porosity levels below 0.01%. These methods are particularly effective for producing transparent ceramics in a timely manner [27, 28].

3.5 AlON

Hot isostatic pressing (HIP) is an effective thermal treatment method for achieving partial densification of transparent ceramics. For instance, the incorporation of $LiAl_5O_8$ as a sintering aid, combined with HIP treatment at 2000°C and a pressure of 2000 bar, has enabled AlON ceramics to achieve a light transmission of 65% in the visible spectrum. Similarly, the addition of Y_2O_3-La_2O_3 with the HIP technique has resulted in achieving 78% transparency in AlON ceramics within the visible light range. Furthermore, employing a two-step sintering process along with HIP and nanoscale additives has increased transparency to 86% under visible light conditions [28, 29].

A critical factor in the performance of optical ceramics is the level of residual stress remaining in transparent ceramic components. To mitigate the effects of residual stress, post-baking treatments such as hot isostatic pressing are employed. Due to the high pressures and temperatures involved in HIP, its effectiveness in reducing residual stresses has been extensively studied. Observations indicate that the HIP process significantly enhances the removal of residual stresses and improves the coefficient of visible light transmission [28].

3.6 $MgAl_2O_4$

Transparent spinel ceramics are widely utilized in military applications due to their excellent thermo-mechanical properties at elevated temperatures, cost-effectiveness, and capability to form complex shapes. As a result, they are among the most commonly used ceramics in the optical industry. Various techniques are employed to fabricate raw spinel components, including slurry casting, cold isostatic pressing, and the sol-gel method. The sintering process varies based on advancements in the field. Some studies have employed single-step sintering techniques under pressure and temperature, such as spark plasma sintering and hot pressing. Conversely, other approaches utilize two-step sintering, involving pre-sintering followed by sintering under pressure and temperature, using methods like gas pressure sintering (GPS) and hot isostatic pressing (HIP). Ultimately, selecting different processing methods is essential to achieve optimal mechanical strength, thermal resistance, and high optical transparency in the visible light spectrum, underscoring the importance of HIP in the production of these ceramics. A variety of additives are used in the manufacturing of spinel ceramics, including LiF, MgF_2, CaO, B_2O_3, Y_2O_3, and SiO_2. However, using these additives in excessive quantities can lead to the formation of a liquid phase and significant grain growth, thereby compromising mechanical strength. In conclusion, the use of the hot isostatic pressing process at temperatures ranging from 1500°C to 1700°C, with pressures between 1500 and 2000 bar and a dwell time of 3 hours, can effectively enhance both the optical transparency and mechanical strength of transparent spinel ceramics [21, 30–35].

3.7 Transparent alumina

In a study, fine-grained aluminum oxide particles with sizes ranging from 0.8 to 2 micrometers were employed. The results indicated that at a sintering temperature of 1400°C, preformed components of various grain sizes achieved a density approaching theoretical density. This was accomplished by maintaining a holding time of 4 hours at the maximum temperature under isostatic pressure of 2000 bar. Transparent polycrystalline alumina (PGA) is subjected to heat treatment at 1800°C in a hydrogen atmosphere. These transparent alumina ceramics have applications in the military, particularly in optical lenses and dielectric ceramics. The PGA production process consists of multiple heat treatment stages, with the final stage of the hot isostatic pressing (HIP) process occurring at 1700°C and a pressure of 2000 bar, maintained for up to 3 hours for optimal clarification. Typically, additives such as MgO are incorporated to facilitate the production of PGA [27, 36, 37].

4. Conclusion

In conclusion, the production of engineering ceramic parts using hot isostatic pressing (HIP) technology offers significant advantages, including enhanced structural integrity through the reduction of porosity and defects, which leads to superior mechanical strength and thermal stability. HIP enables the fabrication of complex geometries and the consolidation of ceramic-based composites, allowing for tailored properties that meet specific application needs. Additionally, the consistent quality and high-performance of HIP-processed ceramics contribute to their cost-effectiveness and reliability across various industries. Overall, HIP technology is crucial for advancing the capabilities and applications of engineering ceramics, making it an invaluable method for producing high-performance components.

Acknowledgements

This research was conducted with financial support from Imam Hossein University, and the authors would like to extend their gratitude to the university's president, Dr. Hasani Ahangar.

Author details

Pouria Dehghani[1], Seyed Salman Seyed Afghahi[1*] and Farshad Soleimani[2]

1 Faculty of Advanced Materials and Nanotechnology, Imam Hussein University, Tehran, Islamic Republic of Iran (IR.Iran)

2 Materials Engineering Department, Malayer University, Hamedan, Islamic Republic of Iran (IR.Iran)

*Address all correspondence to: afghahi@ihu.ac.ir

References

[1] Carlsson R. The shaping of engineering ceramics. Materials and Design. 1989;**10**(1):10-14. DOI: 10.1016/0261-3069(89)90029-0

[2] Koizumi M. Hot Isostatic Pressing—Theory and Applications. Dordrecht: Springer Netherlands; 1992. DOI: 10.1007/978-94-011-2900-8

[3] Loh NL, Sia KY. An overview of hot isostatic pressing. Journal of Materials Processing Technology. 1992;**30**(1):45-65. DOI: 10.1016/0924-0136(92)90038-T

[4] Atkinson HV, Davies S. Fundamental aspects of hot isostatic pressing: An overview. Metallurgical and Materials Transactions A. 2000;**31**(12):2981-3000. DOI: 10.1007/s11661-000-0078-2

[5] Wright JM. Encapsulation method for hot isostatic pressing. US4960550A. 1989

[6] Bocanegra-Bernal MH. Hot Isostatic Pressing (HIP) technology and its applications to metals and ceramics. Journal of Materials Science. 2004;**39**(21):6399-6420. DOI: 10.1023/B:JMSC.0000044878.11441.90

[7] ElRakayby H, Kim K. Effect of glass container encapsulation on deformation and densification behavior of metal powders during hot isostatic pressing. International Journal of Material Forming. 2018;**11**(4):517-525. DOI: 10.1007/s12289-017-1361-8

[8] Richards K. Process for encapsulating a shaped body for hot isostatic pressing by sol-gel method. US5613993A. 1995

[9] Adlerborn J. Method for hot isostatic pressing powder bodies. US4081272A. 1975

[10] Callister WD. Materials Science and Engineering: An Introduction. 4th ed. New York: John Wiley & Sons; 1997

[11] Westman A-K, Larker HT. Interaction of encapsulation glass and silicon nitride ceramic during HIPing. Journal of the European Ceramic Society. 1999;**19**(16):2739-2746. DOI: 10.1016/S0955-2219(99)00059-X

[12] Huusmann O. Glass encapsulation in hip chamber description of methods used in practice. In: Hot Isostatic Pressing—Theory and Applications. Dordrecht: Springer Netherlands; 1992. pp. 517-525. DOI: 10.1007/978-94-011-2900-8_70

[13] Alpha JW. Glass sealing technology for displays. Optics and Laser Technology. 1976;**8**(6):259-264. DOI: 10.1016/0030-3992(76)90039-6

[14] Mouzon J, Maitre A, Frisk L, Lehto N, Odén M. Fabrication of transparent yttria by HIP and the glass-encapsulation method. Journal of the European Ceramic Society. 2009;**29**(2):311-316. DOI: 10.1016/j.jeurceramsoc.2008.03.022

[15] Leriche A, Aleksandrowicz P, Thierry B. Study of glass ceramic diffusion during hot isostatic pressing of encapsulated PZT ceramics. MRS Proceedings. 1991;**251**:265. DOI: 10.1557/PROC-251-265

[16] Kwon O, Messing GL. Gas diffusion during containerless hot isostatic pressing of liquid-phase sintered ceramics. Journal of the American Ceramic Society. 1989;**72**(6):1011-1015. DOI: 10.1111/j.1151-2916.1989.tb06260.x

[17] Ekström T, Käll PO, Nygren M, Olsson PO. Dense single-phase β-sialon ceramics by glass-encapsulated hot

isostatic pressing. Journal of Materials Science. 1989;**24**(5):1853-1861. DOI: 10.1007/BF01105715

[18] Zhilin PL, Gavrilov GN, Ryabtsev AD, Bazhenov EO. The problem of capsule manufacturing for hot isostatic pressing. IOP Conference Series: Materials Science and Engineering. 2020;**971**(2):022049. DOI: 10.1088/1757-899X/971/2/022049

[19] Ekbom R. Application of HIP/PM technique for gas and steam turbines. Metal Powder Report. 1990;**45**(4):284-289. DOI: 10.1016/S0026-0657(10)80056-6

[20] Furong N et al. Fused deposition modeling of Si_3N_4 ceramics: A cost-effective 3D-printing route for dense and high performance non-oxide ceramic materials. Journal of the European Ceramic Society. 2022;**42**(15):7369-7376. DOI: 10.1016/j.jeurceramsoc.2022.08.041

[21] Kim S-Y et al. Optimization of Ca additive and HIP condition in fabrication of the transparent $MgAl_2O_4$ ceramics. International Journal of Refractory Metals and Hard Materials. 2024;**120**:106577. DOI: 10.1016/j.ijrmhm.2024.106577

[22] Huang Z et al. Stereolithography 3D printing of Si3N4 cellular ceramics with ultrahigh strength by using highly viscous paste. Ceramics International. 2023;**49**(4):6984-6995. DOI: 10.1016/j.ceramint.2022.10.137

[23] Bartek A, Ekström T, Herbertsson H, Johansson T. Yttrium α-sialon ceramics by hot isostatic pressing and post-hot isostatic pressing. Journal of the American Ceramic Society. 1992;**75**(2):432-439. DOI: 10.1111/j.1151-2916.1992.tb08198.x

[24] Ekström T, Olsson P-O, Holmström M. O′-sialon ceramics prepared by hot isostatic pressing. Journal of the European Ceramic Society. 1993;**12**(3):165-176. DOI: 10.1016/0955-2219(93)90118-B

[25] Ariff TF et al. Optimizing the synthesis of alumina inserts using hot isostatic pressing (HIP). IOP Conference Series: Materials Science and Engineering. 2018;**290**:012044. DOI: 10.1088/1757-899X/290/1/012044

[26] Bhandhubanyong P, Akhadejdamrong T. Forming of silicon nitride by the HIP process. Journal of Materials Processing Technology. 1997;**63**(1-3):277-280. DOI: 10.1016/S0924-0136(96)02635-0

[27] Yin J, Li X, Lai Y, Zhang X, Yu S, Luo Y. Ultra-high quality factor of transparent Al_2O_3 ceramics fabricated by vacuum sintering and post hot isostatic pressing. Ceramics International. 2023;**49**(22):36879-36884. DOI: 10.1016/j.ceramint.2023.09.018

[28] Jiang N et al. Fabrication of highly transparent AlON ceramics by hot isostatic pressing post-treatment. Journal of the European Ceramic Society. 2017;**37**(13):4213-4216. DOI: 10.1016/j.jeurceramsoc.2017.04.028

[29] Tanaka I, Pezzotti G, Okamoto T, Miyamoto Y, Koizumi M. Hot isostatic press sintering and properties of silicon nitride without additives. Journal of the American Ceramic Society. 1989;**72**(9):1656-1660. DOI: 10.1111/j.1151-2916.1989.tb06298.x

[30] Li J, Zhang B, Tian R, Mao X, Zhang J, Wang S. Hot isostatic pressing of transparent AlON ceramics assisted by dissolution of gas inclusions. Journal of the European Ceramic Society. 2021;**41**(7):4327-4336. DOI: 10.1016/j.jeurceramsoc.2021.02.035

[31] Drdlikova K, Klement R, Rychnovsky D, Maca K, Drdlik D.

Optical properties of Tb3+- and Cr3+-doped MgAl2O4 ceramics prepared by capsule- and carbon-free hot isostatic pressing. Journal of the European Ceramic Society. 2024;**44**(9):5440-5448. DOI: 10.1016/j.jeurceramsoc.2023.11.038

[32] Liu Y, Zhu J, Dai B. Transparent MgAl2O4 ceramics prepared by microwave sintering and hot isostatic pressing. Ceramics International. 2020;**46**(16):25738-25740. DOI: 10.1016/j.ceramint.2020.07.051

[33] Dunaev AA et al. Hot isostatic pressing-induced structural changes in MgAl2O4 ceramics. Inorganic Materials. 2023;**59**(5):526-529. DOI: 10.1134/S0020168523050035

[34] Kim J-M, Kim H-N, Park Y-J, Ko J-W, Lee J-W, Kim H-D. Microstructure and optical properties of transparent MgAl2O4 prepared by Ca-infiltrated slip-casting and sinter-HIP process. Journal of the European Ceramic Society. 2016;**36**(8):2027-2034. DOI: 10.1016/j.jeurceramsoc.2016.02.018

[35] Liu Q et al. Microstructure and properties of MgAl2O4 transparent ceramics fabricated by hot isostatic pressing. Optical Materials. 2020;**104**:109938. DOI: 10.1016/j.optmat.2020.109938

[36] Uematsu K, Itakura K, Uchida N, Saito K, Miyamoto A, Miyashita T. Hot isostatic pressing of alumina and examination of the hot isostatic pressing map. Journal of the American Ceramic Society. 1990;**73**(1):74-78. DOI: 10.1111/j.1151-2916.1990.tb05093.x

[37] Barinov SM, Ponomarev VF, Shevchenko VY. Effect of hot isostatic pressing on the mechanical properties of aluminum oxide ceramics. Refractories and Industrial Ceramics. 1997;**38**(1):9-12. DOI: 10.1007/BF02768225

Chapter 8

Synthesis of Ultra-High Temperature Pyrochlore Ceramics for Extreme Conditions

Radu-Robert Piticescu, Cristina Florentina Ciobota, Mihail Botan and Alexandru Gyorgy Okos

Abstract

Thermal protection systems (TPS) for aerospace industry require the use of coatings made from ultra-high temperature ceramics (UHTCs), a class of materials based on oxides, nitrides, carbides and borides of transitional metals. Yttria-doped zirconia is the actual "gold standard" for aerospace coatings but the high temperature phase transitions limit its use in developing a new generation of TPS. Rare earth zirconia pyrochlores $Ln_2Zr_2O_7$ (Ln = La, Nd, Sm, Gd, Yb) emerged as future candidates for high temperature coatings due to their high thermal stability on a large temperature range. Here, we review the crystalline structures' electronic and thermal properties of these materials, the main methods proposed for their synthesis, their advantages and disadvantages. The use of mixed rare earths as dopant for zirconia ceramics is presented further. Finally, the potential applications as thermal barrier coatings (TBCs) are presented.

Keywords: ultra-high temperature ceramics, pyrochlores, rare earth zirconates, crystalline structures, synthesis, TBCs

1. Introduction

Thermal protective systems (TPS) represent the most effective system to protect structural components of spacecraft vehicles from damage caused by high temperature and corrosive/erosive environments. Ultra-high temperature ceramics (UHTCs) are ceramic-based composites consisting of a matrix of transition metal compounds that can be used around 2000°C and above in an oxidizing or corrosive environment. They are promising candidates for high temperature applications as thermal barrier coatings (TBCs) in hypersonic flight, rocket propulsion, atmospheric re-entry, refractory crucibles, plasma-arc electrodes, etc. [1]. An ideal material for TBC should combine high melting temperature, low density, low thermal conductivity and expansion coefficient close to the metal substrate. A comparison between these properties for oxide and non-oxide UHTCs shows that non-oxide ceramics possess the highest melting temperatures but, in most cases, higher densities and thermal conductivities, as can be seen in **Figure 1**.

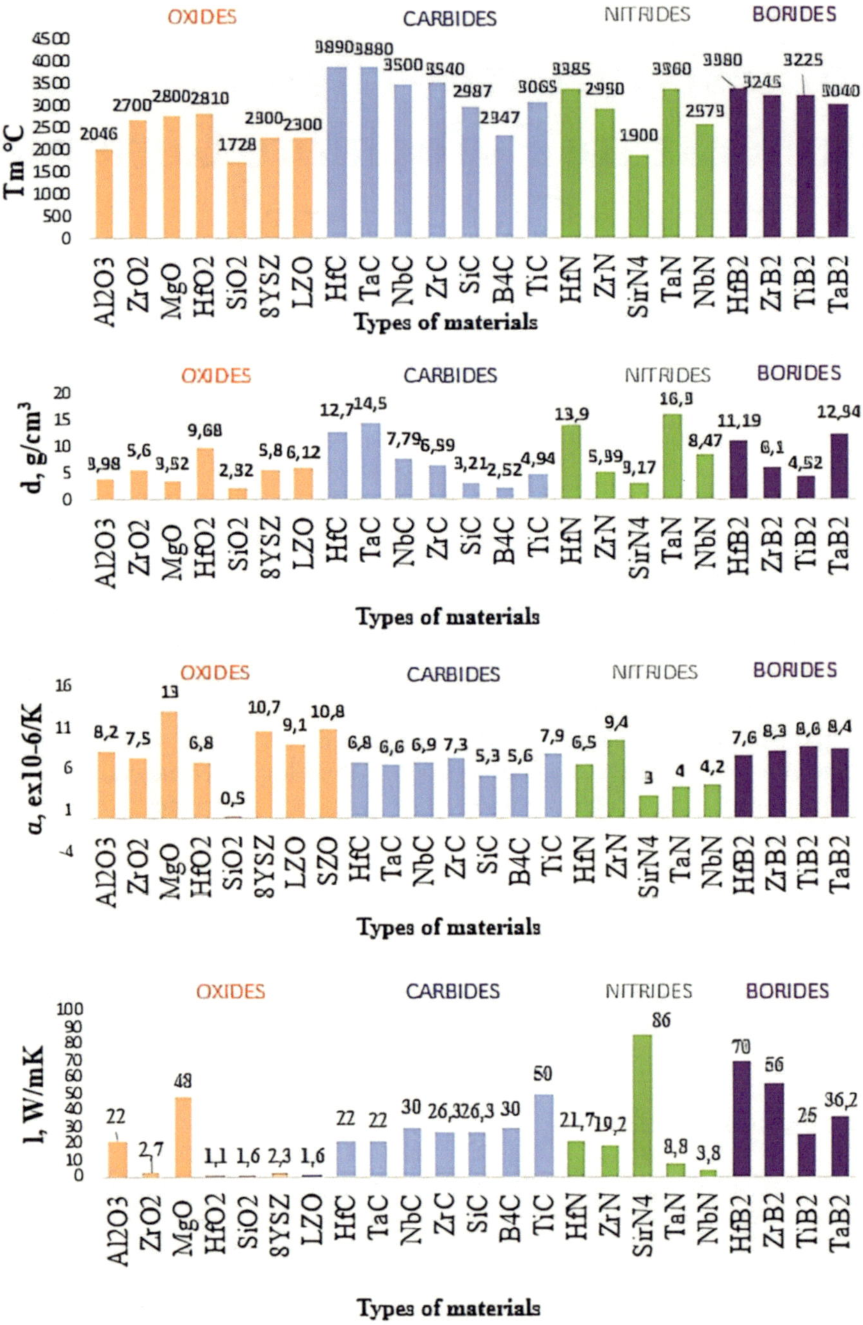

Figure 1.
Properties of oxides, carbides, nitrides and borides required for TBC (Tm—melting temperature; d—density, α—thermal expansion coefficient and k—thermal conductivity) [2].

Non-oxide ceramic coatings also have a lower chemical stability at high temperatures in air and oxidizing atmosphere due to their higher oxygen affinity. Another problem to be solved is the adherence of the coating to the substrate, which needs to find the proper metallic bonding having the capacity to form a thermally grown

Property	Units	Al_2O_3	NiCoCrAlY	8YSZ
Melting temperature	K	~2323	~1863	~2873
Thermal conductivity	$Wm^{-1} K^{-1}$	5.8 (1400 K)	320 (293 K)	~2.5 (298 K)
Thermal expansion coefficient	$X10^{-6} K^{-1}$	9.6 (1273 K)	10.7(293-1273 K)	10.7 (293-1273 K)

Table 1.
Thermal properties of Al_2O3 and NiCoCrAlY alloys and 8YSZ.

protective Al_2O_3 oxide layer (TGO) on the metallic substrate and improve the adhesion force by diffusion of metallic component. While for the bonding of oxide layers, alloys based on NiCoCrAlY perform very well, and for carbides, borides and nitrides, the diffusion is low and different bonding methods must be developed.

The "gold standard" material for ceramic coatings is 8 wt.% yttria-stabilized zirconia (8YSZ). However, it has some main drawbacks for use in higher temperature range: its stability at temperature higher than 1200°C decreases due to phase transformation with volume increase and increasing thermal conductivity for long time exposure (generally greater than 100 h) at high temperatures and susceptibility to CMAS (calcium magnesium aluminum silicates) infiltration leading to delamination [3]. **Table 1** presents the thermal properties of aluminum oxide (Al_2O3) and NiCoCrAlY alloys used as bonding coat and 8YSZ, respectively.

There is a continuous search for alternative candidates, motivated by the industry's demand for more and more performing engines and space components. Numerous candidates have been studied and tested [4]. Here, we discuss the structure, thermal properties and methods for the synthesis of rare earth zirconates with pyrochlore structure as potential materials to improve zirconia coatings in high temperature applications.

2. Crystalline structure and properties of pyrochlore rare earth zirconates

Rare earth zirconate compounds have gained increasing popularity due to their unique features, which make them ideal for different applications, such as radioactive waste immobilization, thermal barrier coatings and photocatalysis [5]. Ceramic materials with general formula $Ln_2Zr_2O_7$ (Ln = La, Nd, Sm, Gd, Yb) with an ionic radius ratio ranging from 1.61 to 1.46 crystallize in cubic pyrochlore structure [6] and have a relatively low oxygen diffusion, thermal conductivity in the range 1.1–1.7 $Wm^{-1} K^{-1}$, higher thermal stability and chemical inertness, making them to be considered among the most promising candidates for TBCs [4]. Pyrochlore zirconates of Promethium (Pm), Terbium (Tb), Dysprosium (Dy), Holmium (Ho), Erbium (Er), Thulium (Tm), Ytterbium (Yb) and Lutetium (Lu) were until now not reported to exist or to be observed experimentally.

Figure 2 shows the typical crystalline structure of $Ln_2Zr_2O_7$ pyrochlores crystallized in the cubic Fd-3 m space group. Ln^{3+} is bonded to eight O^{2-} atoms to form distorted LnO_8 hexagonal bipyramids that share edges with six equivalent LnO_8 hexagonal bipyramids and edges with six equivalent ZrO_6 octahedra. There are two shorter (2.34 Å) and six longer (2.64 Å) Ln-O bond lengths. Zr^{4+} is bonded to six equivalent O^{2-} atoms to form ZrO_6 octahedra that share corners with six equivalent ZrO_6 octahedra and edges with six equivalent LnO_8 hexagonal bipyramids. The corner-sharing octahedral tilt angles are 50°. All Zr-O bond lengths are 2.11 Å. There are two inequivalent O^{2-} sites. In the first O^{2-} site, O^{2-} is bonded to four equivalent Ln^{3+}

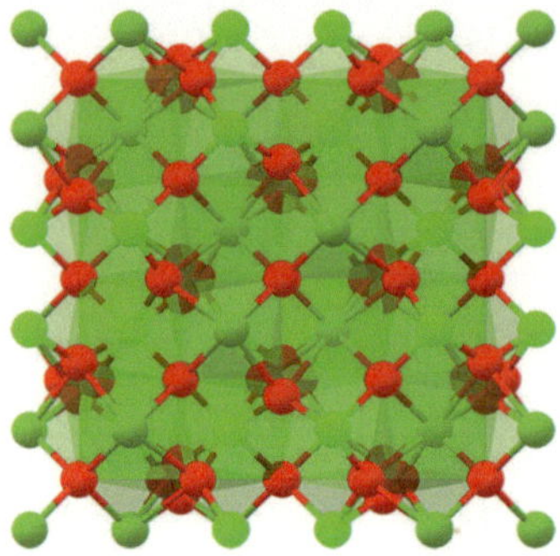

Figure 2.
$Ln_2Zr_2O_7$ crystalline structure [7].

atoms to form corner-sharing OLa_4 tetrahedra. In the second O^{2-} site, O^{2-} is bonded in a 4-coordinate geometry to two equivalent Ln^{3+} and two equivalent Zr^{4+} atoms.

Tables 2 and **3** summarize data related to crystalline structure and electronic properties of pyrochlore based on rare earth zirconates.

Ceramic material	System/space group	Lattice parameters	Bond length	Density [gcm^{-3}]	Ref.
$La_2Zr_2O_7$	Cubic, Fd-3 m	a = 10.81 Å α = 90°	2 short La-O = 2.34 Å 6 long La-O = 2.64 Å Zr-O = 2.11 Å	6.02	[8–15]
	Cubic, Fd-3 m	a = 10.75 Å α = 90°	La-O = 2.38 Å 6 short Zr-O = 2.28 Å 2 long Zr-O = 2.33 Å	6.12	[16, 17]
$Ce_2Zr_2O_7$	Cubic, Fd-3 m	a = 10.68 Å α = 90°	2 short Ce-O = 2.31 Å 6 long Ce-O = 2.47 Å 6 short Zr-O = 2.17 Å 2 long Zr-O = 2.31 Å	6.27	[9, 18–22]
$Pr_2Zr_2O_7$	Cubic, Fd-3 m	a = 10.75 Å α = 90°	2 short Pr-O = 2.33 Å 6 long Pr-O = 2.61 Å Zr-O = 2.1 Å	6.16	[12, 23–29]
$Nd_2Zr_2O_7$	Cubic, Fd-3 m	a = 10.7 Å α = 90°	2 short Nd-O = 2.32 Å 6 long Nd-O = 2.59 Å Zr-O = 2.1 Å	6.31	[9, 10, 23, 30–36]
	Cubic, Fd-3 m	a = 10.69 Å α = 90°	Nd-O = 2.35 Å 6 short Zr-O = 2.28 Å 2 long Zr-O = 2.31 Å	6.35	[37]
$Sm_2Zr_2O_7$	Cubic, Fd-3 m	a = 10.62 Å α = 90°	2 short Sm-O = 2.3 Å 6 long Sm-O = 2.56 Å Zr-O = 2.09 Å	6.59	[9, 12, 38–40]
$Eu_2Zr_2O_7$	Cubic, Fd-3 m	a = 10.62 Å α = 90°	2 short Eu-O = 2.3 Å 6 long Eu-O = 2.56 Å Zr-O = 2.09 Å	6.63	[41–46]
$Gd_2Zr_2O_7$	Cubic, Fd-3 m	a = 10.54 Å α = 90°	Gd-O = 2.29 Å Zr-O = 2.28 Å	6.9	[37, 47–54]

Table 2.
Crystalline structure of pyrochlore based on rare earth zirconates.

Ceramic material	Melting temperature [°C]	Predicted formation energy [eV/atom]	Band gap	Magnetization	Ref.
$La_2Zr_2O_7$	2300	−3.851	3.73	Non-magnetic	[8–14]
	—	−3.468		Non-magnetic	[16, 17]
$Ce_2Zr_2O_7$	> 2000	−3.815	2.53	Ferro-magnetic 2 μB/f.u.	[9]
$Pr_2Zr_2O_7$	—	−3.78	3.46	Non-magnetic	[12, 23, 24]
$Nd_2Zr_2O_7$	> 2000	−3.797	3.33	Non-magnetic	[9, 10, 23, 31, 32]
	—	−3.43		Non-magnetic	[37]
$Sm_2Zr_2O_7$	2000	−3.527		Non-magnetic	[9, 12, 23, 38]
$Eu_2Zr_2O_7$	—	−3.527		Ferro-magnetic 12 μB/f.u.	[41, 42]
$Gd_2Zr_2O_7$	2570	−3.513		Ferro-magnetic 14 μB/f.u.	[37, 47, 48]

Table 3.
Electronic properties of pyrochlore based on rare earth zirconates.

The similarity of the cubic structure in terms of lattice parameters and bond lengths is clearly observed. Density is increasing with the atomic number of rare earth element. Melting temperature for all known rare earth zirconates with pyrochlore structure is higher than 2000°C, with a higher value of 2570°C for $Gd_2Zr_2O_7$. However, research is still needed to have more precise values for the melting point of these compounds.

Pyrochlore materials based on $Ln_2Zr_2O_7$ (Ln = La, Nd, Pr, Sm, Gd, Yb) have relatively low oxygen diffusion, thermal conductivity in the range 1.1–1.7 $W^{-1} K^{-1}$, higher thermal stability and chemical inertness than YSZ, being considered among the most promising candidates for next generation of thermal barrier coating. The absence of phase transitions of these perovskites from room temperature to melting temperature and low thermal conductivities lead to a wider operating temperature range compared to gold standard YSZ.

Thermal conductivity is greatly dependent on temperature, system size, impurities and dislocations. Generally, as the atomic number of a rare earth element increases (from La to Lu), its thermal conductivity decreases. This is frequently linked to enhanced phonon scattering as a result of the higher atomic mass and stronger electron-phonon interaction. Also, thermal conductivity normally decreases with increasing temperature, mainly at higher temperatures, and this is the result of enhanced phonon-phonon scattering.

Yang et al. evaluated thermal conductivity of $La_2Zr_2O_7$ at temperatures ranging from room temperature to 1200°C. At room temperature, thermal conductivity is 3.2 $Wm^{-1} K^{-1}$, drops to 2.1 $Wm^{-1} K^{-1}$ at 400°C and then begins to increase to 2.7 $Wm^{-1} K^{-1}$ at 1200°C [55]. J. M. Zeerati and coworkers claim that they have achieved perfect agreement with experimental results by utilizing the effective harmonic method and homogeneous nonequilibrium molecular dynamics simulations with machine-learning potentials to calculate the thermal conductivity of candidate materials, such as $La_2Zr_2O_7$, $ZrSiO_4$ and $BaZrO_3$, at temperatures up to 1500 K. This work conducted a comparison between the thermal conductivity calculated using the homogeneous nonequilibrium molecular dynamics approach and the experimental data for $La_2Zr_2O_7$. The experimental value was 1.51 $Wm^{-1} K^{-1}$ and the simulated value was 1.47 $Wm^{-1} K^{-1}$; both the values are very close [4].

Other calculated values for pyrochlore compounds based on rare earth elements are presented by Feng et al. Their findings suggest that Cahill's approach is more effective for $Ln_2Zr_2O_7$ molecules. However, for $Nd_2Zr_2O_7$, both approaches provide significantly lower values than the measured ones [56].

3. Synthesis of pyrochlore rare earth zirconates

The synthesis of the pyrochlore materials can be done by different methods, such as solid-state reaction and solution methods, such as co-precipitation, combustion method, hydrothermal synthesis and sol-gel method.

3.1 Solid-state reaction method

According to the solid-state reaction method, the rare earth oxide and zirconia powders mixed at the required stoichiometry are dried in an oven to remove moisture and other impurities at around 110–120°C, until a constant weight is reached. The resultant dried mixture is further ball-milled to form a uniform mixture suspension in distilled water and/or absolute ethanol. The resultant colloidal suspension is dried and then thermally treated (calcined) at above 1500°C in a furnace, until the required pyrochlore structure is obtained. To increase the structural uniformity, the ball milling and thermal treatment processes needed to be repeated many times. The general flowsheet of the solid-state synthesis process is presented in **Figure 3**. Better results are obtained when the final drying step is performed by air spray drying, resulting in powders with better flowability. Zirconia lining and balls are used to avoid impurification with other oxides. The calcination temperatures are in the range 1400–1500°C and milling time 12–72 h [57–59].

3.2 Co-precipitation method

In the co-precipitation method, soluble precursors of lanthanum (Ln) and zirconium (Zr) are used as feedstock material, mainly nitrates with high solubility in water.

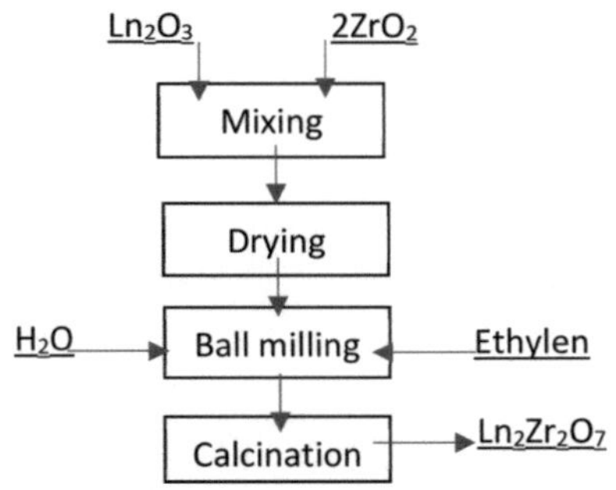

Figure 3.
Schematic flowsheet of the solid-state method.

Rare earth oxides (REOs) are generally easily soluble in nitric acid (HNO_3) solutions and may be used as raw materials. In the next step, the pH is increased to 9–11 by addition of ammonia (NH3) solution or sodium hydroxide (NaOH). The resulting precipitate formed is separated by filtering and washing to remove the alkaline nitrates formed. The use of ammonia is advantageous because ammonium nitrates are more soluble during washing with distilled water. The washed precipitate is further dried and calcined in the temperature range 1000–1300°C for 1 to 5 hours [60–62] to obtain the perovskite powders (**Figure 4**). The process may be described by the reaction

$$2\mathrm{Ln}(NO_3)_3 + 2\mathrm{Zr}(NO_3)_4 + 14NH_4OH = Ln_2Zr_2\mathrm{O}_7 + 14NH_4NO_3 + 7\mathrm{H_2O} \quad (1)$$

The use of a fuel source and oxygen source introduced in the co-precipitated flake may produce auto-ignition and eliminate the thermal treatment stage. The fuel chemical agents used are glycine, hydrazine, citric acid or glycerol. However, a post-heating treatment of the fluffy powder is required to remove the impurities formed during combustion. The combustion method improves process efficiency due to low energy consumption and produces pyrochlore powders with uniform morphology and highly porous structure [63].

3.3 Sol-gel method

The sol-gel processes are based on the controlled hydrolysis of organo-metallic compound precursors, with the formation in the first stage of colloidal suspensions (sols) from which the gel is subsequently formed in the presence of chelating agents.

Unlike the hydrolytic processes of precipitation, where the formation of the new phase takes place through a process of inhomogeneous nucleation, in the case of hydrolysis of organo-metallic compounds the nucleation rate can be more easily controlled and the solid phase is formed through a homogeneous nucleation mechanism. Among the organo-metallic compounds used for the synthesis of ultradisperse and nanocrystalline ceramic powders, metal alkoxides are most often mentioned in the literature. Sol destabilization (formation of a colloidal solid containing the fluid component dispersed in a three-dimensional (3D) lattice) can be achieved by dilution with water or hydrolysis catalyzed by acids or bases. **Figure 5** presents the schematic flowsheet of the sol-gel process used for rare earth zirconates.

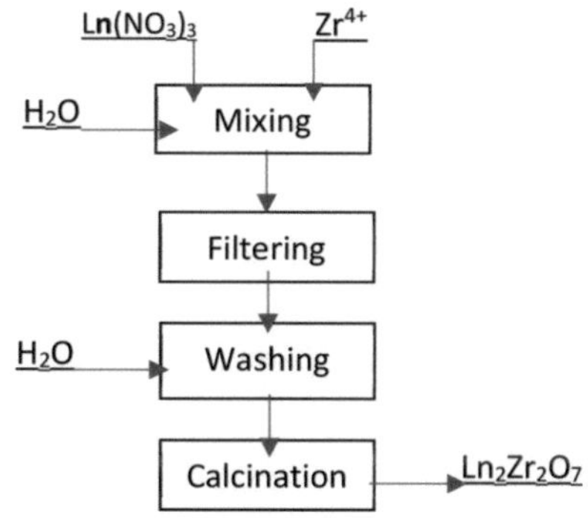

Figure 4.
Schematic flowsheet of the co-precipitation method.

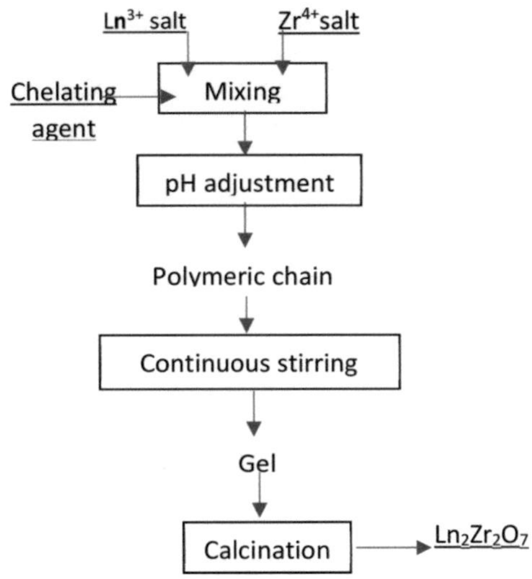

Figure 5.
Schematic flowsheet of the sol-gel method.

$La_2Zr_2O_7$ pyrochlore was prepared using $ZrOCl_2$*$8H_2O$ and $La(NO_3)_3$*$6H_2O$, which were dissolved in a mixed water (H_2O)/ethanol solution. Citric acid was further added to achieve Zr^{4+}:La^{3+}:citric acid ratio = 1:1:0.2. Formamide and polyethylene glycol were added to the above mixture, which was then stirred at room temperature for 2 h to form a uniform transparent sol solution. The resultant solution was placed into a crucible and aged for 20 h at 80°C, which resulted in gel formation. The resultant gel was dried at 110°C for 5 h and calcinated at 1200°C for 2 h in controlled atmosphere to form homogeneous $La_2Zr_2O_7$ coatings [64].

3.4 Hydrothermal method

Hydrothermal methods refer to any heterogeneous reaction in the presence of aqueous solvents or mineralizers under high pressure and temperature conditions. In a hydrothermal process, the dissolution and crystallization processes are enhanced to form new solid phases with controlled stoichiometry. The process involves obtaining a mixed solution with desired composition, adjustment of the pH using mineralizer agents, such as ammonia solution or sodium hydroxide, and transfer of the solution into an autoclave to produce the hydrothermal crystallization of the compound that is separated by filtering and dried to obtain the desired powder. The hydrothermal synthesis of $Ln_2Zr_2O_7$ is described by

$$2\mathrm{Zr}^{+4} + 2\mathrm{La}^{+3} + 14OH^{-} = Ln_2Zr_2\mathrm{O}_7 + 7\mathrm{H_2O} \qquad (2)$$

Figure 6 presents the schematic flowsheet of the hydrothermal process.

The main advantage of the hydrothermal method is the lower temperature of product crystallization, together with faster kinetics and low environmental impact because the process takes place in closed vessels. As an example, $Sm_2Zr_2O_7$ pyrochlore was obtained from SmO_2 and $ZrOCl_2{\cdot}8H_2O$ precursors dissolved in nitric acid

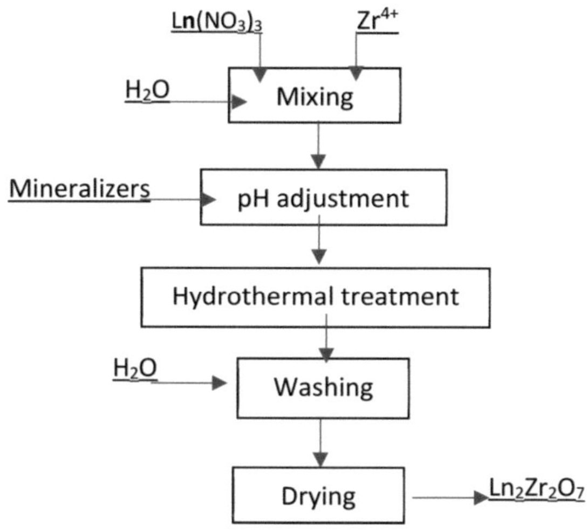

Figure 6.
Schematic flowsheet of the hydrothermal process.

(HNO_3), addition of aqueous ammonia solution to obtain a pH of 5–9, washing and addition of potassium hydroxide (KOH) solution as mineralizer under vigorous stirring and hydrothermal treatment into a Teflon-lined autoclave at 200°C. The powder obtained was filtered, washed and dried in vacuum at 70°C for 18 h [40]. Similarly, $La_2Zr_2O_7$ was obtained using La and Zr nitrates and NaOH as mineralizer to reach pH = 11, followed by hydrothermal reaction at 200°C for 1 h [65].

Table 4 presents the main advantages and disadvantages of the solid-state reactions, co-precipitation, sol-gel and hydrothermal methods used to obtain rare earth zirconium perovskites and zirconia co-doped with mixed rare earth oxides.

Main parameters	Synthesis method			
	Solid-state	**Co-precipitation**	**Sol-gel**	**Hydrothermal**
Process reaction	Temperature Time	pH Time	pH Temperature Surfactants	Temperature Pressure pH
Precursors	Oxides	Soluble salts/RE oxides	Metallic alkoxides	Soluble salts
Nucleation	Non-homogeneous	Non-homogeneous	Homogeneous	Homogeneous
Duration	High	Moderate	Moderate	Low
Particle size/ shape	Non-uniform/ irregular	Uniform size/ irregular shapes	Non-uniform/ irregular	Uniform sizes and shapes
Heat treatment	Required	Required	Required	No
Kinetic enhancement	No	pH control	pH and surfactants	Microwaves

Table 4.
Advantages and disadvantages of perovskite synthesis.

4. Synthesis of zirconia ceramics doped with mixed rare earth oxides

Modern TBCs have a multilayer structure, with the first layer made by NiCoCrAlY bonding layer forming the alumina TGO, and successive ceramic coatings with the top coating from the ceramic material with the higher melting temperature and thermal stability. Despite the advantages of rare earth zirconates as UHTCs provided by higher melting points and lower thermal conductivity compared to YSZ, their use as intermediate layers is limited due to the chemical, structural and thermal expansion coefficient mismatches. A potential solution to improve the thermal properties of TBC is to use as intermediate layer ZrO_2 doped with mixed rare earth oxides (REOs).

In such complex systems, the study of the phase equilibrium diagrams was conducted and the crystalline phases formed with temperature analyzed.

In Piticescu et al. [66], the authors performed a first study on the crystalline phase modifications of hydrothermally synthesized YSZ doped with 4, 6 and 8 wt.% La, Nd, Sm and Gd oxides, respectively, during thermal treatment in the temperature range 400–1400°C and their thermal conductivities.

The phase evolution of powders during thermal treatment in the range 400–1400°C evidenced the progressive formation of different solid solutions by isomorphic substitution of Zr^{4+} with Ln^{3+}. A cubic ZrO_2 solid solution is the major phase and secondary phases with generic formula $(Ln_{0.07}Y_{0.14}Z_{r0.79})O_{1.90}$; $(Ln_{0.11}Y_{0.14}Z_{r0.75})O_{1.88}$; and $(Ln_{0.14}Y_{0.14}Zr_{0.72})O_{1.86}$ were formed depending on the dopant concentration. At temperatures 1200°C and 1400°C, the formation of cubic solid solutions with pyrochlore structures ($RE_2Zr_2O_7$ where RE = Y and Ln) is observed for all compositions, except for samples co-doped with 4, 6 and 8 wt.% Nd, 6 and 8 wt.% Sm and 8 wt.% Gd, when only the cubic solution $(ZrO_2)_{SS}$-$(Ln_{0.14}Y_{0.14}Zr_{0.72})O_{1.86}$ phase was observed.

Low values of the thermal conductivities were measured by the hot plate method at room temperature (**Table 5**). These values are an order of magnitude lower than the values predicted and measured for rare earth zirconates in **Table 6** and closer to thermal conductivity values given in Ref. [67], respectively, 0.7 $Wm^{-1}\,K^{-1}$ for co-precipitated $Pr_2Zr_2O_7$, 0.4 $Wm^{-1}\,K^{-1}$ for $Sm_2Zr_2O_7$ and 0.38 $Wm^{-1}\,K^{-1}$ for $(Pr, Sm)_2Zr_2O_7$, suggesting that both the synthesis method and the complex defect structures induced by co-doping have a beneficial influence on reducing thermal conductivities of perovskites.

Sample	Thermal conductivity [$Wm^{-1}\,K^{-1}$]	Thermal diffusivity [$Mm2s^{-1}$]	Specific heat [$MJ^*m^{-3}\,K^{-1}$]
ZrO_2-$8Y_20_3$-$6La_2O_3$	0.3049 ± 0.0079	0.2603 ± 0.0075	0.4048 ± 0.0036
ZrO_2-$8Y_20_3$-$6Nd_2O_3$	0.3276 ± 0.0004	0.2899 ± 0.0018	0.3502 ± 0.0023
ZrO_2-$8Y_20_3$-$6Sm_2O_3$	0.3923 ± 0.0019	0.2766 ± 0.0053	0.3987 ± 0.0060
ZrO_2-$8Y_20_3$-$6Gd_2O_3$	0.3799 ± 0.0012	0.2799 ± 0.0084	0.3809 ± 0.0106

Table 5.
Thermal conductivities, thermal diffusivities and specific heat of zirconia co-doped with rare earth oxides by hydrothermal method.

Material	Units	Debye model	Cahill model	Experimental
$La_2Zr_2O_7$ (LZO)	$Wm^{-1}K^{-1}$	1.2	1.3	1.5
$Nd_2Zr_2O_7$ (NZO)	$Wm^{-1}K^{-1}$	1.2	1.1	1.6
$Sm_2Zr_2O_7$ (SZO)	$Wm^{-1}K^{-1}$	1.3	1.5	1.5
$Gd_2Zr_2O_7$ (GZO)	$Wm^{-1}K^{-1}$	1.2	1.2	1.6
$Pr_2Zr_2O_7$ (PZO)	$Wm^{-1}K^{-1}$	1.2	1.4	1.5
ZrO2 cubic	$Wm^{-1}K^{-1}$	2.3	2.4	2.5

Table 6.
Predicted and experimental values for thermal conductivities of rare earth zirconates.

In a next step to study more complex systems, nanostructured materials (1-x) $ZrO_{2-x}(Ln_2O_3)$ and ZrO_2 - x(La,Sm,Gd,Yb,Nd)O_3 (x = 0.2) encoded as ZrO_2-LSGYN were synthesized in a one-step process by hydrothermal method (**Figure 7**) starting from water-soluble salts of Zr and oxides: $ZrCl_4$ (Merck, New Jersey, USA, p.a. 98%), Sm_2O_3 (Alfa Aesar, Massachusetts, USA, p.a. >99.9%), Nd_2O_3 (Alfa Aesar, Massachusetts, USA, p.a. >99.9%), Yb_2O_3 (Alfa Aesar, Massachusetts, USA, p.a. >99.9%), Gd_2O_3 (Thermo Scientific, Massachusetts, USA, p.a. >99.9%) and La_2O_3 (Thermo Scientific, Massachusetts, USA, p.a. >99.9%). Ammonia solution (25 wt.% NH_3) was used as a mineralizer to control the solution pH. The amounts of Sm_2O_3, La_2O_3, Nd_2O_3, Gd_2O_3, Sm_2O_3 and $ZrCl_4$ were established in agreement with the theoretical molar formula ZrO_2- xLn_2O_3 (x = 0.2, Ln = La, Sm, Gd, Yb, Nd). The final pH was in the range of 9.5–11. The suspension thus obtained was subjected to hydrothermal processing (pressure 100 atm, temperature 200°C and time 2 h). The suspension was introduced into a Teflon-lined vessel of a sealed hydrothermal autoclave reactor (5 L, Berghof Products + Instruments GmbH, Berghof, Germany) endowed with a

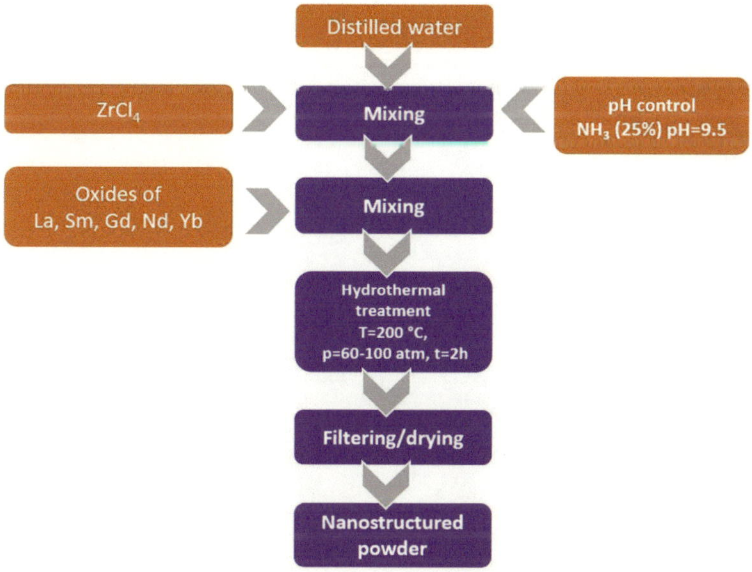

Figure 7.
Schematic flowsheet for hydrothermal synthesis of ZrO_2-xLn_2O_3 (x = 0.2, Ln = La, Sm, Gd, Yb, Nd).

cooling coil inserted into the autoclave vessel to avoid grain growth. To control the reaction pressure, argon gas was purged inside the autoclave. Powders were sintered in air at 1200°C to obtain homogeneous crystalline structures.

Powders were characterized from the chemical point of view by the Inductively Coupled Plasma Atomic Emission Spectroscopy (ICP-OES) technique, using ICP-OES 725 (AGILENT, Santa Clara, California, USA) equipment, according to ASTME E1479-16 standards (**Table** 7).

Micromeritics® TriStar II Plus (Norcross-Atlanta, USA) using the program for data collecting and calculation was used to analyze the surface proper¬ties of powders.

Powders were characterized from the chemical point of view by the Inductively Coupled Plasma Atomic Emission Spectroscopy (ICP-OES) technique, using ICP-OES 725 (AGILENT, Santa Clara, California, SUA) equipment, according to ASTME E 1479–16 standards.

Micromeritics® TriStar II Plus (Norcross-Atlanta, USA) using the program for data collecting and calculation TriStar II Plus was used to analyze the surface properties of powders.

The chemical compositions of the powders obtained are presented in **Table** 5 and confirm that the composition is conform with the initially designed one.

Figures 8–11 present the specific surface areas, pore volumes and average pore sizes. The analysis of the surface properties of as-synthesized powders is compared with the analysis of the surface properties of powders thermally treated at 1200°C and aims to optimize the sintering process for further obtaining pellets used for characterization

Compound	Chemical analysis, wt.%					
	Zr	La	Sm	Gd	Yb	Nd
(1-x)ZrO_2-xSm_2O_3 x = 0.2	44.1	—	35	—	—	—
(1-x)ZrO_2-La_2O_3 x = 0.2	40.36	34	—	—	—	—
(1-x)ZrO_2-Gd_2O_3 x = 0.2	38.63	—	—	37.7	—	—
(1-x)ZrO_2-Yb_2O_3 x = 0.2	41.4	—	—	—	37.9	—
(1-x)ZrO_2-Nd_2O_3 x = 0.2	43.1	—	—	—	—	33.5
ZrO_2-LSGYN	41.8	7.44	6.60	7.85	8.74	6.32

Table 7.
Chemical compositions of ZrO_2-xLn_2O_3 (x = 0.2) powders.

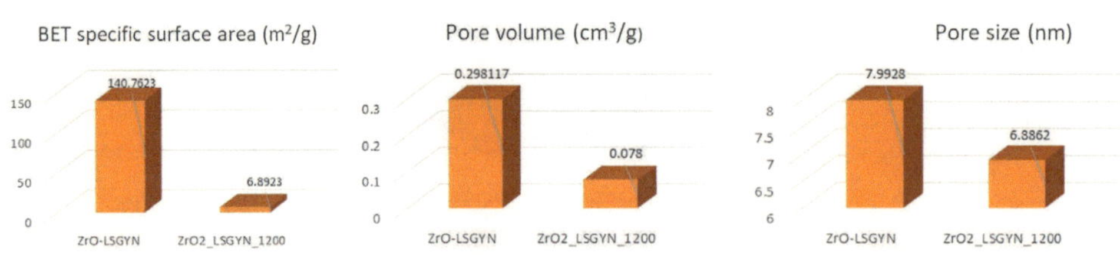

Figure 8.
Brunauer-Emmett-Teller (BET) surface area analysis results for ZrO2-LSGYN and ZrO2-LSGYN heat-treated at 1200°C: specific surface area (left), pore volume (center) and pore size (right).

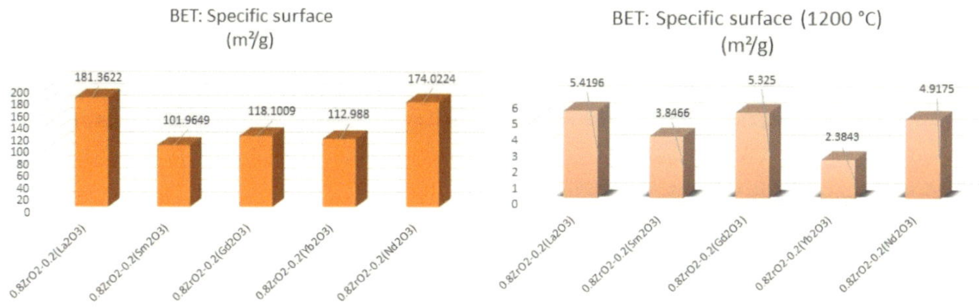

Figure 9.
BET specific surface area analyzed for the five specimens. Left: As obtained 0.8ZrO2-0.2(Ln2O3); right: 0.8ZrO2-0.2(Ln2O3) heat-treated at 1200°C, where Ln = La, Nd, Yb, Sm.

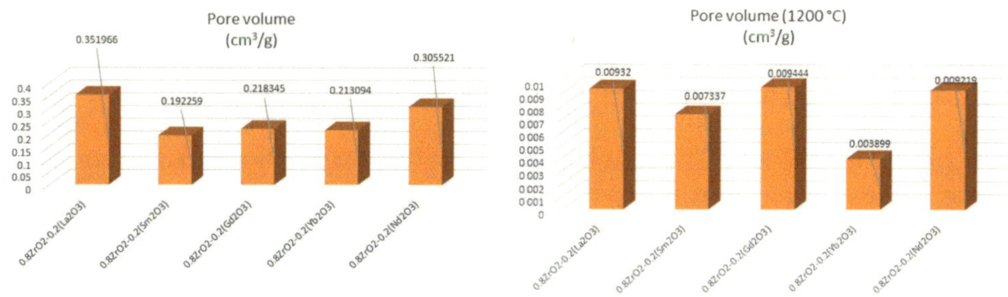

Figure 10.
*Pore volumes analyzed for the five specimens. Left: As obtained 0.8ZrO2-0.2(Ln2O3); right: 0.8ZrO2*0.2(Ln2O3) heat-treated at 1200°C, where Ln = La, Nd, Yb, Sm.*

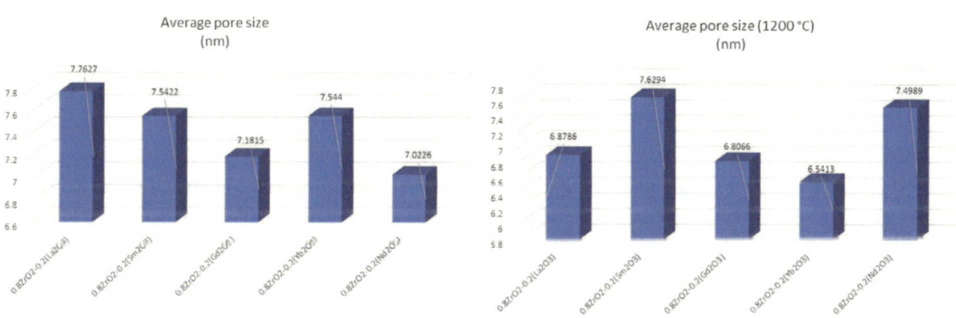

Figure 11.
Average pore size analyzed for the five specimens. Left: As obtained 0.8ZrO2-0.2(Ln2O3); right: 0.8ZrO2-0.2(Ln2O3) heat-treated at 1200°C, where Ln = La, Nd, Yb, Sm.

of thermal properties. The combination of grain growth, doping effects and porosity reduction at high temperatures results in a decrease in the surface area of ZrO_2 doped with rare earth elements. This phenomenon is crucial in understanding the behavior of these materials in various applications, such as thermal barrier coatings.

In terms of pore size, ZrO_2 samples doped with Sm and Yb oxides treated at 1200°C show that the pore size rises with temperature rather than decreasing, as seen in the other samples. The pore radius in the ZrO_2-Nd_2O_3 and ZrO_2-Sm_2O_3 powders increases when subjected to heat treatment at 1200°C compared to those that were not heat-treated. This can be attributed to various phenomena, including

i) the rearrangement of particles during calcination; it is possible that the particles agglomerate differently from their original structure, leading to the formation of larger pores, (ii) crystallite growth, (iii) partial sintering of the particles, (iv) phase transformations that can change the density and structure of the original material, and (v) microstructure development—calcination influences the microstructure of the material, creating a more open porous structure compared to that of uncalcined materials, which is a more compact structure [68].

5. Obtaining thermal barrier coatings based on rare earth zirconates

The increasing demand for new functional coatings has been a strong incentive for research toward not only understanding the fundamentals and technical aspects of film nucleation and growth, but also developing new deposition techniques that allow for better control of the deposition process. Electron Beam Physical Vapor Deposition (EB-PVD), Powder Flame Spraying, Plasma Thermal Spray and Cold Gas Dynamic Spray Coating are techniques suited for creating different types of TBCs. The EB-PVD process takes place in a high vacuum chamber, ensuring a relatively high deposition rate up to 150 nm/minute. The adhesion of the coating may be improved by controlled heating of the substrate during deposition. Combining these process's merits, denser coatings could be achieved. The EB-PVD process may be successfully used for selective deposition of multi-layered films based on refractory metal, oxides, carbides, nitrides, etc., for the parts to be used in extreme conditions. A specific aspect of EB-PVD coatings is related to their columnar microstructure determining the behavior of the coating during its service life [69]. The control of substrate temperature on coatings adhesion remains a major issue for optimizing final properties.

Five different TBC systems, YSZ, $La_2Zr_2O_7$, $Gd_2Zr_2O_7$, YSZ/, $La_2Zr_2O_7$ and YSZ/ $Gd_2Zr_2O_7$, were produced and exposed to furnace thermal cyclic oxidation tests. The deposition of protective ceramic top coats was performed with EB-PVD, single-layer coatings having a thickness of approximately 200 μm, while the thickness of each two-layer coating was approximately 100 μm. It was found that TBC double layers have a higher lifetime compared to TBC's with single layer. As a result of the performed cycling tests, the best performance was exhibited by the YSZ/$Gd_2Zr_2O_7$ coating system [70].

To prove the effect of the $La_2Zr_2O_7$ layer deposited on the YSZ layer on the thermal conductivity and oxygen penetration, two systems were exposed to isothermal and thermal cyclic oxidation tests. The two systems, YSZ and YSZ / $La_2Zr_2O_7$ top coats, were deposited using EB-PVD technique. The stress distribution showed that the YSZ / LZ (lanthanum zirconate) system has a longer life span than the YSZ TBC. Although the oxidation and thermal cyclic tests showed that double-layer YSZ/LZ TBC exhibited better performance than single-layer YSZ TBC, there is no drastic difference between the thermal performances of TBCs [71].

Another study suggested the use of $La_2Zr_2O_7$ in the design of TBC for its low thermal conductivity. Although YSZ has been used as a top coat for TBC due to its thermal diffusivity, the major disadvantage of its use is due to its limited long-term operating temperature. Thus, two systems of ceramic coatings were obtained Al_2O_3 /$La_2Zr_2O_7$/ YSZ, functionally graded (a 5-layer system— Al_2O_3/75%Al_2O_3 + 25%$La_2Zr_2O_7$ /25%Al_2O_3 + 25%YSZ + 50%$La_2Zr_2O_7$/50%YSZ + 50%$La_2Zr_2O_7$ /75%YSZ + 25%$La_2Zr_2O_7$, and a 6-layer system—Al_2O_3/75%Al_2O_3 + 25%$La_2Zr_2O_7$/25%Al_2O_3 + 25%YSZ + 50%$La_2Zr_2O_7$/50%YSZ + 50%$La_2Zr_2O_7$/75%YSZ + 25%$La_2Zr_2O_7$/100%YSZ) using atmospheric plasma spray (APS) equipment.

The researchers used alumina directly as a starting layer, near the metal substrate, because it acts as an oxygen diffusion barrier that protects the metal substrate from oxidation. $La_2Zr_2O_7$ powder was prepared using the sol-gel technique. It was found that samples coated with YSZ over 75% YSZ + 25% $La_2Zr_2O_7$ possess higher resistance to the attack of a Na_2SO_4 + V_2O_5 molten mixture up to 50 h of treatment over those coated with 75%YSZ + 25% $La_2Zr_2O_7$ top coats [72].

Compared to a conventional ceramic double ceramic layers coating (DCLC) of two ceramics linked with different thermal expansion coefficients, a functionally graded coating (FGC) is expected to be an interesting and effective approach to improve TBC performance. To this end, C. Wang and colleagues proposed for comparison two such systems obtained by suspension plasma spraying (SPS): a FGC, with gradual compositional variation along the thickness direction from pure 8YSZ (8 wt% Y2O3-stabilized ZrO2) to pure LZ on the outer surface and a LZ/8ZZ DCLC. Nanosized LZ power (30–50 nm) was prepared by hydrothermal method. The thermal cycling lifetime of optimized $La_2Zr_2O_7$/8YSZ functionally graded coatings is obviously longer than those of DCLC and LZ SCLC at 1000°C and 1200°C. The thermal cycling times of FGC (67 cycles) is increased by 55% than that of DCLC at 1000°C, and that (12 cycles) is increased by 50% than that of DCLC at 1200°C. The graded coatings reduce the thermal mismatch between two ceramics and integrate them into composite coatings, which help improve the high temperature reliability of TBCs [73].

A comparison was made between YSZ as a conventional ceramic top coating material, $Gd_2Zr_2O_7$ and YSZ/$Gd_2Zr_2O_7$7 as new generation coating materials with rare earth zirconate content, deposited as ceramic top coatings with EB-PVD method onto the CoNiCrAlY bond coat. These samples were subjected to hot corrosion tests by spreading mixtures of 55% V2O5 and 45% Na2SO4 salt at 5 h intervals at 1000° C. It turned out that double coatings of YSZ/ $Gd_2Zr_2O_7$ proved to be more resistant to hot corrosion than other single coatings. After the hot corrosion tests, the monoclinic zirconia and YVO4 long rod-shaped crystal structures are formed in YSZ coating, whereas the monoclinic zirconia and GdVO4 crystal structure are formed in coatings with $Gd_2Zr_2O_7$ and YSZ/$Gd_2Zr_2O_7$ as a hot corrosion product. $Gd_2Zr_2O_7$coatings, a good alternative coating material for TBC, have a relatively low fracture strength compared to YSZ, causing extension stresses and early cracking in the deterioration of hot corrosion that occurs at high temperatures. On the other hand, the low thermal conductivity and the high thermal expansion coefficient of the $Gd_2Zr_2O_7$ coating materials require greater stability at high temperatures [74].

Another YSZ / $Gd_2Zr_2O_7$ system was developed but this time by the APS technique. Plasma grade flowable 8 wt% YSZ and gadolinium zirconate (GZO) powders were prepared by a single-step co-precipitation technique. Field emission scanning electron microscopy (FESEM) cross-sectional analysis of the YSZ / GZO bilayer coating after corrosion test affirms the effectiveness of the bilayer design in preventing the penetration of corrosive salts to the YSZ layer. Also, YSZ/GZO bilayer TBC exhibited a higher thermal cyclic life (300 cycles) than the single-layer 8YSZ (175 cycles) coatings at 1100° C [75]. **Table 8** summarizes the results obtained regarding thermal behavior of different coatings using rare earth zirconates.

In another application, the bonding alloy powders NiCrAlY, ceramic 8YSZ, $La_2Zr_2O_7$ and $Gd_2Zr_2O_7$ are fed into the four dense graphite crucibles inserted in water-cooled copper slots of a carousel system that rotates around the electron beam generating source (**Figure 12a**), with a maximum filling degree of 80% of the crucible's volume. The slab or bar substrates on which the thermal barrier is deposited are cleaned of traces of organic substances with alcohol in an ultrasonic bath, then

Nr.	Material for TBC	Method	Properties
1	YSZ top coat	EB-PVD	106 ± 10 cycles
	Gd2Zr2O7 top coat		81 ± 7 cycles
	La2Zr2O7 top coat		coat 23 ± 4 cycles
	YSZ/Gd2Zr2O7 top coat		139 ± 13 cycles
	YSZ/La2Zr2O7 top coat		131 ± 12 cycles
2	YSZ	EB-PVD	
	YSZ / La 2Zr2O7		
3	5 layers - Al_2O_3/75%Al_2O_3 + 25%$La_2Zr_2O_7$/25% Al_2O_3 + 25%YSZ + 50%$La_2Zr_2O_7$/ 50%YSZ + 50%$La_2Zr_2O_7$/75%YSZ + 25%$La_2Zr_2O_7$	APS	The samples that have on the last layer YSZ have a higher resistance to the attack of a Na2SO4 + V_2O_5.
	6 layers - Al_2O_3 / 75%Al_2O_3 + 25%$La_2Zr_2O_7$ / 25%Al_2O_3 + 25%YSZ + 50%$La_2Zr_2O_7$ / 50%YSZ + 50%$La_2Zr_2O_7$ /75%YSZ + 25%$La_2Zr_2O_7$/ 100% YSZ		
4	$La_2Zr_2O_7$/ 8YSZ FGC / 8YSZ FGC	SPS	The thermal cycling lifetime of FGC is obviously longer than that of DCLC; OFCG helps improve the high temperature reliability of TBCs.
5	YSZ / Gd_2Zr_2O	EB-PVD	It is more resistant to hot corrosion than other single coatings; relatively low fracture toughness compared to YSZ
6	YSZ/GZO	APS	preventing the penetration of corrosive salts to the YSZ layer; higher thermal cyclic life (300 cycles) than the single-layer 8YSZ (175 cycles) coatings at 1100°C

Table 8.
Thermal behavior of different coatings using rare earth zirconates.

mounted in the rotating clamping place located above the crucibles with the material to be deposited (**Figure 12b**), which can be preheated by means of a mobile device with infrared (IR) lamps (**Figure 12c**). Close the front sealing door of the installation and start the vacuum pumps. After the vacuum reaches a value of approx. 10^{-6} torr, the electron cannon is turned on and the deposition of the bonding layer begins by melting and evaporating the NiCrAlY powder from the first crucible, with the help of the electron flux focused on the surface of the powder in the crucible. After consuming, the crucible carousel is rotated and the first ceramic layer is deposited by melting and evaporating the 8YSZ powder in electron flow. The crucible carousel is rotated again and the second ceramic layer is deposited by melting and evaporating the $La_2Zr_2O_7$ powder in electron beam. The crucible carousel is then rotated and the top ceramic layer is deposited by melting and evaporating the $Gd_2Zr_2O_7$ top coating layer. After the deposition operation is completed, the vacuum pumps are stopped, the installation is ventilated until the pressure inside reaches atmospheric pressure

DOI: http://dx.doi.org/10.5772/intechopen.1007904

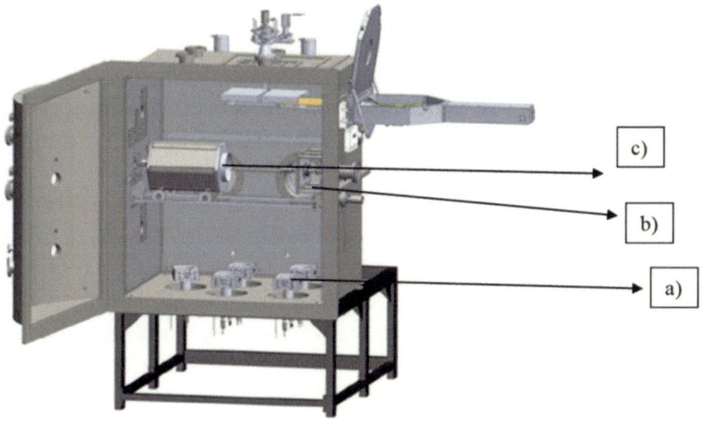

Figure 12.
Combinatorial EB-PVD coating system: (a) electron beam source; (b) device for fixing the substrates to be covered; and (c) mobile IR system for heating the substrates.

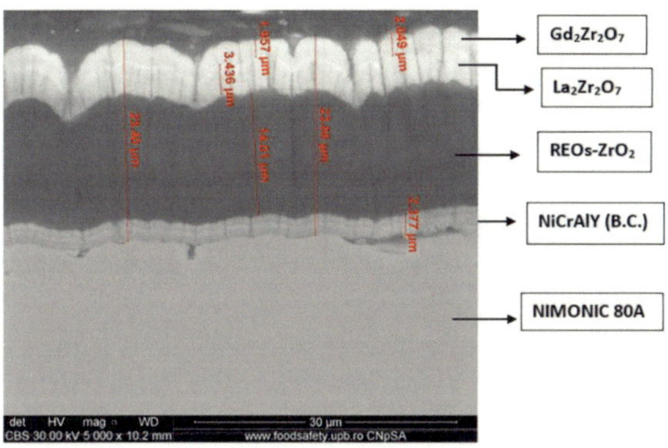

Figure 13.
SEM micrograph of the EB-PVD coatings with rare earth perovskite layers.

and the coated substrates in the clamping device are disassembled. Scanning electron microscopy (SEM) analysis is mentioned in the section in **Figure 13**, highlighting the four layers of the material deposited. The thermal shock resistance check shows that the architecture deposited on the NIMONIC 80 commercial refractory alloy substrate withstands at least 100 heating and cooling cycles at a test temperature of 1200°C without detachment and degradation.

6. Conclusions

Ultra-high temperature ceramics are promising candidates for high temperature applications as thermal barrier coatings (TBCs) in hypersonic flight, rocket propulsion, atmospheric re-entry, refractory crucibles, plasma-arc electrodes, etc., to protect structural components of spacecraft vehicles from damage caused by high temperature and corrosive/erosive environments due to their high melting temperatures and low thermal conductivities.

Non-oxide ceramic coatings have highest melting temperatures but lower chemical stability at high temperatures in air and oxidizing atmosphere due to their higher oxygen affinity and lower adhesion forces to the base metallic substrate.

The "gold standard" material for ceramic coatings is yttria-stabilized zirconia (YSZ). However, it has some main drawbacks for use in higher temperature range due to phase transitions at temperatures higher than 1200°C decreasing thermal stability and susceptibility to (calcium magnesium aluminum silicates) infiltration leading to delamination.

Rare earth zirconate compounds with general formula $Ln_2Zr_2O_7$ (Ln = La, Nd, Sm, Gd, Yb) have unique features, which make them ideal for different applications, such as radioactive waste immobilization, thermal barrier coatings and photocatalysis. Their crystalline cubic pyrochlore structures have a relatively low oxygen diffusion, low thermal conductivity, high thermal stability and chemical inertness, making them among the most promising candidates for TBCs. Melting temperature for all known rare earth zirconates with pyrochlore structure is higher than 2000°C, with higher value 2570°C for $Gd_2Zr_2O_7$. The thermal conductivity according to modeling and some experimental data is in the range 1.1–1.7 $Wm^{-1}\,K^{-1}$; however, lower values of the thermal conductivities in the range 0.38–0.7 $Wm^{-1}\,K^{-1}$ were reported for co-precipitated rare earth zirconates and 0.31–0.38 for ZrO2-8Y2O3-6Ln2O3 co-doped powders, suggesting that both the synthesis method and the complex defect structures induced by co-doping have a beneficial influence on reducing thermal conductivities of perovskites. Solid-state reactions, co-precipitation, sol-gel and hydrothermal methods have been used for the synthesis of $Ln_2Zr_2O_7$, with hydrothermal being the process with lower crystallization temperatures.

Electron Beam Physical Vapor Deposition (EB-PVD), Powder Flame Spraying, Plasma Thermal Spray and Cold Gas Dynamic Spray Coating are the main techniques used for creating different types and architectures of TBCs from rare earth zirconates and zirconia co-doped with mixed rare earth zirconates. In all cases, the bonding of the coatings to the metallic substrate to be thermally protected by ceramic coatings needs the deposition of a NiCoCrAlY thin bonding layer, enabling the formation of a thermal growth alumina layer reducing oxygen diffusion and ensuring the adhesion energy of the ceramic layers to the substrates.

Different architectures containing Al_2O_3, YSZ, YSZ-co-doped with rare earth oxides and rare earth zirconates were proposed. The parameters defining the TBCs architectures are the number of layers, the kind and sequence of ceramic materials' and layers' thicknesses.

With the actual rare earth zirconate materials and configurations used, thermal protection systems were able to support a minimum of 100 heating-cooling cycles.

Fundamental and applied research are further needed to model the thermal and adhesion properties of different rare earth zirconates and oxides co-doped with multiple rare earth oxides, including the development of high entropy oxides based on rare earth oxides.

Acknowledgements

The authors thank the Romanian Ministry of Research, Innovation and Digitization in the frame of Contract No. PN 23 25 01 01, project "Integration of Combinatorial EB-PVD Technology in Developing Materials for Applications in Energy Co-generation."

DOI: http://dx.doi.org/10.5772/intechopen.1007904

Conflict of interest

The authors declare no conflict of interest.

Author details

Radu-Robert Piticescu[1]*, Cristina Florentina Ciobota[1], Mihail Botan[2] and Alexandru Gyorgy Okos[1]

1 National Research and Development Institute for Non-ferrous and Rare Metals, Pantelimon-Ilfov, Romania

2 National Institute for Aerospace Research "Elie Carafoli", Bucharest, Romania

*Address all correspondence to: rpiticescu@imnr.ro

References

[1] Li D. Materials. 2013;**6**:551-564. DOI: 10.3390/ma6020551

[2] Grilli ML, Valerini D, Slobozeanu AE, Postolnyi BO, Balos S, Rizzo A, et al. Critical raw materials saving by protective coatings under extreme conditions: A review of last trends in alloys and coatings for aerospace engine applications. Materials. 2021;**14**:1656. DOI: 10.3390/ma14071656

[3] Tailor S, Mohanty RM, Doub AV. Development of a new TBC system for more efficient gas turbine engine application. Materials Today Proceedings. 2016;**3**:2725-2734. DOI: 10.1016/j.matpr.2016.06.019

[4] Zeraati M, Oganov AR, Fan T. Searching for low thermal conductivity materials for thermal barrier coatings: A theoretical approach. Physical Review Materials. 2024;**8**:033601. DOI: 10.1103/PhysRevMaterials.8.033601

[5] Gupta M, Dwivedi G, Nylen P, Vackel A, Sampath S. An experimental study of microstructure-property relationships in thermal barrier coatings. Journal of Thermal Spray Technology. 2013;**22**(5):659-670

[6] Menelaou M. Rare-Earth zirconates (Ln2Zr2O7): A family of ceramic (nano) materials. In: Nanotechnology (Materials and Applications). India: KD Publications; 2022. pp. 84-92. ISBN: 978-93-94570-00-9

[7] Materials Project Database. Available from: https://next-gen.materialsproject.org/materials/mp-4974?chemsys=La-Zr-O

[8] Tabira Y, Withers RL, Yamada T, Ishizawa N. Annular dynamical disorder of the rare earth ions in a La2Zr2O7 pyrochlore via single crystal synchroton X-ray diffraction. Zeitschrift für Kristallographie (1979-2010). 2001;**216**:92-98

[9] Uno M, Kosuga A, Okui M, Horisaka K, Muta H, Kurosaki K, et al. Photoelectrochemical study of lanthanide zirconium oxides, Ln_2 Zr_2 O_7 (ln = La, Ce, Nd and Sm). Journal of Alloys and Compounds. 2006;**420**:291-297

[10] Harvey EJ, Whittle KR, Lumpkin GR, Smith RI, Redfern SAT. Solid solubilities of $(La, Nd)_2$ (Zr, Ti)$2O_7$ phases deduced by neutron diffraction. Journal of Solid State Chemistry. 2005;**178**:800-810

[11] Bjoerketun ME, Knee CS, Nyman BJ, Wahnstroem G. Protonic defects in pure and doped La_2 Zr_2 O_7 pyrochlore oxide. Solid State Ionics. 2008;**178**:1642-1647

[12] Tabira Y, Withers RL, Minervini L, Grimes RW. Systematic structural change in selected rare earth oxide pyrochlores as determined by wide-angle CBED and a comparison with the results of atomistic computer simulation. Journal of Solid State Chemistry. 2000;**153**:16-25

[13] Pruneda JM, Artacho E. First-principles study of structural, elastic and bonding properties of pyrochlores. Physical Review B. 2005;**72**:085107. DOI: 10.1103/PhysRevB.72.085107

[14] Blanchard PER, Clements R, Kennedy BJ, Ling CD, Reynolds E. Does local disorder occur in the pyrochlore zirconates? Inorganic Chemistry. 2012;**51**:13237-13244

[15] Niu J, Wu X, Zhang H, Qin S. Pressure-induced phase transition of

$La_2Zr_2O_7$ and $La_{0.5}Gd_{1.5}Zr_2O_7$ pyrochlore. Royal Society of Chemistry Advances. 2019;**9**:18954, DOI: 10.1039/c9ra03438b

[16] Jinhua Z, Jishun Y, Xun C, Shuen H. Thermal expansion and solubility limits of cerium-doped lanthanum zirconates. Journal of Alloys and Compounds. 2012;**525**:78-81

[17] Whittle KR, Swainson IP, Cranswick LMD, Redfern SAT, Lumpkin GR. Lanthanum pyrochlores and the effect of yttrium addition in the systems $La_{(2-x)}$ Y_x Zr_2 O_7 and $La_{(2-x)}$ $Y_{(x)}$ Hf_2 O_7. Journal of Solid State Chemistry. 2009;**182**:442-450

[18] Sasaki T, Ukyo Y, Kuroda K, Arai S, Muto S, Saka H. Crystal structure of $Ce_2Zr_2O_7$ and beta-$Ce_2Zr_2O_{7.5}$. Journal of the Ceramic Society of Japan. 2004;**112**:440-440

[19] Raison PE, Pavel CC, Jardin R, Suard E, Haire RG, Popa K. Thermal expansion behavior of ce2 zr2 o7 up to 898 k in conjunction with structural analyses by neutron diffraction. Physics and Chemistry of Minerals. 2010;**37**:555-559

[20] Urban S, Dolcet P, Möller M, Chen L, Klar PJ, Djerdj I, et al. Synthesis and full characterization of the phase-pure pyrochlore $Ce_2Zr_2O_7$ and the -$Ce_2Zr_2O_8$ phases. Applied Catalysis B: Environmental. 2016;**197**:23-34

[21] Popa K, Konings RJM, Wastin F, Colineau E, Magnani N, Raison PE. A re-evaluation of the heat capacity of cerium zirconate ($Ce_2Zr_2O_7$). Journal of Physics and Chemistry of Solids. 2008;**69**:70-75

[22] Otobe H, Nakamura A, Yamashita T, Minato K. Oxygen potential and defect structure of oxygen-excess pyrochlore $Ce_2Zr_2O_{(7+x)}$. Journal of Physics and Chemistry of Solids. 2005;**66**:329-334

[23] Yang M, Rodgers JA, Middler LC, Fuertes A, Attfield JP. Direct solid-state synthesis at high pressures of new mixed-metal Oxynitrides: $RZrO_{2n}$ (R = Pr, Nd, and Sm). Inorganic Chemistry. 2009;**48**:11498-11500

[24] Koohpayeh SM, Wen JJ, Trump BA, Broholm CL, McQueen TM. Synthesis, floating zone crystal growth and characterization of the quantum spin ice Pr_2 Zr_2 O_7 pyrochlore. Journal of Crystal Growth. 2014;**402**:291-298

[25] Matović B, Maletaškić J, Yoshida K, Yano T. Synthesis, characterization and sintering of fluorite and pyrochlore-type compounds: $Pr_2Zr_2O_7$, $Sm_2Zr_2O_7$ and $PrSmZr_2O_7$. Materials Today: Proceedings. 2019;**16**:156-162

[26] Fang H, Yang S, Ye W, Zhong F, Luo Y, Wang S, et al. Mechanism insight into enhanced oxygen reduction reaction over heterovalent ion incorporated pyrochlore $Pr_2Zr_2O_7$ for direct ammonia solid oxide fuel cells. Chemical Engineering Science. 2024;**290**:119778

[27] Alam J, Jana YM, Biswas AA. Crystal-field study of magnetization and specific heat properties of frustrated pyrochlore $Pr_2Zr_2O_7$. Journal of Magnetism and Magnetic Materials. 2016;**416**:391-400

[28] dos Santos Veiga EL, Morte MF, MIR HB, Cordoncillo E. Effect of the oxidation states on the electrical properties of Fe-doped $Pr_2Zr_2O_7$ pyrochlore. Journal of Materials Research and Technology. 2022;**16**:201-215

[29] Zhong F, Yang S, Chen C, Fang H, Chen K, Zhou C, et al. Defect-induced pyrochlore $Pr_2Zr_2O_7$ cathode rich in oxygen vacancies for direct ammonia solid oxide fuel cells. Journal of Power Sources. 2022;**520**:230847

[30] Hagiwara T, Yamamura H, Nomura K, Igawa M. Relationship

between crystal structure and oxide-ion conduction in Ln2Zr2O7 (ln = Eu, Nd and La) system deduced by neutron and X-ray diffraction. Journal of the Ceramic Society of Japan. 2013;**121**:205-205

[31] Clements R, Hester JR, Kennedy BJ, Ling CD, Stampfl APJ. The fluorite - pyrochlore transformation of $Ho_{(2-y)} Nd_{(y)} Zr_2 O_7$. Journal of Solid State Chemistry. 2011;**184**:2108-2113

[32] Mandal BP, Krishna PSR, Tyagi AK. Order-disorder transition in the nd(2-y) y(y) zr2 o7 system: Probed by X-ray diffraction and Raman spectroscopy. Journal of Solid State Chemistry. 2010;**183**:41-45

[33] Liu X, Lu Y, Sun J, Liu G, Liu W, Zhang P, et al. NdAlO3-Nd2Zr2O7 composite with eutectic composition for advanced nanostructured thermal insulation coatings. Surface & Coatings Technology. 2024;**478**:130502

[34] Migas D, Moskal G, Jucha S. Hot corrosion behavior of double-phase $Nd_2Zr_2O_7$-YSZ thermal barrier coatings. Surface & Coatings Technology. 2022;**449**:128955

[35] Hagiwara T, Nomura K, Yamamura H. Relationship between crystal structure and oxide-ion conduction in $Nd_2Zr_2O_7$ and $La_2Zr_2O_7$ deduced by high-temperature neutron diffraction. Solid State Ionics. 2014;**262**:551-554

[36] Ajabshir SZ, Ghasemian N, Kamazani MM, Niasari MS. Effect of zirconia on improving NO_x reduction efficiency of $Nd_2Zr_2O_7$ nanostructure fabricated by a new, facile and green sonochemical approach. Ultrasonics Sonochemistry. 2021;**71**:105376

[37] Mandal BP, Banerji A, Sathe V, Deb SK, Tyagi AK. Order-disorder transition in $Nd_{2-y} Gd_y Zr_2 O_7$ pyrochlore solid solution: An x-ray diffraction and Raman spectroscopic study. Journal of Solid State Chemistry. 2007;**180**:2643-2648

[38] Sayed FN, Grover V, Bhattacharyya AA, Jain D, Arya A, Pillai CGS, et al. $Sm_{(2-x)} Dy_{(x)} Zr_2 O_7$ pyrochlores: Probing order-disorder dynamics and multifunctionality. Inorganic Chemistry. 2011;**50**:2354-2365

[39] Moskal G, Jasik A, Mikuśkiewicz M, Jucha S. Thermal resistance determination of $Sm_2Zr_2O_7$ + 8YSZ composite type of TBC. Applied Surface Science. 2020;**515**:145998

[40] Wang Q, Cheng X, Li J, Jin H. Hydrothermal synthesis and photocatalytic properties of pyrochlore $Sm_2Zr_2O_7$ nanoparticles. Journal of Photochemistry and Photobiology A: Chemistry. 2016;**321**:48-54

[41] Hart G, Forcade R. Generating derivative structures from multilattices: Algorithm and application to hcp alloys. Physical Review B. 2009;**80**:014120

[42] Shlyakhtina AV, Knotko AV, Boguslavskii MV, Stefanovich SY, Kolbanev IV, Larina LL, et al. Effect of non-stoichiometry and synthesis temperature on the structure and conductivity of $Ln_{2+x} M_{2-x} O_{7-x/2}$ (ln = Sm - Gd; M = Zr, Hf; x = 0 - 0.286). Solid State Ionics. 2007;**178**:59-66

[43] Hagiwara T, Nomura K, Kageyama H. Differences in local structures around zirconium atoms in $Eu_2Zr_2O_7$ and $La_2Zr_2O_7$. Solid State Ionics. 2019;**335**:32-37

[44] Fabrichnaya O, Kriegel MJ, Pavlyuchkov D, Seidel J, Dzuban A, Savinykh G, et al. Heat capacity for the $Eu_2Zr_2O_7$ and phase relations in

the ZrO_2–Eu_2O_3 system: Experimental studies and calculations. Thermochimica Acta. 2013;**558**:74-82

[45] Boujnah M, Chavira E. Enhanced the physical properties through the oxygen vacancy in $(Eu_{1-x}Sc_x)_2Zr_2O_7$ Zirconate pyrochlore. Optical Materials. 2020;**110**:110499

[46] Hagiwara T, Nomura K, Kageyama K. Crystal structure analysis of Ln2Zr2O7 (ln = Eu and La) with a pyrochlore composition by high-temperature powder X-ray diffraction. Journal of the Ceramic Society of Japan. 2017;**125**:65-65

[47] Mandal BP, Pandey M, Tyagi AK. Gd_2 Zr_2 O_7 pyrochlore: Potential host matrix for some constituents of thoria based reactor's waste. Journal of Nuclear Materials. 2010;**406**:238-243

[48] Kennedy BJ, Qingdi Z, Avdeev M. Neutron diffraction studies of $Gd_2Zr_2O_7$ pyrochlore. Journal of Solid State Chemistry. 2011;**184**:1695-1698

[49] Doleker KM, Karaoglanli AC, Ozgurluk Y, Kobayashi A. Performance of single YSZ, $Gd_2Zr_2O_7$ and double-layered YSZ/$Gd_2Zr_2O_7$ thermal barrier coatings in isothermal oxidation test conditions. Vacuum. 2020;**177**:109401

[50] Lei M, Weimin M, Xudong S, Jianan L, Lianyong J, Han S. Structure properties and sintering densification of $Gd_2Zr_2O_7$ nanoparticles prepared via different acid combustion methods. Journal of Rare Earths. 2015;**33, 2**:195

[51] Deng M, Huang Z, Wang H. Amorphization of $Gd_2Zr_2O_7$ induced by mechanical deformation in nanoindentation. Scripta Materialia. 2024;**242**:115945

[52] Koao LF, Motloung SV, Smith JW, Motaung TE, Malevu TD, Pawade VB, et al. The influence of annealing temperature on the structure and optical properties of $Gd_2Zr_2O_7$ nanostructures prepared by chemical bath deposition method. Journal of Alloys and Compounds. 2021;**863**:158367

[53] Ahmad S, Gouadria S, Jabbour K, Naz A, Manzoor S, Abdullah M, et al. Iron doped $Gd_2Zr_2O_7$ hierarchical nanoflakes arrays as robust electrodes materials for energy storage application. Journal of Energy Storage. 2023;**60**:106687

[54] Zuo S, Shu X, Fu Y, He Y, Zhao G, Wei G, et al. Ultra-fast immobilization of simulated TRPO waste into as-prepared $Gd_2Zr_2O_7$ by flash sintering. Ceramics International. 2024;**50**:28867-28876

[55] Yang C, Wan M, Zhao M, Shahid M, Pan W. Effective blocking of radiative thermal conductivity in La2Zr2O7/LaPO4 composites for high temperature thermal insulation applications. Journal of the European Ceramic Society. 2016;**36**:3809-3814

[56] Feng B, Xiao CL, Wan ZX, Qu ZC, Huang JC, Chen R, et al. Electronic structure, mechanical properties and thermal conductivity of Ln2Zr2O7 (ln=La, Pr, Nd, Sm, Eu and Gd) pyrochlore. Acta Materialia. 2011;**59**(4):1742-1760. DOI: 10.1016/j.actamat.2010.11.041

[57] Li JY, Dai H, Zhong XH, Zhang YF, Ma XF, Meng J, et al. Lanthanum zirconate ceramic toughened by BaTiO3 secondary phase. Journal of Alloys and Compounds. 2008;**452**(2):406-409

[58] Zhong X, Zhao H, Zhou X, Liu C, Wang L, Shao F, et al. Thermal shock behavior of toughened gadolinium zirconate/YSZ double-ceramic-layered

thermal barrier coating. Journal of Alloys and Compounds. 2014;**593**:50-55

[59] Hu Q, Zeng J, Wang L, Shu X, Shao D, Zhang H, et al. Helium ion irradiation effects on neodymium and cerium co-doped Gd2Zr2O7 pyrochlore ceramic. Journal of Rare Earths. 2018;**36**(4):398-403

[60] Zhang C, Zhao J, Yang L, Zhou Y, Wang Q, Chen H, et al. Preparation and corrosion resistance of nonstoichiometric lanthanum zirconate coatings. Journal of the European Ceramic Society. 2020;**40**(8):3122-3128

[61] Rodriguez JT, Cano VG, Menelaou M, Kaštyl J, Cihlář J, Tkachenko S, et al. Rare-earth zirconate Ln2Zr2O7 (ln: La, Nd, Gd, and Dy) powders, xerogels, and aerogels: Preparation, structure, and properties. Inorganic Chemistry. 2019;**58**(21):14467-14477

[62] Kaliyaperumal C, Sankarakumar A, Palanisamy J, Paramasivam T. Fluorite to pyrochlore phase transformation in nanocrystalline Nd2Zr2O7. Materials Letters. 2018;**228**:493-496

[63] Matovic B, Maletaskic J, Zagorac J, Pavkov V, Maki RS, Yoshida K, et al. Synthesis and characterization of pyrochlore lanthanide (Pr, Sm) zirconate ceramics. Journal of the European Ceramic Society. 2020;**40**(7):2652-2657

[64] Wang S, Li W, Wang S, Zhang J, Chen Z. Deposition of SiC/La2Zr2O7 multi-component coating on C/SiC substrate by combining sol-gel process and slurry. Surface and Coatings Technology. 2016;**302**:383-388

[65] Hongming Z, Danqing Y. Effect of rare earth doping on thermo-physical properties of lanthanum zirconate ceramic for thermal barrier coatings. Journal of Rare Earths. 2008; **26**:770-774

[66] Piticescu R-R, Slobozeanu AE, Valsan SN, Ciobota CF, Ghita AN, Motoc AM, et al. Hydrothermal synthesis of Nanocrystalline ZrO_2-$8Y_2O_3$-xLn_2O_3 powders (ln = La, Gd, Nd, Sm): Crystalline structure. Thermal and Dielectric Properties, Materials. 2021;**14**:7432. DOI: 10.3390/ma14237432

[67] Pokhrel M, Gupta SK, Wahid K, Mao Y. Pyrochlore rare-earth hafnate RE2Hf2O7 (RE= La and Pr) nanoparticles stabilized by molten-salt synthesis at low temperature. Inorganic Chemistry. 2019;**58**(2):1241-1251

[68] Loong C-K, Thiyagarajan P, Richardson JW, Ozawa M, Suzuki S. Microstructural evolution of zirconia nanoparticles caused by rare-earth modification and heat treatment. Journal of Catalysis. 1997;**171**(2):498-505

[69] Piticescu R-R, Urbina M, Rinaldi A, Cuesta-Lopez S, Sobetkii A. Development of novel material systems and coatings for extreme environments: A brief overview. JOM. 2019;**71**(2):683-690. DOI: 10.1007/s11837-018-3273-6

[70] Karaoglanli AC, Doleker KM, Ozgurlu Y. Interface failure behavior of yttria stabilized zirconia (YSZ), La2Zr2O7, Gd2Zr2O7,YSZ/La2Zr2O7 and YSZ/Gd2Zr2O7 thermal barrier coatings (TBCs) in thermal cyclic exposure. Materials Characterization. 2020;**159**:110072

[71] Doleker KM, Ozgurluk Y, Karaoglanl AC. Isothermal oxidation and thermal cyclic behaviors of YSZ and double layered YSZ/La2Zr2O7 thermal barrier coatings (TBCs). Surface & Coatings Technology. 2018;**351**:78-88

[72] Naga SM, Awaad M, El-Maghraby HF, Hassan AM, Elhoriny M, Killinger A, et al. Effect of La2Zr2O7 coat on the hot corrosion of multi-layer

thermal barrier coatings. Materials and Design. 2016;**102**:1-7

[73] Wang C, Wang Y, Fan S, You Y, Wang L, Yang C. Optimized functionally graded La2Zr2O 8YSZ thermal barrier coatings. Journal of Alloys Compounds. 2015;**649**:1182-1190

[74] Ozgurluk Y, Doleker KM, Karaoglanli AC. Hot corrosion behavior of YSZ, Gd2Zr2O7 and YSZ/Gd2Zr2O7 thermal barrier coatings exposed to molten sulfate and vanadate salt. Applied Surface Science. 2018;**438**:96-113

[75] Wang Y, Ma Z, Liu L, Liu Y. Influence of the grain size on CMAS attack of $Sm_2Zr_2O_7$ ceramic. Ceramics International. 2021;**47**:24453-24457

Chapter 9

Polymer-Derived Advanced Engineering Ceramics

Jinxue Ding and Wei Li

Abstract

Over the past few decades, considerable research efforts and progress have been made concerning processing strategies of advanced ceramics as well as their structural/functional applications. Moreover, there are emerging research activities related to developing synthetic pathways to advanced ceramics with tunable composition, controllable morphologies, or improved sinterability. The polymer-derived ceramic (PDC) route is a relatively young technology for ceramic manufacturing compared with conventional ceramic powder technology, which brings a significant technological breakthrough for the development of ceramic science and technology. As the PDC route allows the processing and manufacturing of advanced ceramics from the liquid/ solid polymeric precursors, they are highly interesting, for example, for the fabrication of near-net shape ceramics, ceramic matrix composites (CMCs), additive manufacturing of advanced ceramics, and so on. The main objective of the present chapter is related to the recent developments of PDCs, to their processing strategies for ceramic components, and to the potential applications of PDCs.

Keywords: polymer-derived ceramics, molecular design and engineering, manufacturing techniques, structural/functional properties, potentail applications

1. Introduction

Molding and sintering fine ceramic powders constitute the typical process for ceramic manufacturing [1]. However, as the applications of ceramics expand, a variety of advanced technologies are being developed and utilized for their production, catering to a range of requirements [2–4]. The polymer-derived ceramic (PDC) route, pioneered in the 1960s, is one of the advanced technologies for ceramic production [5, 6]. Over the last half-century, extensive research on polymer-derived ceramics (PDCs) has predominantly concentrated on silicon-based systems, given their foundation in silicon chemistry [7–9]. From a ceramic composition perspective, the initial focus was on the first two non-oxide binary carbides, SiC and nitrides Si_3N_4 [10, 11]. Following this, ternary compounds such as oxy-carbides SiOC and carbonitrides SiCN, as well as quaternary systems like SiBCN and Si*M*CN (with *M* representing a transition metal), were developed and explored [12–15]. The compositional evolution from binary systems to more complex ternary and quaternary systems is to fulfill diverse requirements for structural/functional properties [16, 17]. Furthermore, these

PDC systems provide a pathway to prepare carbides, nitrides, or carbonitrides within a thermally stable amorphous matrix, broadening both their functionality and structural applications. Besides, other systems without silicon, such as BCN, TiC, and AlN, have also been investigated for various purposes [18, 19]. The PDC route, as indicated by its name, involves the process of acquiring ceramic materials through the controlled thermal decomposition, or thermolysis, of inorganic or organic polymeric precursors. These precursors, typically synthesized through various chemical reactions, undergo controlled pyrolysis to eliminate organic components and transform into inorganic ceramic structures. Indeed, starting from liquid or solid polymeric precursors enables the PDC route to be highly versatile [7, 15]. For example, the viscoelastic properties of polymers allow for the fabrication of ceramics with near-net shapes and complex geometries, which are not always achievable through traditional ceramic processing methods [20]. Consequently, the PDC products are very diverse, such as fibers, coatings, foams, as well as dense monoliths, to suit different applications [21–23]. Most importantly, the PDC route, as an emerging technology, offers advantages in creating intricate and complex ceramic structures with high precision and tailored properties, making it particularly valuable in applications requiring advanced ceramic materials with specific shapes or forms.

2. Polymer-derived ceramic technology

Polymer-derived ceramic technology is a relatively recent development in ceramic manufacturing compared with ceramic powder technology, offering great possibilities for processing advanced ceramic materials. More importantly, the composition, microstructure, and corresponding properties of PDCs can be controllably designed at the molecular level [7, 15]. As aforementioned, it begins with the synthesis of polymeric precursors, in which the molecular structure is carefully designed to impart specific ceramic chemical/physical properties. The next critical step is to shape the precursors into desired forms, followed by crosslinking to stabilize their structure. Subsequently, the organic polymer is heat treated under controlled conditions, such as atmospheres (Ar, N_2, NH_3, etc.), temperatures, and pressures, to remove the organic components and form an amorphous ceramic matrix. Finally, controlled crystallization serves to refine the phase composition of ceramics, thereby tailoring properties such as mechanical strength and chemical and thermal stability. In this section, each stage of the process will be comprehensively elaborated upon.

2.1 Polymeric precursor synthesis

The phase composition and microstructure of the polymer-derived ceramic products are closely related to the molecular type and structure of the polymeric precursors used. Therefore, the design of polymer synthesis plays a crucial role in the field of PDCs. Several organosilicon polymers have proven suitable as precursors for Si-based ceramics. As presented in **Figure 1**, the backbone X defines the class of the polymer, such as polysilanes with X = Si, polysiloxanes with X = O, polycarbosilanes with X = CH_2, polysilazanes with X = NH, and polysilylcarbodiimides with X = [N=C=N]. The functional groups R1 and R2 attached at the silicon atoms are usually hydrogen, aliphatic, or aromatic groups, which provide opportunities to design polymer precursors at the nanoscale level [7]. For example, transition metals or other organic substituents can be introduced onto the side groups R1 and R2, allowing precise

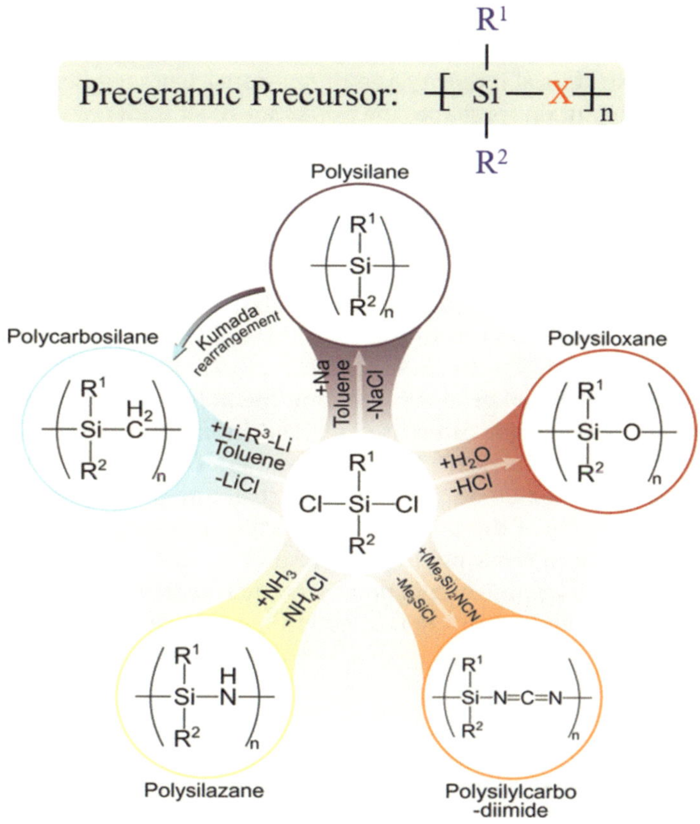

Figure 1.
General synthesis routes of commonly used organosilicon polymers [24].

control over the ceramic yield and the carbon content in the final ceramics [25, 26]. Moreover, the solubility and rheological properties of polymers, which are crucial features for subsequent processing, can also be finely controlled by the side groups [27].

The chlorosilanes (R_xSiCl_{4-x}, x = 0–3, R = organic group), generally formed as byproducts in the silicone industry, are the most frequently used raw materials for the synthesis of Si-based precursors because of their high reactivity, commercial availability, and low cost [2]. As shown in **Figure 1**, the synthetic route for Si-based precursors typically involves catalytic dechlorination of organic chlorosilanes, assisted by catalysts such as alkali metals (like lithium, sodium, potassium), ammonia, amines, or water [24].

2.2 Shaping and crosslinking

Due to the polymeric nature of preceramic precursors, they can be readily shaped and fabricated into various forms, including one-dimensional (1D), two-dimensional (2D), and three-dimensional (3D) ceramic products [2, 7]. It should be noted that the polymeric precursor can exist in various forms: as a cross-linkable liquid, a meltable and curable solid, or an unmeltable but soluble solid [28]. This inherent flexibility in processing allows for a wide range of forming techniques, including extrusion, molding, and casting, making it possible to produce complex ceramic components. One of the advantages is that pre-ceramization machining prevents issues such as tool wear and brittle fracture during component finishing, ensuring a higher-quality final product with improved structural integrity and dimensional accuracy [29].

Importantly, to maintain its shape, the precursor must undergo crosslinking to transform into a thermoset after shaping, which helps to minimize weight loss during ceramization [30]. Thermal cross-linking is the predominant mechanism, typically occurring within the temperature range of 200 to 400°C [31]. During this process, cross-linking reactions take place between functional groups, including Si–H, Si–OH, N–H and vinyl/allyl substituents, leading to the formation of a three-dimensional network structure [32]. Other techniques like oxidative processes, γ-radiation, electron-beam irradiation, and reactive gases or plasma can also be used to initiate cross-linking reactions [21, 33]. In particular, UV-crosslinking can occur when there are photo-sensitive functional groups present on the polymeric precursors [34]. Besides, selective laser curing is often used in 3D printing processes, enabling the fabrication of complex and intricate 3D structures [35]. One can select the appropriate method for cross-linking based on the molecular structure of the polymeric precursor and the desired properties of the final ceramic derived therefrom. Noteworthy, careful control of the extent of cross-linking is crucial because it significantly influences the rheological behavior of preceramic polymers, such as their spinnability for fiber production and uniformity in coating fabrication [36, 37].

2.3 Pyrolysis and crystallization

Generally, the thermal treatment (pyrolysis) occurs after crosslinking to realize the polymer-to-ceramic transition, which is typically completed at temperatures between 400 and 1100°C [7, 15]. During the pyrolysis, the rearrangement and radical reactions occurring above 400°C lead to the cleavage of chemical bonds and the release of hydrogen, low-weight oligomers, and hydrocarbons, consequently forming amorphous ceramic materials. A flow of inert (Ar, N_2) or reactive (NH_3, H_2, CO_2, Ar + H_2O) gas is often used. On the one hand, the inert gas can protect the material from oxidation or other possible reactions. On the other hand, the pyrolysis atmosphere plays an important role in adjusting the phase composition of resultant ceramics [38]. For instance, reactive gas NH_3 is commonly used in the Si–N, Si–B–N, and Si–*M*–N systems, as it can effectively remove residual carbon introduced by the solvent and increase the nitrogen content in the final product [39]. The thermal decomposition accompanied by gas release results in a weight loss typically of 10–30%. It is noteworthy that high weight loss can cause substantial shrinkage in the final ceramic components. To enhance ceramic yield, efforts can focus on optimizing the elemental composition and thermal stability of the preceramic polymer, adjusting the structural arrangement (branched, ring, or linear) within the polymer backbone, improving the effectiveness of crosslinking, and incorporating suitable fillers [40]. Recently, ceramic yields as high as 90% have been reported in the literature [41, 42].

Figure 2 schematically summarized the temperature range for the transformation of amorphous to crystalline in PDC technology [15]. At temperatures exceeding 1100°C, the amorphous phase undergoes crystallization and decomposition, leading to the formation of multiphase crystalline structures [43]. The stability of the amorphous phase is primarily influenced by the addition of other chemical elements such as carbon, boron, aluminum, and transition metals (e.g., Hf, Zr, Ta) [44, 45]. These added elements form a percolation network within the amorphous matrix, acting as diffusion barriers that prevent local crystallization and maintain the size of nuclei

Figure 2.
The processing temperature range for PDC technology [15].

below the critical radius. The nucleation and growth of the crystalline grains from the amorphous matrix occur at higher temperatures (above 1300°C) due to the increase in atomic mobility and the lower energy state of the crystalline phase compared to the amorphous phase. In addition, carbothermal reduction of SiO_2 and Si_3N_4 takes place at elevated temperatures due to the existence of free carbon in SiOC and SiCN ceramics, which leads to the formation of SiC and a further weight loss [46].

3. Microstructure of PDCs

As discussed in Section 2.3, when exposed to high temperatures, PDCs undergo profound microstructural changes due to their intrinsic complexity, comprising an amorphous matrix, free carbon, and nanodomains or a combination thereof. The segregated carbon from the amorphous phase presents a unique characteristic distinct from ceramic products prepared through powder technology. The remarkable resistance of PDCs to crystallization has been attributed to the nature of nanodomains, as proposed. Therefore, understanding the temperature-dependent microstructural evolution in PDCs from amorphous ceramics to nanocomposites is crucial for comprehending their properties, such as mechanical strength, thermal stability, chemical resistance, and electrical and optical properties, to name a few. This evolution affects the overall performance and suitability of PDCs in various high-demand applications, making it a key area of study. To provide a basic introduction, we will elaborate on the microstructure and phase composition of two well-researched systems: polymer-derived SiOC and SiCN.

3.1 Polymer-derived SiOC

SiOC is characterized as an amorphous material, with a significant amount of carbon incorporated into its silica glass structure [47]. The silicon atoms are tetrahedrally coordinated by oxygen and carbon atoms, forming SiO_4C_{4-x} units. The substitution of divalent oxygen atoms with tetravalent carbon atoms increases the bond densities in the system, which explains the enhanced physical properties such as viscosity, elastic modulus, and hardness when compared to vitreous silica counterparts [48]. At pyrolysis temperatures approximately 1100°C, polymer-derived SiOC ceramics remain largely amorphous. However, when advanced, high-resolution analytical techniques are applied, polymer-derived SiOC is found to exhibit nanoscale heterogeneity, due to its typical non-stoichiometric nature and inherent presence of free carbon phases within its structure. Nanosized phases or nanodomains that can exist within the amorphous SiOC matrix include amorphous silica (a-SiO_2), free carbon (C_{free}), amorphous or crystalline silicon carbide (SiC), and interfaces of SiO_4C_{4-x} [49].

It is noteworthy that the carbon content in polymer-derived SiOC can be easily adjusted by designing the molecular structure of the polymer precursor. For instance, varying the ratio of triethoxysilane (*T*H) and methyldiethoxysilane (*D*H) monomers during polymer synthesis allows for precise control over the carbon content [47]. Differences in microstructure can also be observed when varying the carbon content, as displayed in **Figure 3a** [50]. The microstructure exhibits graphitic carbon nanodomains represented by dark-shaded circles, surrounded by C-rich SiO_4C_{4-x}

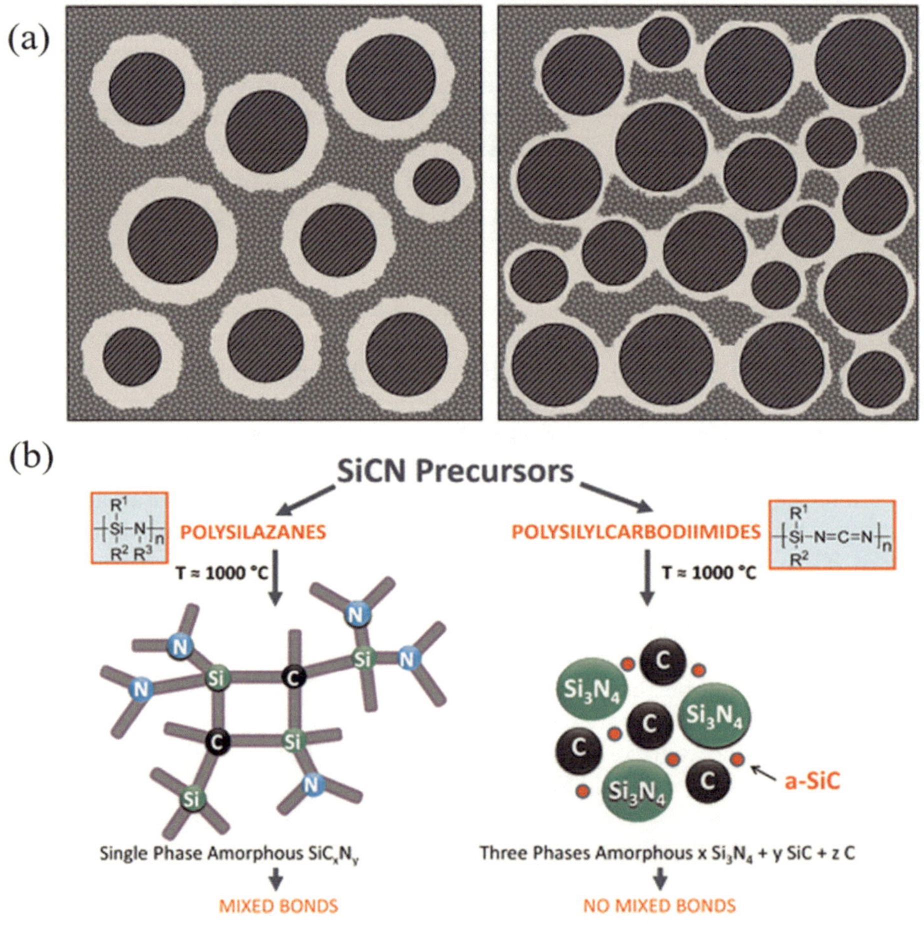

Figure 3.
Schematic microstructure of polymer-derived (a) SiOC and (b) SiCN [7, 50].

structural units depicted in light gray regions. The oxygen-rich SiO_4C_{4-x} structural units are shown as the gray patterned matrix. It is noted that a high content of carbon in the pre-ceramic polymer leads to the percolation of the graphitic carbon regions and the C-rich structural units within the ceramics. Moreover, by controlling the carbon content to low levels (i.e., $\leq$ 8 wt%), it is possible to achieve nearly stoichiometric SiOC, which facilitates the investigation of microstructural evolution of polymer-derived SiOC [51]. It is reported that the phase separation starts when the annealing temperature is up to 1300°C, due to the redistribution of the Si–O and Si–C bonds. Upon further annealing at higher temperatures (1400–1600°C), nanosized β-SiC phases form and grow due to carbothermal reactions, leading to a decrease in the free carbon content [52].

3.2 Polymer-derived SiCN

In the SiCN system, a completely different microstructure is observed depending on the backbone structure of the polymeric precursor used: polysilazanes (X = N) or polysilylcarbodiimides (X = [N=C=N]), as presented in **Figure 3b** [53]. Polysilazanes produce a single-phase amorphous SiC_xN_y structure characterized by mixed Si–C–N bonding and localized areas of free carbon [54]. In this structure, silicon maintains its tetrafunctional coordination, allowing for mixed bonding with nitrogen and carbon atoms to form SiC_4N_{4-x} configurations. Despite some variations in microstructural evolution caused by the presence of linear or cyclic units in polysilazanes, the microstructure after pyrolysis at 1000°C can generally be depicted by the left picture in **Figure 3b**. However, SiCN ceramics derived from polysilylcarbodiimides form a multiphase amorphous system primarily consisting of amorphous Si_3N_4 and carbon clusters, with little to no presence of SiC_4N_{4-x} [55]. The formation of the amorphous SiC phase requires a sufficiently high carbon content. Compared to polysilazane-derived SiCN, the development of particular amorphous structures inhibits the processes of crystallization and phase separation. In addition, the backbone structure of polysilylcarbodiimides (whether branched or linear) influences the structural evolution during pyrolysis. For example, a branched poly(phenylsilsequicarbodiimide) forms a nanostructure with regions containing free carbon and amorphous Si_3N_4 (a-Si_3N_4), alongside an interfacial region characterized by C–N bonds. Instead, a linear poly (vinylsilylcarbodiimide) produces a nanostructure with dispersed a-Si_3N_4 surrounded by interconnected free carbon phase [54].

Similarly, after pyrolysis at 1100°C, polymer-derived SiCN ceramics remain mainly amorphous with some nanodomain structures. The onset of crystallization occurs within the temperature range of 1300–1500°C, varying depending on the C/Si ratio in the system. It has been reported that the thermal stability against crystallization can be enhanced with an increase in the C/Si ratio [56]. In this stage, the coarsening of nanodomains and the transformation of amorphous Si_3N_4 into α-Si_3N_4 can also be observed. The presence of the free carbon phase has been demonstrated to restrict the formation of the α-Si_3N_4 phase by acting as a diffusion barrier [57]. Upon thermal treatment at higher temperatures (1600–2000°C), many of the Si_3N_4 phase can undergo transformation into SiC phase due to carbothermal reaction, accompanied by the release of N_2 gas [58]. Above 1700°C, all Si-based phases become fully crystalline, typically with a carbon matrix embedded with α- and β-SiC phases. Besides, many researchers have recently been interested in introducing boron or transition metals (*M* = Hf, Zr, Ta) into polysilazanes or polysilylcarbodiimides to form quaternary SiBCN or Si*M*CN systems, in order to increase the nitrogen content in the final ceramic products and improve high-temperature properties. Results show that the formed intergranular nanodomains BC_xN_y and MC_xN_y contribute a lot to the thermal stability, oxidation resistance, and high-temperature mechanical properties [59].

4. Advanced engineering applications of PDCs

The ability to tune the chemical and physical properties of PDCs endows them with multifunctional capabilities. This versatility allows PDCs to be customized for a wide range of applications, from high-temperature structural materials to components in electronic and optical devices, as well as in energy storage and conversion systems [15, 23]. Here, we will summarize selected applications based on the types of final ceramic products.

4.1 Dense ceramic bulks

Polymer-derived ceramic monoliths based on SiC and Si_3N_4 exhibit significant potential for applications in high-temperature structural components, as well as in grinding and cutting operations [60]. Compared with conventional sintering using binders to promote densification of green pellets, which requires burning out during processing and inevitably introduces impurities, preceramic polymer binders offer several advantages [61]. They enhance green density significantly, particularly with powders that have low packing density. Upon pyrolysis, these binders transform into ceramics with extremely small particle sizes, which facilitate sintering without the need of external additional aids. Besides, incorporating inert or reactive fillers into the polymer matrix is effective in minimizing pyrolysis shrinkage. This approach allows for achieving linear dimensional changes of less than 0.1% through precise control of time and temperature during partial pyrolysis of complex components. This process results in ceramic products, especially with complex shapes, exhibiting enhanced density and thus improved mechanical properties [2, 20].

Specifically, to produce dense, crack-free, and stoichiometric SiC bulks using the PDC route, four main steps are typically followed in PDCs: obtaining amorphous powder through pyrolysis; shaping using warm-pressing techniques; ceramization of the shaped green bodies through heat treatment; and liquid precursor infiltration and pyrolysis (PIP). Recently, dense SiC-based monoliths have also been prepared using hot-pressing (HP), spark plasma sintering (SPS), and 3D printing techniques [62–64]. Wen et al. [64] sintered boron-containing HfC/SiC-based ceramics using SPS. Incorporating boron results in the formation of low-viscous borosilicate, which facilitates the rapid formation of a continuous, dense scale upon oxidation, thereby providing excellent oxidation resistance over a wide temperature range (from 20 to 1500°C). Li et al. [41] prepared additive-free amorphous bulk polymer-derived SiHfN ceramics by using warm-pressing followed by ammonolysis and annealing. The critical issues related to gas evolution and crystallization, which can lead to bloating and cracking, were addressed through controlled thermolysis and pressure management. The SiHfN ceramic achieved a hardness of up to 19.7 GPa through warm-pressing at 120 MPa, followed by annealing the green sample at 1300°C. Xiong et al. [62] successfully prepared crack-free dense monolith and lattice skeleton structural polymer-derived

SiOC ceramics by vat photopolymerization additive manufacturing. As displayed in **Figure 4**, the pyrolyzed samples exhibited relatively uniform shrinkage and crack-free structures because the introduction of phenolic resin (PR) enabled a smooth gas-release process by establishing ordered channels within the sample bodies during pyrolysis. The introduction of PR can enhance not only high-temperature stability but also mechanical properties.

4.2 Ceramic fiber and fiber-reinforced ceramic matrix composites (CMCs)

Ceramic fibers, particularly those based on SiC, have stood out as the most successful commercial application of PDCs since the last century. Ceramic fibers have paved the way for the development of fiber-reinforced ceramic matrix composites (CMCs), which have attracted considerable attention for their applications in high-temperature environments where the metallic materials suffer serious corrosion and mechanical property deterioration [23]. Therefore, ceramic fibers must possess superior mechanical and chemical properties to withstand the demanding operational environments of turbine engines, rocket nozzles, and furnaces. The polymer-derived ceramic fiber process has enabled the successful production of advanced ceramic fibers spanning from nanometers to micrometers in diameter [36].

To produce fibers, typically, a spinning step, either melting spinning or electrospinning, needs to be incorporated. Therefore, suitable rheology of the polymer precursor, both in its molten state and in solution, is essential for processes. Maintaining green fibers under tension during the polymer-to-ceramic conversion is

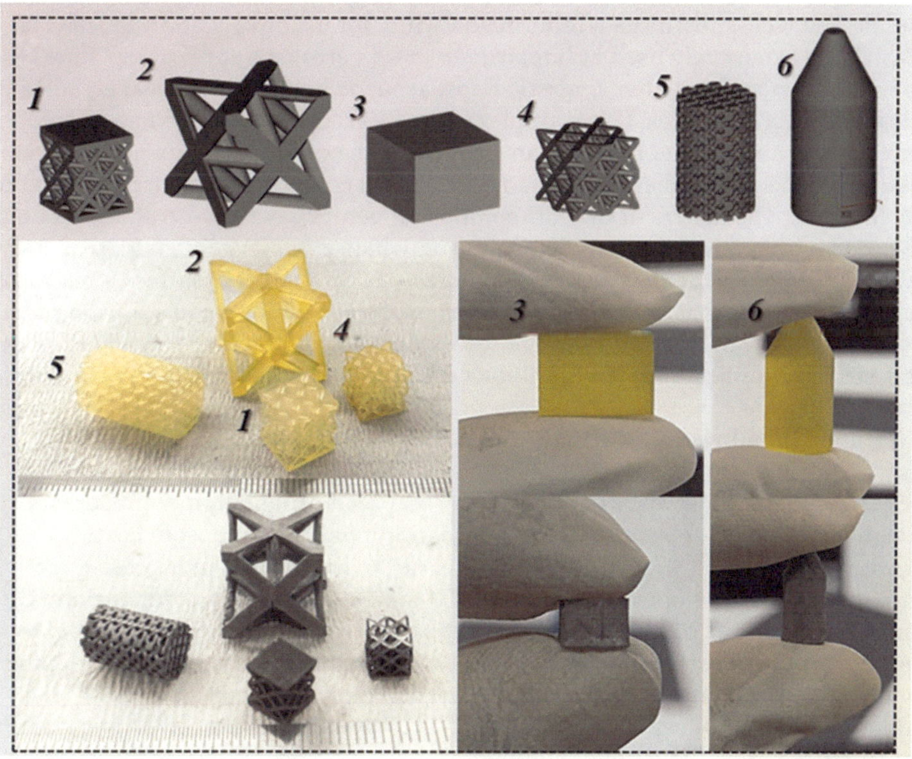

Figure 4.
Various digitally designed 3D models with different shapes [62].

crucial to prevent crimping of the fibers due to thermal shrinkage, and all stages of the process are carried out in a controlled atmosphere (either inert or reactive). For example, SiC-based ceramic fibers are commercially manufactured by melt spinning and widely employed as heat-resistant materials, as well as for reinforcing ceramic matrix composites (CMCs). SiC-based ceramic fibers are derived from polycarbosilane or polymetallic ocarbosilane precursors and have undergone the stages of development according to their chemical composition, oxygen content, and C/Si atomic ratio [65]. Amorphous SiBCN fibers have also been successfully prepared using polyborosilazane precursors [66]. As-prepared amorphous SiBCN fibers with a dense and smooth texture show high thermal stability in a N_2 atmosphere of up to 1600°C. Electrospinning is also extensively studied for producing fibers with precise control over size and morphology, such as novel BN fibers and core-shell nanofibers [67]. This method offers a versatile and controllable approach to creating nanoscale ceramic fibers, facilitating the integration of functional materials with fibers and enabling new applications to be explored.

Fiber-reinforced CMCs are fabricated by encapsulating carbon fibers (C_f) or ceramic fibers within a dense ceramic matrix. The matrix phase can consist of a single or multiple phases with compositional variations. The dense matrix is typically formed using three primary techniques: chemical vapor infiltration (CVI), polymer infiltration and pyrolysis (PIP), and reactive melt infiltration (RMI) or liquid silicon infiltration (LSI) [23]. Among them, the PIP method offers several advantages, including simplified and cost-effective equipment requirements, lower processing temperatures that maintain the strength of ceramic fibers, shorter cycle times, enhanced infiltration depth, improved uniformity, and the capability to produce large and intricate components. The well-known fiber-reinforced CMCs, for example, C/SiC, SiC/SiC, and C/C-SiC, are extensively used as hot structures for aerospace applications, like aero-engine combustor liners, ducts, nozzle flaps, acoustic liners, turbine vanes, turbine blades, and aircraft brakes. For example, using C/SiC composites as materials for combustion chambers and nozzles can elevate the maximum combustion temperature to above 1650°C [68]. Fiber-reinforced SiOC-based composites are also explored due to its low cost. The matrix in C/SiOC composites comprises nano-crystallites of SiC embedded within an amorphous SiOC matrix, which exhibits good oxidation resistance [69]. However, at elevated temperatures, the SiOC matrix suffers from poor creep resistance, primarily due to the presence of silica nanodomains. Introducing boron to the SiOC matrix can improve the oxidation resistance, as the SiBOC matrix forms stable borosilicate glass and inhibits crystallization during oxidization [70].

4.3 Ceramic membranes, coatings, and adhesives

Ceramic membranes and coatings can be prepared using polymer precursors through liquid phase deposition (e.g., dip coating, spray coating, spin coating) followed by pyrolysis [46]. This method is easier to handle and more cost-effective compared to physical vapor deposition (PVD) and chemical vapor deposition (CVD). Generally, thin layers of membranes are deposited by coating them onto tubular asymmetric porous ceramic substrates. Polymer-derived SiOC membranes demonstrate high permeability to hydrogen over carbon dioxide, showcasing molecular sieving capabilities for separating H_2/CO_2 [71]. Polymer-derived SiBCN membranes offer significant advantages over commercial alumina membranes, including ease of manufacturing, high thermal stability, and excellent oxidation resistance, making them suitable for high-temperature separation applications [72].

Polymer-derived ceramic coatings are extensively utilized as protective layers on metals or CMCs to prevent oxidation when used as hot structural components in the aerospace industry [23]. For example, ceramic-like gradient Si(C)N coatings can form on stainless steel sheets, which effectively shield it from oxidation up to 1000°C [73]. The gradient Si(N)CrO layers act as a diffusion barrier against oxygen, resulting in a significant reduction of both weight gain and the oxidation rate by up to two orders of magnitude. However, during the pyrolysis of precursors into ceramics, porosity and residual stress may occur due to shrinkage and decomposition. To tackle these issues, passive fillers such as BN, ZrO_2, and Al_2O_3, as well as active fillers like $ZrSi_2$ and $TiSi_2$, can be incorporated to minimize shrinkage during the pyrolysis of ceramic precursors. This approach enables the creation of high-performance environmental barrier coatings capable of achieving substantial thicknesses, up to 100 μm [74]. Alternatively, another effective solution is the design of a double-layer coating system with a PDC bond coat and a glass/ceramic filled PDC topcoat [75]. The PDC bond coat plays a crucial role by shielding the metallic surface, enhancing adhesion between the surface and the topcoat, and reducing the coefficient of thermal expansion (CTE) mismatch between the metal and the topcoat. Apart from metallic components, it is also important to develop protective coatings for CMC components [76]. For example, C/C composites undergo degradation above 500°C in an oxidizing environment. Despite C/SiC, C/C-SiC, and SiC/SiC composites exhibiting good oxidation resistance, they still necessitate protective coatings to prolong their service life. The SiC-coated C/C composite experiences approximately one-fourth the mass loss compared to the uncoated C/C composite, after 50 minutes of exposure to an oxidizing environment [77].

In many cases, it is more cost-effective to construct complex shapes by assembling simple geometric forms. Ceramic adhesive bonding is the most cost-effective method of joining without significant energy wastage, and it does not cause severe damage to the substrates [78]. Therefore, high-temperature adhesives are crucial structural materials for aerospace and nuclear power applications, where they must offer strong bonding capabilities in elevated temperature conditions. NASA has developed an adhesive called Non-Oxide Adhesive eXperimental (NOAX), which blends a sophisticated formulation of preceramic sealant (hydridopolycarbosilane) with silicon carbide powder and additional additives. This adhesive meets NASA's stringent requirements for space missions [79]. It is applied to fill cracks or gaps in protective coatings measuring up to 0.5 mm wide and 100 mm long. Following exposure to temperatures exceeding several hundred degrees, NOAX achieves a semi-rigid state and reaches complete hardening only upon encountering the extreme heat during reentry into Earth's atmosphere.

4.4 Microelectromechanical systems (MEMSs) and sensor

While silicon-based MEMSs have been applied in sensors and actuators, their performance is frequently constrained by poor thermal stability above 250°C and mechanical stability above 600°C [80]. Certain PDCs, such as SiOC, SiCN, and SiCNO, also exhibit semiconducting properties, primarily attributed to the presence of graphitic carbon and other nanodomains within the amorphous matrix. Reportedly, both SiCN and SiOC ceramics exhibit remarkable piezoresistive properties, with exceptionally high coefficients observed along both longitudinal and transverse directions, even under elevated temperatures [81, 82]. According to Ryu et al., semiconductive properties of SiCNO PDCs were measured and confirmed up to 1300°

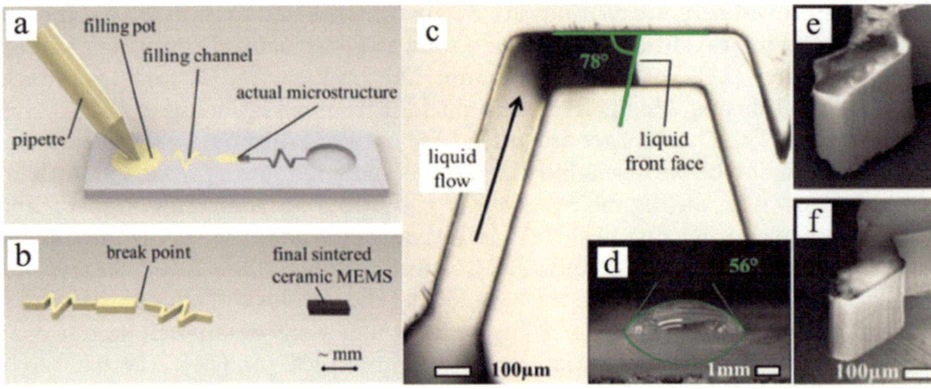

Figure 5.
Illustration of the shaping process of PDCs [80].

C without dopants. The band-gap of SiCNO PDCs decreased from 2.2 to 0.1 eV as the annealing temperature was raised from 1100 to 1400°C [83].

Additionally, the semiconductive properties of PDCs can be readily tuned by controlling their microstructure. SiC, with its wide bandgap semiconductor properties, has been proposed as a solution to address temperature limitations in Si MEMSs for high-temperature sensing applications. Its superior resistance to corrosion, excellent thermal stability, chemical stability, and mechanical performance have sparked considerable interest in high-temperature semiconductor applications [84]. **Figure 5** illustrates the shaping process and microscopic features of PDC MEMSs, demonstrating how the polymer route utilizes micropatterning to shape liquid precursors [80].

Due to their temperature-dependent electrical resistances, PDCs have shown promise for use as flux sensors or temperature sensors in harsh environments. As an example, Nagaiah et al. engineered a high-temperature heat flux sensor for gas turbine engines utilizing polymer-derived SiCN ceramics [85]. This advanced sensor shows a temperature coefficient of resistance of 4000 ppm/°C and maintains extended functionality at 1400°C, surpassing conventional heat flux sensors. In addition, amorphous SiCN-based ceramics have demonstrated effectiveness in hydrogen (H_2) sensing at temperatures as high as 500°C, attributed to their semiconducting nature. Thinner layers of SiCN ceramics exhibit heightened sensitivity in measuring resistivity upon exposure to adsorbed hydrogen molecules [86].

4.5 Porous ceramics

Porous ceramics, such as cellular ceramics/ceramic foams, ceramic sandwich structures, and ceramic aerogels, are ideal for lightweight thermal protection systems (TPS) [87]. Cellular ceramics are ceramic materials known for their substantial porosity, with hollow cells that can vary in repetition and shape. These cells can be regular in size and form, akin to a honeycomb structure, or irregular, resembling a foam structure. Using preceramic polymers to produce cellular ceramics offers several benefits, such as controlling the composition of ceramics, adapting to new processing techniques, and easily machining the precursor foam followed by post-curing. The fabrication of ceramic foams using preceramic polymers can be achieved through three different methods: (i) replica method, (ii) sacrificial template method, and (iii) direct foaming, as illustrated in **Figure 6** [88]. For example, SiOC foams typically

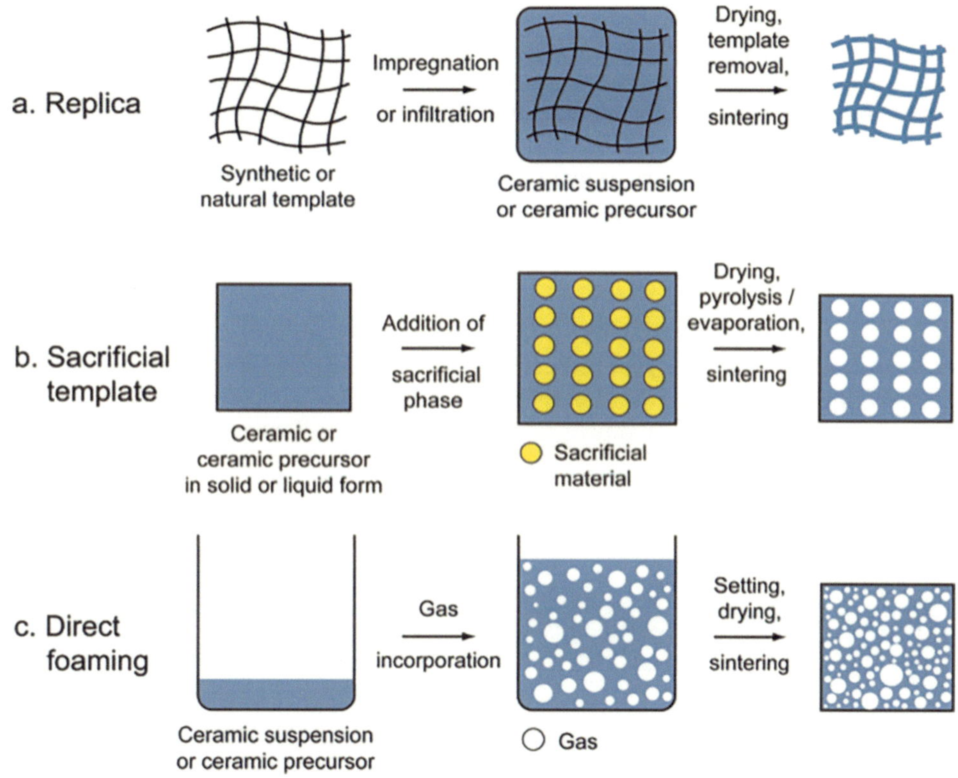

Figure 6.
Scheme of possible processing routes used for the production of macroporous ceramics [88].

undergo a decrease in compressive strength at elevated temperatures, mainly due to carbothermal reaction [89]. In contrast, SiC foams can maintain their strength under similar conditions due to the absence of carbothermal reaction [90]. Derived from polycarbosilane, SiC foams demonstrate an enhanced load-bearing capability through increased strut density, attributable to a reduction in pore size. Besides that, ceramic aerogel, with its ultra-low density and low thermal conductivity, is highly sought after as a thermal protection system (TPS) material for space applications [91].

5. Conclusions

Advanced ceramics, being among the most sought-after technological materials, find extensive applications across various industries and engineering fields. The polymer-derived ceramic route paves a disruptive way for the ability to realize high-performance ceramic components, even with complex shapes. The use of polymer precursors enables design at the molecular scale, employing highly versatile shaping methods and fine-tuning mechanical and functional properties. Careful control of the pyrolysis process (e.g., temperature and atmosphere) allows for the preparation of polymer-derived ceramics with controlled composition, tailored microstructure, and desired properties. The versatility in shaping makes it possible to produce ceramics in all kinds of forms, such as fibers, coatings, and dense and porous monoliths. Due to their excellent high-temperature stability, oxidation resistance, and semiconducting properties, polymer-derived ceramics have been employed in various applications including hot structural components, protective coatings, sensors, and thermal protection systems. In recent years, researchers have shown a growing interest in enhancing properties through microstructural design, such as by introducing transition metal elements, and through advanced processing techniques like additive manufacturing. With continuous efforts in this field, it is anticipated that more potential applications for polymer-derived ceramics will be explored.

Acknowledgements

Jinxue Ding and Wei Li gratefully acknowledge the financial support from the ECR Publication Fund provided by the Technical University of Darmstadt. Additionally, Wei Li extends appreciation for the support from the TU Darmstadt Career Bridging Grant.

Conflict of interest

The authors declare no conflict of interest.

Author details

Jinxue Ding[1] and Wei Li[1,2]*

1 Technische Universität Darmstadt, Institut für Materialwissenschaft, Darmstadt, Germany

2 Department of Mechanical and Materials Engineering, University of Alabama, Birmingham, AL, USA

*Address all correspondence to: wli3@uab.edu

References

[1] Rahaman MN. Ceramic Processing. 1st ed. Boca Raton: CRC Press; 2017

[2] Greil P. Polymer derived engineering ceramics. Advanced Engineering Materials. 2000;**2**(6):339-348

[3] Richerson DW, Lee WE. Modern Ceramic Engineering: Properties, Processing, and Use in Design. 4th ed. Boca Raton: CRC Press; 2018

[4] Lakhdar Y, Tuck C, Binner J, Terry A, Goodridge R. Additive manufacturing of advanced ceramic materials. Progress in Materials Science. 2021;**116**:100736

[5] Ainger FW, Herbert JM. The preparation of phosphorus-nitrogen compounds as non-porous solids. Special Ceramics. 1960:168-182

[6] Chantrell PG, Popper P. Inorganic polymers and ceramics. Special Ceramics. 1965:87-103

[7] Colombo P, Mera G, Riedel R, Soraru GD. Polymer-derived ceramics: 40 years of research and innovation in advanced ceramics. Journal of the American Ceramic Society. 2010;**93**(7): 1805-1837

[8] Shengyang F, Zhu M, Zhu Y. Organosilicon polymer-derived ceramics: An overview. Journal of Advanced Ceramics. 2019;**8**(4):457-478

[9] Xia A, Yin J, Chen X, Liu X, Huang Z. Polymer-derived si-based ceramics: Recent developments and perspectives. Crystals. 2020;**10**(9):824

[10] Schilling Jr CL. Polymeric routes to silicon carbide. British Polymer Journal. 1986;**18**(6):355-358

[11] Schmidt WR, Sukumar V, Hurley Jr WJ, Garcia R, Doremus RH, Interrante LV, et al. Silicon nitride derived from an organometallic polymeric precursor: Preparation and characterization. Journal of the American Ceramic Society. 1990;**73**(8): 2412-2418

[12] Bujalski DR, Wieber GM, Zank GA, et al. Stoichiometry control of sioc ceramics by siloxane polymer functionality. Journal of Materials Chemistry. 1998;**8**(6):1427-1433

[13] Galusek D, Reschke S, Riedel R, Dreßler W, Šajgaík P, Lenčéš Z, et al. In-situ carbon content adjustment in polysilazane derived amorphous sicn bulk ceramics. Journal of the European Ceramic Society. 1999;**19**(10):1911-1921

[14] Lee J, Butt DP, Baney RH, Bowers CR, Tulenko JS. Synthesis and pyrolysis of novel polysilazane to sibcn ceramic. Journal of Non-Crystalline Solids. 2005;**351**(37–39):2995-3005

[15] Wen Q, Yeping X, Binbin X, Fasel C, Guillon O, Buntkowsky G, et al. Single-source-precursor synthesis of dense sic/ hfc x n 1- x-based ultrahigh-temperature ceramic nanocomposites. Nanoscale. 2014;**6**(22):13678-13689

[16] Colombo P. Polymer Derived Ceramics: From Nano-Structure to Applications. Pennsylvania, USA: DEStech Publications, Inc; 2010

[17] Ren Z, Mujib SB, Singh G. High-temperature properties and applications of si-based polymer-derived ceramics: A review. Materials. 2021;**14**(3):614

[18] Jiang Z, Rhine WE. Synthesis and pyrolysis of novel polymeric precursors to tic/al$_2$o$_3$, tin/al$_2$o$_3$, and aln/tin nanocomposites. Chemistry of Materials. 1994;**6**(7):1080-1086

[19] Zhang T, Zhang J, Wen G, Zhong B, Xia L, Xiaoxiao Huang H, et al. Ultra-light h-bcn architectures derived from new organic monomers with tunable electromagnetic wave absorption. Carbon. 2018;**136**:345-358

[20] Greil P. Near net shape manufacturing of polymer derived ceramics. Journal of the European Ceramic Society. 1998;**18**(13):1905-1914

[21] Ichikawa H. Polymer-derived ceramic fibers. Annual Review of Materials Research. 2016;**46**(1):335-356

[22] Vakifahmetoglu C, Zeydanli D, Colombo P. Porous polymer derived ceramics. Materials Science and Engineering: R: Reports. 2016;**106**:1-30

[23] Packirisamy S, Sreejith KJ, Devapal D, Swaminathan B. Polymer-derived ceramics and their space applications. In: Handbook of Advanced Ceramics and Composites: Defense, Security, Aerospace and Energy Applications. Cham, Switzerland: Springer; 2020. pp. 975-1080

[24] Hotza D, Nishihora RK, Machado RAF, Geffroy P-M, Chartier T, Bernard S. Tape casting of preceramic polymers toward advanced ceramics: A review. International Journal of Ceramic Engineering and Science. 2019;**1**(1): 21-41

[25] Widgeon S, Mera G, Gao Y, Sen S, Navrotsky A, Riedel R. Effect of precursor on speciation and nanostructure of sibcn polymer-derived ceramics. Journal of the American Ceramic Society. 2013;**96**(5):1651-1659

[26] Thor N, Bernauer J, Petry N-C, Ionescu E, Riedel R, Pundt A, et al. Microstructural evolution of si (hf_xta_{1-x}) (c) n polymer-derived ceramics upon high-temperature anneal. Journal of the European Ceramic Society. 2023;**43**(4): 1417-1431

[27] Balan C, Riedel R. Rheological investigations of a polymeric precursor for ceramic materials: Experiments and theoretical modeling. Journal of Optoelectronics and Advanced Materials. 2006;**8**(2):561

[28] Ackley BJ, Martin KL, Key TS, Clarkson CM, Bowen JJ, Posey ND, et al. Advances in the synthesis of preceramic polymers for the formation of silicon-based and ultrahigh-temperature non-oxide ceramics. Chemical Reviews. 2023; **123**(8):4188-4236

[29] Maria R, da Rocha P, Greil JC, Bressiani, and Ana Helena de Almeida Bressiani. Complex-shaped ceramic composites obtained by machining compact polymer-filler mixtures. Materials Research. 2005;**8**:191-196

[30] Dong S, Li Y-L, An H-J, Liu X, Hou F, Li J-Y, et al. Pyrolytic transformation of liquid precursors to shaped bulk ceramics. Journal of the European Ceramic Society. 2010;**30**(6): 1503-1511

[31] Schulz M, Börner M, Haußelt J, Heldele R. Polymer derived ceramic microparts from x-ray lithography —Cross-linking behavior and process optimization. Journal of the European Ceramic Society. 2005;**25**(2–3):199-204

[32] Choong Kwet Yive NS, Corriu RJP, Leclercq D, Mutin PH, Vioux A. Silicon carbonitride from polymeric precursors: Thermal cross-linking and pyrolysis of oligosilazane model compounds. Chemistry of Materials. 1992;**4**(1): 141-146

[33] Zhiming S, Zhang L, Li Y, Li S, Chen L. Rapid preparation of sic fibers

using a curing route of electron irradiation in a low oxygen concentration atmosphere. Journal of the American Ceramic Society. 2015;**98**(7): 2014-2017

[34] He W, Chen L, Peng F. Coating formed by sibcn single source precursor via uv-photopolymerization. Materials Letters. 2017;**206**:121-123

[35] Li S, Duan W, Zhao T, Han W, Wang L, Dou R, et al. The fabrication of sibcn ceramic components from preceramic polymers by digital light processing (dlp) 3d printing technology. Journal of the European Ceramic Society. 2018;**38**(14):4597-4603

[36] Miele P, Bernard S, Cornu D, Toury B. Recent developments in polymer-derived ceramic fibers (pdcfs): Preparation, properties and applications–a review. Soft Materials. 2007;**4**(2–4):249-286

[37] Bernauer J, Kredel SA, Ionescu E, Riedel R. Polymer-derived ceramic coatings with excellent thermal cycling stability. Advanced Engineering Materials. 2024;**26**:2301820

[38] Sorarù GD, Tavonatti C, Kundanati L, Pugno N, Biesuz M. Effect of the pyrolysis atmosphere on the mechanical properties of polymer-derived sioc and sicn. Journal of the American Ceramic Society. 2020; **103**(11):6519-6530

[39] Li W, Widenmeyer M, Ding J, Jiang T, Feldmann L, Liu J, et al. Phase evolution and oxidation resistance of $si_3n_4/hfb_xc_yn^{1-x-y}$ ceramic nanocomposites prepared from tailored preceramic polymers. Ceramics International. 2023;**49**(21):34164-34172

[40] Zhuang K, Lin S, Huang W, Liao L, Zheng Y, Li L, et al. Realizing high ceramic yield and low shrinkage of in-situ formed lightweight 3d-sic (rgo) px polymer-derived ceramics with excellent fracture toughness. Ceramics International. 2020;**46**(17):27426-27436

[41] Li W, Li F, Zhaoju Y, Wen Q, Fan B, Feng Y, et al. Polymer-derived sihfn ceramics: From amorphous bulk ceramics with excellent mechanical properties to high temperature resistant ceramic nanocomposites. Journal of the European Ceramic Society. 2022;**42**(11): 4493-4502

[42] Li W, Hanzi D, Tian C, Jiang T, Bernauer J, Widenmeyer M, et al. Single-source-precursor derived bulk si_3n_4/hfb_xn_{1-x} ceramic nanocomposites with excellent oxidation resistance. Zeitschrift für Anorganische und Allgemeine Chemie. 2022;**648**(21):e202200203

[43] Poerschke DL, Braithwaite A, Park D, Lauten F. Crystallization behavior of polymer-derived si–o–c for ceramic matrix composite processing. Acta Materialia. 2018;**147**:329-341

[44] Viard A, Fonblanc D, Lopez-Ferber D, Schmidt M, Lale A, Durif C, et al. Polymer derived si–b–c–n ceramics: 30 years of research. Advanced Engineering Materials. 2018;**20**(10):1800360

[45] Yang N, Kathy L. Effects of transition metals on the evolution of polymer-derived sioc ceramics. Carbon. 2021;**171**:88-95

[46] Barrios E, Zhai L. A review of the evolution of the nanostructure of sicn and sioc polymer derived ceramics and the impact on mechanical properties. Molecular Systems Design and Engineering. 2020;**5**(10):1606-1641

[47] Sorarù GD, D'andrea G, Campostrini R, Babonneau F, Mariotto G. Structural characterization

and high-temperature behavior of silicon oxycarbide glasses prepared from sol-gel precursors containing si-h bonds. Journal of the American Ceramic Society. 1995; **78**(2):379-387

[48] Sorarù GD, Kundanati L, Santhosh B, Pugno N. Influence of free carbon on the young's modulus and hardness of polymer-derived silicon oxycarbide glasses. Journal of the American Ceramic Society. 2019;**102**(3):907-913

[49] Chaney H, Zhou Y, Kathy L. Understanding sioc atomic structures via synchrotron x-ray and reactive force field potential studies. Materials Today Chemistry. 2023;**29**:101429

[50] Widgeon SJ, Sen S, Mera G, Ionescu E, Riedel R, Navrotsky A. 29si and 13c solid-state nmr spectroscopic study of nanometer-scale structure and mass fractal characteristics of amorphous polymer derived silicon oxycarbide ceramics. Chemistry of Materials. 2010;**22**(23):6221-6228

[51] Bois L, Maquet J, Babonneau F, Mutin H, Bahloul D. Structural characterization of sol-gel derived oxycarbide glasses. 1. Study of the pyrolysis process. Chemistry of Materials. 1994;**6**(6):796-802

[52] Bréquel H, Parmentier J, Walter S, Badheka R, Trimmel G, Sylvie Masse J, et al. Systematic structural characterization of the high-temperature behavior of nearly stoichiometric silicon oxycarbide glasses. Chemistry of Materials. 2004;**16**(13):2585-2598

[53] Gao Y, Mera G, Nguyen H, Morita K, Kleebe H-J, Riedel R. Processing route dramatically influencing the nanostructure of carbon-rich sicn and sibcn polymer-derived ceramics. Part i: Low temperature thermal transformation. Journal of the European Ceramic Society. 2012;**32**(9):1857-1866

[54] Widgeon S, Mera G, Gao Y, Stoyanov E, Sen S, Navrotsky A, et al. Nanostructure and energetics of carbon-rich sicn ceramics derived from polysilylcarbodiimides: Role of the nanodomain interfaces. Chemistry of Materials. 2012;**24**(6):1181-1191

[55] Bill J, Schuhmacher J, Müller K, Schempp S, Seitzy J, Dürr J, et al. Investigations into the structural evolution of amorphous si–c–n ceramics from precursors. International Journal of Materials Research. 2000;**91**(4):335-351

[56] Iwamoto Y, Völger W, Kroke E, Riedel R, Saitou T, Matsunaga K. Crystallization behavior of amorphous silicon carbonitride ceramics derived from organometallic precursors. Journal of the American Ceramic Society. 2001; **84**(10):2170-2178

[57] Klausmann A, Morita K, Johanns KE, Fasel C, Durst K, Mera G, et al. Synthesis and high-temperature evolution of polysilylcarbodiimide-derived sicn ceramic coatings. Journal of the European Ceramic Society. 2015;**35**(14): 3771-3780

[58] Mera G, Tamayo A, Nguyen H, Sen S, Riedel R. Nanodomain structure of carbon-rich silicon carbonitride polymer-derived ceramics. Journal of the American Ceramic Society. 2010;**93**(4): 1169-1175

[59] Yuan J, Hapis S, Breitzke H, Yeping X, Fasel C, Kleebe H-J, et al. Single-source-precursor synthesis of hafnium-containing ultrahigh-temperature ceramic nanocomposites (uhtc-ncs). Inorganic Chemistry. 2014; **53**(19):10443-10455

[60] Degenhardt U, Günter Motz W, Krenkel FS, Berroth K, Harrer W, Danzer R. Si_3n_4/sic materials based on preceramic polymers and ceramic

powder. Ceramic Transactions. 2010; **209**:379-387

[61] Riedel R, Seher M, Mayer J, Vinga D, Szabó. Polymer-derived si-based bulk ceramics, part i: Preparation, processing and properties. Journal of the European Ceramic Society. 1995;**15**(8):703-715

[62] Xiong S, Liu J, Cao J, Li Z, Idrees M, Lin X, et al. 3d printing of crack-free dense polymer-derived ceramic monoliths and lattice skeletons with improved thickness and mechanical performance. Additive Manufacturing. 2022;**57**:102964

[63] Esfehanian M, Oberacker R, Fett T, Hoffmann MJ. Development of dense filler-free polymer-derived sioc ceramics by field-assisted sintering. Journal of the American Ceramic Society. 2008;**91**(11): 3803-3805

[64] Wen Q, Zhaoju Y, Liu X, Bruns S, Yin X, Eriksson M, et al. Mechanical properties and electromagnetic shielding performance of single-source-precursor synthesized dense monolithic sic/hfcxn$_{1-x}$/c ceramic nanocomposites. Journal of Materials Chemistry C. 2019;7(34): 10683-10693

[65] Flores O, Bordia RK, Nestler D, Krenkel W, Motz G. Ceramic fibers based on sic and sicn systems: Current research, development, and commercial status. Advanced Engineering Materials. 2014;**16**(6):621-636

[66] Viard A, Miele P, Bernard S. Polymer-derived ceramics route toward sicn and sibcn fibers: From chemistry of polycarbosilazanes to the design and characterization of ceramic fibers. Journal of the Ceramic Society of Japan. 2016;**124**(10):967-980

[67] Salles V, Bernard S, Brioude A, Cornu D, Miele P. A new class of boron nitride fibers with tunable properties by combining an electrospinning process and the polymer-derived ceramics route. Nanoscale. 2010;**2**(2):215-217

[68] Liu C, Chen J, Han H, Wang Y, Zhang Z. A long duration and high reliability liquid apogee engine for satellites. Acta Astronautica. 2004;**55**(3–9):401-408

[69] de Omena Pina SR, Pardini LC, Yoshida IVP. Carbon fiber/ceramic matrix composites: Processing, oxidation and mechanical properties. Journal of Materials Science. 2007;**42**:4245-4253

[70] Vijay V, Siva S, Sreejith KJ, Prabhakaran PV, Devasia R. Effect of boron inclusion in sioc polymer derived matrix on the mechanical and oxidation resistance properties of fiber reinforced composites. Materials Chemistry and Physics. 2018;**205**:269-277

[71] Jüttke Y, Richtera H, Voigta I, Prasadb RM, Bazarjanib MS, Gurlob A, et al. Polymer derived ceramic membranes for gas separation. Chemical Engineer. 2013;**32**:1891-1896

[72] Ionescu E, Riedel R. Polymer processing of ceramics. In: Ceramics and Composites Processing Methods. New Jersey, USA: Wiley-American Ceramic Society; 2012. pp. 235-270

[73] Günthner M, Kraus T, Dierdorf A, Decker D, Krenkel W, Motz G. Advanced coatings on the basis of si (c) n precursors for protection of steel against oxidation. Journal of the European Ceramic Society. 2009;**29**(10): 2061-2068

[74] Günthner M, Schütz A, Glatzel U, Wang K, Bordia RK, Greißl O, et al. High performance environmental barrier coatings, part i: Passive filler loaded sicn

system for steel. Journal of the European Ceramic Society. 2011;**31**(15):3003-3010

[75] Parchoviansky M, Petríková I, Barroso GS, Svancarek P, Galuskova D, Motz G, et al. Corrosion and oxidation behavior of polymer derived ceramic coatings with passive glass fillers on aisi 441 stainless steel. Ceramics Silikáty. 2018;**62**:146-157

[76] Bill J, Heimann D. Polymer-derived ceramic coatings on c/c-sic composites. Journal of the European Ceramic Society. 1996;**16**(10):1115-1120

[77] Devapal D. Studies on inorganic and organometallic polymers [PhD thesis]. Kottayam: Mahatma Gandhi University; 2007

[78] Luan X'g, Chang S, Riedel R, Cheng L. An air stable high temperature adhesive from modified sibcn precursor synthesized via polymer-derived-ceramic route. Ceramics International. 2018;**44**(7):8476-8483

[79] Riedell JA, Easler TE. Ceramic Adhesive and Methods for on-Orbit Repair of re-Entry Vehicles. Technical Report. New York, USA: Joe Pramberger-NASA Tech Briefs; 2013

[80] Grossenbacher J, Gullo MR, Bakumov V, Blugan G, Kuebler J, Brugger J. On the micrometre precise mould filling of liquid polymer derived ceramic precursor for 300-µm-thick high aspect ratio ceramic mems. Ceramics International. 2015;**41**(1):623-629

[81] Zhang L, Wang Y, Wei Y, Weixing X, Fang D, Zhai L, et al. A silicon carbonitride ceramic with anomalously high piezoresistivity. Journal of the American Ceramic Society. 2008;**91**(4):1346-1349

[82] Toma L, Kleebe H-J, Müller MM, Janssen E, Riedel R, Melz T, et al. Correlation between intrinsic microstructure and piezoresistivity in a sioc polymer-derived ceramic. Journal of the American Ceramic Society. 2012; **95**(3):1056-1061

[83] Ryu H-Y, Wang Q, Raj R. Ultrahigh-temperature semiconductors made from polymer-derived ceramics. Journal of the American Ceramic Society. 2010;**93**(6): 1668-1676

[84] Chowdhury MAR, Wang K, Jia Y, Chengying X. Semiconductor-conductor transition of pristine polymer-derived ceramics sic pyrolyzed at temperature range from 1200 c to 1800 c. Journal of the American Ceramic Society. 2020; **103**(4):2630-2642

[85] Nagaiah NR, Kapat JS, An L, Chow L. Novel polymer derived ceramic-high temperature heat flux sensor for gas turbine environment. Journal of Physics: Conference Series. 2006;**34**:458

[86] Ren X, Ebadi S, Chen Y, An L, Gong X. High-temperature characterization of sicn ceramics for wireless passive sensing applications up to 500ºc. In: WAMICON 2011 Conference Proceedings. Florida, USA: IEEE; 2011. pp. 1-5

[87] Colombo P. Engineering porosity in polymer-derived ceramics. Journal of the European Ceramic Society. 2008;**28**(7): 1389-1395

[88] Studart AR, Gonzenbach UT, Tervoort E, Gauckler LJ. Processing routes to macroporous ceramics: A review. Journal of the American Ceramic Society. 2006;**89**(6):1771-1789

[89] Wolff F, Nicolat BC, Fey T, Greil P, Münstedt H. Extrusion foaming of a preceramic silicone resin with a variety

of profiles and morphologies. Advanced Engineering Materials. 2012;**14**(12): 1110-1115

[90] Wang J, Oschatz M, Biemelt T, Lohe MR, Borchardt L, Kaskel S. Preparation of cubic ordered mesoporous silicon carbide monoliths by pressure assisted preceramic polymer nanocasting. Microporous and Mesoporous Materials. 2013;**168**:142-147

[91] Zera E, Perolo A, Campostrini R, Li W, Sorarù GD, et al. Synthesis and characterization of polymer-derived sicn aerogel. Journal of the European Ceramic Society. 2015;**35**(12):3295-3302

Chapter 10

Effect of Calcination on Alumina Ball-Milled Powders toward Lead-Free KNN-Type Ceramics Synthesis

Alina Pruna, Ashley Bonilla, Rut Benavente, Maria D. Salvador-Moya, David Busquets-Mataix and Amparo Borrell

Abstract

The synthesis of lead-free $(Na_{0.5}K_{0.5})NbO_3$ (KNN) ceramics for potential piezoelectric applications is reported by conventional solid-state reaction between alkaline carbonates and Nb_2O_5. Prior to the synthesis, the reactant powders and their corresponding stoichiometric mixture are alumina ball-milled to homogenize the particle size and as pre-activation treatment, respectively. The synthesis of the KNN-based ceramics was investigated systematically with the duration of the ball milling and calcination conditions in terms of mass change evolution at involved temperature steps. The properties of the obtained ceramics including phase structure, morphology, composition, relative density and microhardness were assessed by Field Emission Scanning Electron microscopy, Energy Dispersive Spectroscopy and X-ray diffraction. The obtained results indicate that longer ball milling duration is detrimental to the synthesis of KNN ceramics while tailoring of the KNN properties can be achieved by adjustment of calcination conditions including calcination rate, calcination temperature stage and calcination dwell duration.

Keywords: KNN, piezoelectric, ceramics, alumina ball-milling, calcination

1. Introduction

The piezoelectric materials are widely applied in several fields including piezoelectric actuators, piezoelectric transformers, and piezoelectric energy harvesters [1–3]. The corresponding market has been dominated by lead zirconium titanate (commonly abbreviated with PZT) ceramics, but these contain over 60 wt% lead and so, they are ineluctably toxic. Therefore, lead-free alternatives are sought [1]. Lately, the $(Na_{0.5}K_{0.5})NbO_3$-based system (KNN) has received much attention thanks to its outstanding properties, including considerable piezoelectric coefficient d_{33} of 300–400 pC N^{-1} and high Curie temperature beyond 400°C, besides its lead-free content [4, 5].

Given the high complexity in the composition of KNN ceramics, many challenges are encountered during the synthesis, which would result in poorly reproducible functional properties [6]. The conventional route to synthesize KNN ceramics is based on the solid-state reaction between sodium and potassium carbonates and Nb_2O_5 powders. Despite many advantages, an important drawback of the conventional solid-state reaction consists in the hygroscopic nature of the carbonate precursors that leads to water content and thus, light stoichiometric changes that are able to induce important structural changes or appearance of secondary phase [7]. The non-homogeneous distribution of Nb and varying diffusion rates of Na and K ions could also lead to local heterogeneities, incorrect stoichiometry, and the presence of intermediate and secondary phases.

It has been shown that the ferroelectric and piezoelectric properties of KNN materials can be tailored by proper control of alkaline content and the K/Na ratio and by adjusting the content of intermediate and secondary phases. The optimal piezoelectric output in KNN systems was indicated for a K/Na ratio critical value of 1 [8]. On the other hand, during the synthesis, particle coarsening seldom takes place, which in turn diminishes the powder sinterability and lowers the piezoelectric response. Thus, a densification control is also required to improve the piezoelectric properties of KNN ceramics.

Regarding the synthesis, mechanochemical activation has been proposed as a suitable pre-treatment approach. Mechanochemical reactions are promoted via energy transfer between milling bodies and the powder being milled. Thus, the reaction between the powder precursors can be activated during the ball milling process given that the corresponding energetic barrier is reached by the cumulative kinetic energy provided by the high-energy employed in the milling [19]. The mechanochemical activation has important advantages, namely, it decreases the calcination temperature and overall synthesis duration, and it allows the control over the volatilization of alkaline species by reducing it. Thus, it enables the synthesis of KNN ceramics with enhanced chemical homogeneity and improved crystal structure [9, 10]. The common ball milling conditions refer to ball milling of precursor mixture with zirconia bodies for as long as 24 h, on/off time as 2/8 minutes. Following the ball milling mechanochemical activation of the precursor mixture, calcination is applied. Common calcination temperatures refer to 900°C and dwelling for as long as 5 h. Most of the works report a synthesis approach based on ball milling of the mixture followed by calcination and sintering or a second milling step after the calcination and another calcination step or varying sintering steps [11].

During the sintering, densification takes place together with grain growth and coalescence. In a first stage, a liquid phase is formed due to excess Nb_2O_5 and it helps to initiate the sintering when necking appears together with particle downsizing, thus inducing a light densification [12, 13]. Further, the densification continues as the grains get interconnected and the pores reduce their section area. In a last stage, the pores diminish greatly up to their complete disappearance.

The sintering temperature in conventional method was reported to be between 1000 and 1150°C. While a higher sintering temperature and dwelling were indicated to be useful in diminishing the intermediate phase and the residual porosity, an excess temperature or dwelling would result in grain growth and evaporation of the alkaline elements due to their low fusion points (892 and 760°C for Na and K), thus, in composition variation [14–17].

The maximum densification in KNN structure of about 97% was observed at 1100°C. By increasing the sintering dwelling, the improvement of densification required for the enhancement of electrical properties of KNN ceramics can be achieved [13, 18]. An optimal dwelling at maximum sintering temperature was reported as 2 h [19]. Nevertheless, the dwell time and heat rate must be adjusted to completely remove the intermediate phase and impede the evaporation of alkaline elements by a prolonged dwelling.

Despite many reports on similar processing conditions, the significantly different properties of the obtained KNN ceramics show that more studies are required to define proper fabrication protocols that allow enhanced reproducibility of the final KNN properties. The mechanical energy supply during grinding and thermal energy delivery during calcination and densification have been indicated as key parameters to focus on due to the direct relation between particle sizes and powders' reactivity [20]. In this work, a novel fabrication approach is proposed for KNN systems to be applied in the conventional solid-state route, namely by mechanochemical activation of downsized KNN precursors powders, and adjusting the calcination conditions. Both the downsizing of precursors and the mechanochemical activation of their mixture have been achieved by ball milling with alumina bodies as the cheaper alternative. The obtained results indicate the proposed synthesis approach is a valuable tool in the fabrication of reliable KNN ceramics.

2. Materials and methods

2.1 Materials

Commercial niobium (V) oxide (Nb_2O_5, 325 mesh, <44um), K_2CO_3, and Na_2CO_3 (anhydrous) were purchased from Sigma-Aldrich (UK) Co., Ltd.

2.2 Synthesis

The precursors were homogenized individually by ethanol wet milling in a planetary ball mill in a zirconia jar in a Fritsch Pulverisette 7 machine at 400 rpm for eight cycles of 10 minutes on time and 5 minutes idle time. The ball material employed in the attrition milling was Al_2O_3 of 5 mm diameter. The milling energy was adjusted by employing a ball-to-powder ratio (BPR) of 10:1. The homogenized powders were dried at 100°C for 24 hours and further mixed according to the stoichiometric formula, namely in the mass fractions 68.5247, 13.6604 and 17.8150% for Nb_2O_5, Na_2CO_3, and K_2CO_3, respectively.

The lead-free systems were prepared using the conventional solid-state reaction method reaction in three steps, namely, mixing, calcination and sintering. First, the KNN mixtures were mixed by employing the same conditions of ball milling as for the individual precursors, namely by employing Al_2O_3 balls of 5 mm diameter in 10:1 BPR. To assess the effect of precursor mixing duration on the reactivity and formation of KNN ceramics, the ball milling duration was varied from 2 to 3 h. After the mixing, the powder mixtures were dried at 100°C for 24 hours and sieved.

The sieved mixtures were subjected to calcination in the air by using a Carbolyte HTF1800 furnace. To assess the effect of calcination conditions, the calcination rate, temperature and dwelling were varied. The calcination conditions that were employed are 700°C for 2 h, with a rate of 3°C min^{-1} (method CA); 700°C for 5 h, with a rate of 10°C min^{-1} (method CB); 800°C for 2 h, with a rate of 10°C min^{-1} (method CC). Method CA and CB take up similar processing time and were applied with the purpose of lowering the calcination temperature, by proper adjustment of calcination rate, dwelling and mixing. The method CC was employed to confirm the requirement of higher calcination temperature, as a function of mixing duration for the powders. After calcination, the mixtures were left to cool naturally overnight, sieved again and compacted uniaxially under 15 MPa pressure for 2 minutes.

The sintering step was further performed by temperature increase up to 1100°C with a heat rate of 10°C min^{-1} and 2 h dwelling followed by furnace-cooling to room temperature overnight.

2.3 Characterization

The particle size of the precursors and mixtures was obtained by light scattering in water as dispersing liquid with a Mastersizer Hydro 2000, Malvern Instruments analyzer. The volumetric particle size distribution was averaged from three measurements of 12 s each at 2000 rpm pumping.

The mass changes upon heating and the temperatures where they take place were observed on both precursors and mixtures with a thermogravimetric analyzer TGA Q50, TA Instruments in air atmosphere to simulate the calcination environment. The heat rate employed was 10°Cmin^{-1}.

The fracture surface of the sintered samples has been studied using a Field Emission Gun Scanning Electron Microscope (FE-SEM, GEMINI ULTRA 55 MODEL, ZEISS). Elemental analysis was performed on the same microscope equipped with an Energy Dispersive Spectroscopy analyzer.

An X-ray diffractometer (D2 Phaser, Bruker, Karlsruhe, Germany) was used to determine the room temperature phase structure. A Bruker D2 Phaser diffractometer using radiation of Cu-kα with a wavelength λ = 1.54 Å was employed. The XRD peak at about 45° was deconvoluted using Voigt function and the coexistence of orthorhombic and tetragonal phases was evaluated.

The bulk density of the sintered samples was measured using Archimedes' principle by employing distilled water as immersing liquid. At least five measurements were averaged for each sample. The relative density was calculated by dividing the bulk density by the theoretical density of the powder mixture (4.506 g cm^{-3}) [21].

Vickers microhardness assessments were carried out by indentation with a microhardness tester (400 A Innovatest, Maastricht, The Netherlands)) employing a conventional diamond pyramid indenter and a 0.1 kg for 10 seconds. The diagonals of each indentation were measured using an optical microscope. The value of HV is the relationship between the applied load and the surface area of the diagonals of indentation. At least five measurements were obtained per sample and further averaged. Prior to microhardness measurement, the samples were embedded in epoxy resin (TransOptic, Buehler) by using a pressing unit (LaboPress-3, Struers, Ballerup, Denmark). A compression force of 15 kN, heating at 180°C for 7 minutes and cooling for 7 minutes were the parameters set for the embedding process. Polishing with decreasing particle size up to 1 μm was achieved by employing a Struers, model Roto-Pol-31 equipment.

3. Results and discussion

The homogenized precursors employed in the synthesis of KNN ceramics were analyzed first in terms of particle size. All precursors have been downsized with respect to the as-received precursors of about 95%. **Figure 1** presents the differential size distribution for the three reactants upon homogenization. As it can be observed,

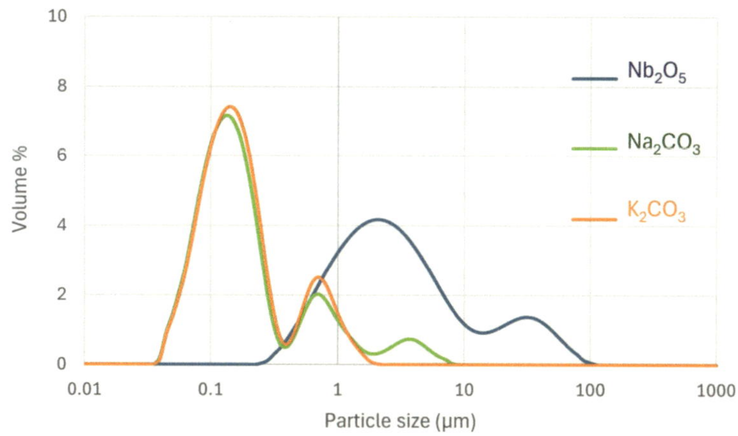

Figure 1.
Particle size distribution for the KNN precursors.

all reactants exhibit a bimodal distribution, except for Na_2CO_3 that presents a third one, which is attributed to agglomerates due to its hygroscopic nature.

The evolution of percentile size values below 10, 50 and 90% of sample content were retrieved according to the cumulative particle size distribution. The results obtained for the individual precursors upon ball milling homogenization are shown in **Table 1**. As it can be observed, the carbonates show similar small size distribution, with submicron average size, while the Nb_2O_5 average size particles have about 3 microns.

The individual precursors were further weighed in the corresponding stoichiometric values and mixed by ball milling. The evolution of percentile size values for the mixtures obtained with varying milling duration is depicted in **Table 2**. As it can be observed, the mixtures show intermediate values with respect to the individual precursors. The median particle size of the mixture is lower after the ball milling mixing, which is indicative of further downsizing of the reactants in the mixing step. On the other hand, it is observed that prolonged mixing by ball milling results in increasing size distribution across the percentiles, which could be due to limited downsizing followed by agglomeration.

The next step in the synthesis of the KNN ceramics is represented by the calcination of precursor mixture. To observe the effect of the milling duration for the mixing of the reactants on the behavior to solid-state reaction in the calcination step, the percentile size values were obtained. For exemplification, the results for the calcination at 700°C for 2 h, with a rate of 3°Cmin^{-1} (method CA) are depicted in **Table 3.**

It was observed that the particle size values increased with respect to the mixtures prior to calcination, which is attributed to the decomposition and coalescence of the reactant particles during the solid-state reaction in the calcination step. Moreover,

Precursor	d10 (µm)	d50 (µm)	d90 (µm)
Nb_2O_5	0.755	3.017	24.199
Na_2CO_3	0.072	0.169	1.898
K_2CO_3	0.071	0.149	0.666

Table 1.
Cumulative size distribution for the homogenized precursors.

Mixing duration	d10 (μm)	d50 (μm)	d90 (μm)
2 h	0.395	1.448	5.131
3 h	0.428	1.776	17.050

Table 2.
Cumulative size distribution for the mixtures obtained with different mixing duration.

Mixing duration	d10 (μm)	d50 (μm)	d90 (μm)
2 h	0.985	2.848	34.823
3 h	0.973	3.478	40.339
PF 2 h	2.494	1.967	6.787
PF 3 h	2.273	1.958	2.366

Table 3.
Effect of mixing duration on the particle size upon calcination with CA method.

the mixture milled for 3 h resulted in particle size values greater than the one milled for 2 h, which might be explained by the larger particle size of the mixture prior to calcination. The solid-state reaction was further monitored by employing a parameter described as particle size increase factor (PF) obtained by dividing the percentile values obtained upon calcination by those prior to calcination as shown in **Table 3**. The obtained values show similar PF for up to 50% of the sample, irrespective of the mixing duration. On the other hand, the PF for the particles up to 90% of the sample is three times higher for the mixture obtained for 2 h mixing, which can be attributed to improved solid-state reaction due to the smaller particle size of the mixture subjected to calcination.

To get an insight into the effect of the calcination conditions on the solid-state reaction, the percentile values were obtained and analyzed by the PF with respect to the particle size values corresponding to the mixture prior to calcination (as depicted in **Table 2**). For exemplification, the evolution of particle size values for the mixture obtained by 2 h mixing is shown in **Table 4**.

As it can be observed, the particle size values increase with calcination conditions in the order CA < CC < CB. The particle size increase factor (PF) for each percentile exhibits the same trend, that is PF CA < PF CC < PF CB. The obtained particle size and PF values indicate that for the same calcination temperature value, as low as

Calcination condition	d10 (μm)	d50 (μm)	d90 (μm)
CA	0.985	2.848	34.823
CB	1.212	4.56	69.617
CC	1.093	3.815	42.656
PF CA	2.494	1.967	6.787
PF CB	3.068	3.149	13.568
PC CC	2.767	2.635	8.313

Table 4.
Effect of calcination conditions on the particle size for powders obtained by 2 h mixing.

700°C, a longer dwelling induces improved reaction despite a higher calcination rate (calcination condition CB) with respect to a shorter dwelling and slower calcination rate (calcination condition CA). The calcination CB also appears to induce improved coalescence with respect to condition CC, despite the lower temperature (700 vs. 800°C), which is explained by the increased dwelling.

Following the calcination step, the mixtures were sieved, uniaxially compacted and subjected to sintering step in the same conditions, namely 1100°C for 2 h and 10°C min^{-1} rate. The solid-state reaction was further monitored by recording the weight loss (%) upon individual synthesis steps, that is calcination step and sintering one, by reporting the final weight to the initial weight of the mixture subjected to each step, as well as total weight loss obtained by summing the individual ones. **Table 5** depicts the evolution of weight losses with the calcination conditions and milling duration of the reactant mixing.

It is observed that the weight loss values are higher for the mixture obtained by 3 h mixing than 2 h mixing, indicating that the particles undergo more decomposition and evaporation and less formation reaction, due to less contact surface of the larger particles. After the calcination step, the weight loss is observed to be the lowest for the CA calcination condition, attributed to the lower calcination temperature. The weight loss is similar for the CB and CC calcination conditions, and higher than CA condition.

The weight loss after the sintering step has an opposite trend, that is it decreases from condition CA to CB=CC, which is attributed to a larger extent of decomposition due to the high temperature applied in the sintering step. As the CB and CC mixtures already initiated the formation reaction in the calcination step, they lose less weight upon sintering one, where the formation reaction is completed.

The total weight loss recorded after the two steps (calcination and sintering) decreased in the order CA > CB > CC, indicating a higher yield of KNN formation in calcination conditions CB and CC and that proper control of the three parameters such as calcination rate, temperature and dwelling time is needed to adjust the synthesis of KNN ceramics, besides the control of reactant particle size.

The solid-state reaction synthesis of KNN ceramics was further analyzed using thermogravimetric measurements (TGA). The weight changes of the individual precursors indicated that Nb_2O_5 is stable over the whole temperature range while Na_2CO_3 and K_2CO_3 dehydrate around 100°C and start to decompose from 625°C, with temperature peaks at 820°C Na_2CO_3 and 830°C for K_2CO_3, in agreement with the melting points [22]. The thermograms obtained for the mixtures indicate varying behavior as a function of the ball milling mixing duration, as shown in **Figure 2**. The weight losses recorded below 100°C are attributed to water evaporation. The water loss from 2 h mixture amounts 1.9% while the 3 h one reaches double percentage. Another weight loss is recorded at about 280°C and it is assigned to water loss from hydrated

Calcination condition	Weight loss % after calcination		Weight loss % after sintering		Total weight loss %	
	2 h	3 h	2 h	3 h	2 h	3 h
CA	11.94	13.34	17.03	17.65	28.98	30.99
CB	15.30	15.91	11.43	11.88	26.73	27.79
CC	15.34	15.93	10.67	11.55	26.01	27.48

Table 5.
Effect of calcination and precursor mixing on the weight loss in the synthesis steps.

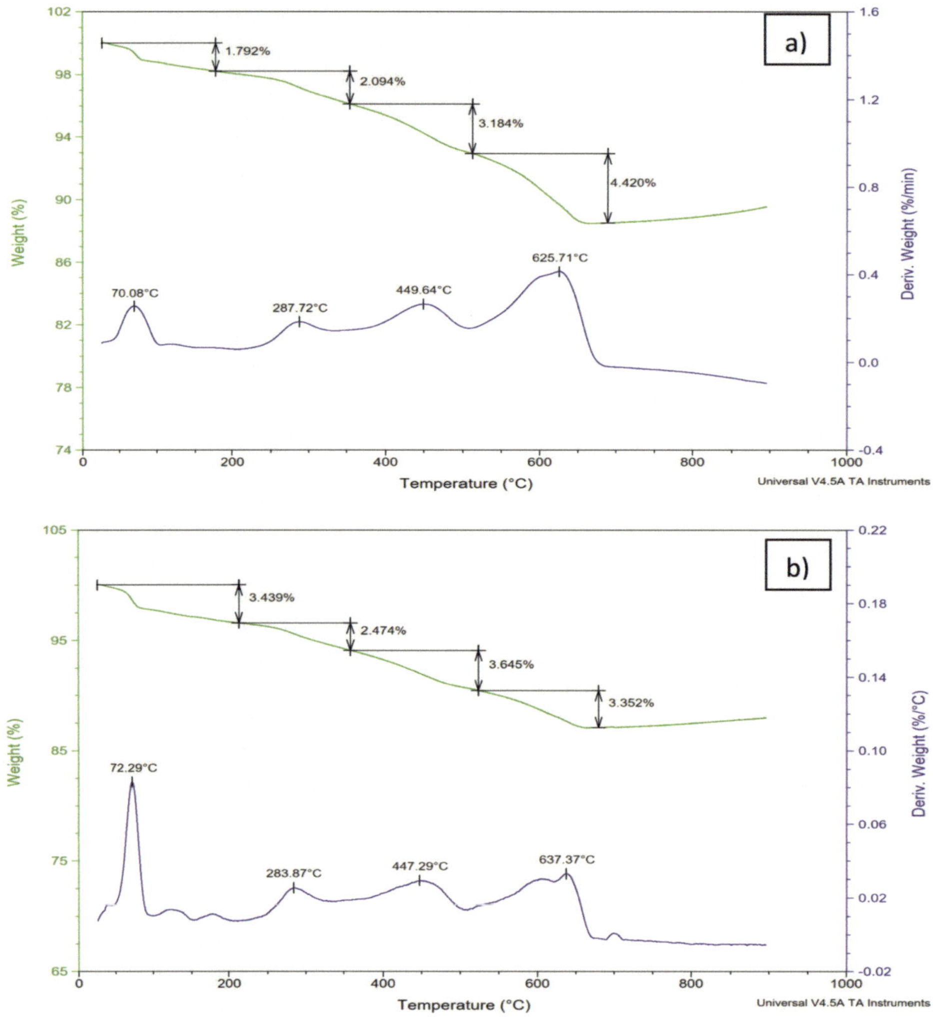

Figure 2.
Thermograms (TG, green) and Differential Thermograms (DTG, blue) of the mixture obtained by 2 h mixing (a) and 3 h mixing (b).

carbonates together with CO_2 release [7]. This weight loss increased from 2% for the 2 h mixture to 2.5% for the 3 h one. The formation of an intermediate phase $(K,Na)_2Nb_4O_{11}$ at the Nb_2O_5 surface upon decomposition of the carbonates together with the initiation of the formation of stoichiometric $(K_{0.50}Na_{0.50})NbO_3$ at the surface of the intermediate phase [23] is indicated by the wide peak centered at 447°C. The associated weight loss is about 3.2 and 3.6% for the 2 h and 3 h mixture, respectively. Another peak at about 630°C is assigned to the completion of the formation of KNN and diffusion of alkali elements toward the unreacted Nb_2O_5 [23].

While all temperature peaks were located at similar values, irrespective of the mixing duration, the weight losses (%) were higher for the mixture obtained for 3 h than for 2 h, indicating higher particle experience evaporation due to their larger size. Considering the weight loss observed up to 700°C, the mixture obtained for 2 h reached a value of 11.46%, which is closer to the theoretical weight loss for the KNN precursors, that is, 11.34% [23], than the one recorded for the mixture obtained for

3 h, namely 12.84%. Moreover, the smaller weight change at the temperature assigned to the KNN formation completion, that is, 630°C, for the mixture obtained for 3 h is indicative of the lower contact surface between the particles which affects the balance between Nb_2O_5 and carbonate particles and thus KNN formation completion.

The morphology of the sintered KNN was investigated by FESEM measurements and the typical morphologies are depicted in **Figure 3** as a function of precursor mixing duration and calcination conditions. It is observed that the grain shape changed toward typical cube-like one [24, 25] by varying the calcination conditions in the order CA < CB < CC for both cases of mixing durations.

In general, the size of the grains was larger for the mixture obtained by 3 h mixing. The low calcination rate in CA condition results in decomposition and evaporation of material and in combination with a low contact surface due to larger particle size as the 3 h mixture exhibits, some of the calcined grains appear smaller than in the case of 2 h calcined mixture. By a faster calcination rate as in CB condition, although the large dwell allows for the formation reaction between the smaller particles, the bigger ones do not have enough time to decompose with the same degree; therefore, the grains exhibited by 3 h calcined mixture appear larger than the 2 h one. Finally, the

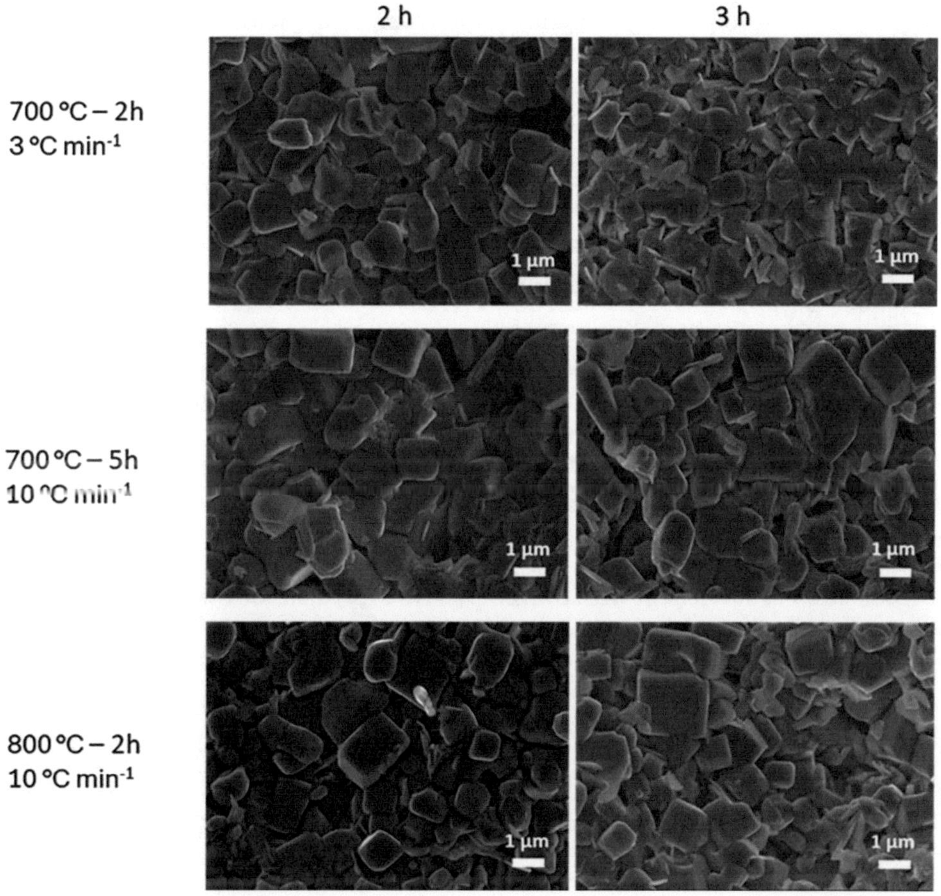

Figure 3.
FESEM images for sintered KNN as a function of precursor mixing duration (2 or 3 h) and calcination conditions.

combination of the higher calcination rate and temperature in CC condition allows for a more homogeneous grain size distribution within the sample, while the 3 h mixture still presents larger grains. The grain size ranged between 1 and 2 μm, and it agrees with other reports [26].

The elemental analysis by energy dispersive spectroscopy (EDS) was performed to obtain further information regarding the solid-state reaction and formation of KNN ceramics. K ions have a lower diffusion rate; thus, it is highly affecting the final properties of KNN ceramics. For this reason, the total alkaline content, $A/(A + \text{Nb})$ and K/A (where A is the alkaline content as K + Na), were investigated, as they reflect the formation completion for KNN structure due to particle size induced reactivity. **Table 6** presents the evolution of such indicators with the mixing duration and calcination conditions. The theoretical values for the KNN and the intermediate phase, $K_2Na_2Nb_4O_{11}$ for $A/(A + \text{Nb})$, are 0.50 and 0.33, respectively, while the value of K/A is 0.50 for both.

As it can be observed in **Table 6**, the values for the $A/(A + \text{Nb})$ and K/A are closer to the theoretical values corresponding to the intermediate phase $(K,Na)_2Nb_4O_{11}$. The $A/(A + \text{Nb})$ increases in the order of calcination conditions CA < CB = CC, while the K/A decreases in the same order, for both mixture cases, toward the theoretical values obtained for perovskite KNN phase. The evolution of these two parameters is indicative of the increasing formation of KNN structure with proper calcination conditions achieved in CB and CC for the 2 h mixture and in CC for the 3 h mixture. Nevertheless, the two indicators are closer to the theoretical KNN phase for the 2 h mixture than the 3 h one. The calcination conditions CB and CC appear to result in similar A and K contents for the 2 h mixture, indicating that the smaller particle size allow more reactivity and KNN formation due to a larger contact surface and it can be achieved also at lower calcination temperature with further adjustment of dwelling time. On the other hand, for the larger particles in the 3 h mixture, only calcination condition CC appears mostly favorable for KNN formation due to the higher calcination temperature.

The crystal structure of KNN ceramics was further analyzed using XRD measurements. The calcined mixtures were first investigated with respect to the individual precursors. Knowing that high calcination temperature induces good KNN formation reaction, to observe the effect of the precursor mixing duration in this aspect, the calcination conditions such as the lowest calcination temperature and dwelling time were considered as relevant. Thus, **Figure 4** depicts the XRD spectra of the mixtures calcined in condition CA (700°C, 2 h dwelling, 3°C min^{-1} rate) vs. precursors. The precursors show the typical peaks [26]. The temperature of 700°C was selected for the calcination in agreement with the temperature revealed for the synthesis by TGA results (above 650°C). Indeed, both spectra of the mixtures depict the typical perovskite peaks and low content of the intermediate phase (intermediate phase

EDS analysis		2 h			3 h	
	CA	CB	CC	CA	CB	CC
A/(A + Nb)	0.45	0.48	0.48	0.46	0.48	0.50
K/A	0.56	0.51	0.51	0.57	0.54	0.52

Table 6.
Evolution of alkaline elements and K from total alkaline content with the precursor mixing duration and calcination condition.

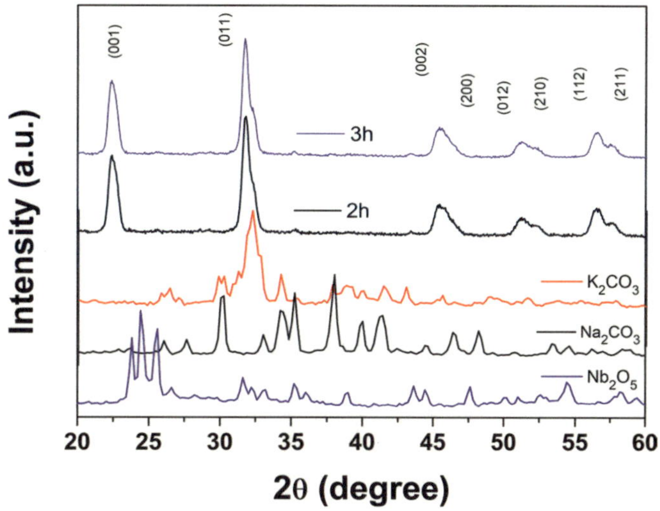

Figure 4.
XRD spectra for the mixtures calcined in condition CA.

$(K,Na)_2Nb_4O_{11}$ prior to the formation of KNN). Both spectra present a sharp (110) plane at about 32° of maximum intensity, indicative of the perovskite formation. The coexistence of both orthorhombic and tetragonal phases with dominancy from orthorhombic one is shown by the separation in two of the peaks [27].

The peak at 45° corresponding to the orthorhombic and tetragonal coexistence [28] was found to increase in intensity and show a marked split upon sintering as shown in **Figure 5a** and **b**, for both mixtures, obtained by 2 and 3 h, respectively. The smaller full-width half maximum of the peak located at 32° is indicative of the reaction formation where the particles fuse and result in larger structure. The insets in **Figure 5** depict a different behavior during the synthesis depending on the precursor mixing duration. On one hand, the sintered 2 h mixture shows a decrease in the angle f with the calcination condition or the 45° peak, namely in the order CC < CB < CA. This aspect indicates changes in the composition, and they are attributed to improved reactivity and diffusion of alkaline elements toward the formation of pure perovskite KNN structure, in agreement with the EDS results in **Table 6** for 2 h mixture. On the other hand, the sintered 3 h mixture exhibits an opposite behavior, that is, the peak at 45° shifts toward increasing angle CA < CB < CC. The peak intensity increases for both mixtures from CA to CB and CC calcination conditions.

The peak at 45° is further depicted in a magnified image in **Figure 5c**, to show the difference between the mixtures at the highest calcination temperature. It is observed that the angle corresponding to the 2 h mixture is lower than the 3 h one, attributed to improved K^+ diffusion (larger cation radius than Na^+) [29] in the KNN structure thanks to the larger contact surface conferred by smaller particles in the 2 h mixture.

The density measurements indicated the densification of the KNN ceramics depended greatly on the calcination conditions as well as on the reactant mixing duration, as shown in **Table 7**. The relative density ranged between 92.10 and 96.31%, in agreement with other reports. The relative density values of the KNN ceramics were higher for the 2 h mixture, which can be explained by the smaller average particle size of reactant mixture subjected to solid-state reaction as induced by proper precursor mixing duration by ball milling. On the other hand, the relative density increased in

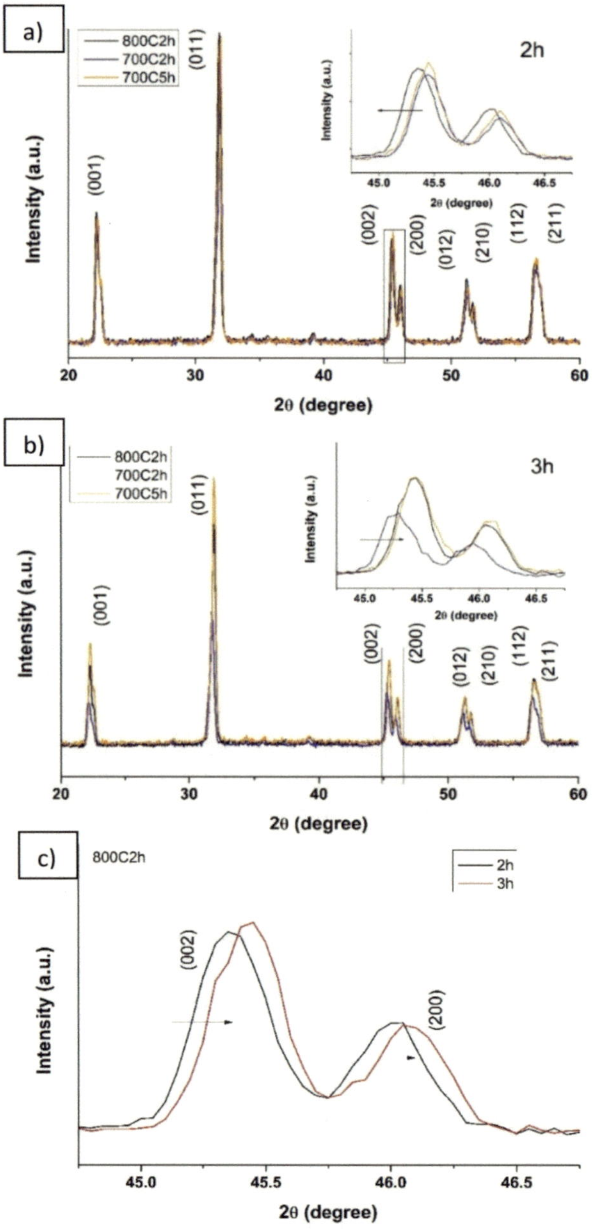

Figure 5.
XRD spectra for the sintered mixture 2 h (a), 3 h (b) with the calcination conditions and magnified XRD spectra in condition CC (c).

the order CA < CB < CC of calcination conditions, indicating that a higher calcination temperature is most determining for the densification, most probably due to the reducing porosity between the grains.

Despite the interest raised by KNN ceramics, there are only a few reports on their mechanical properties [30–32]. Here, the mechanical properties of the KNN ceramics were evaluated by microhardness measurements. The HV0.1 evolution with the average particle size induced by reactant mixing by ball milling and with calcination

Calcination condition/mixing duration	Relative density %	
	2 h	3 h
CA	92.10	91.53
CB	95.36	94.98
CC	96.31	95.70

Table 7.
Evolution of relative density for sintered KNN with calcination and precursor mixing.

Calcination condition/mixing duration	HV0.1	
	2 h	3 h
CA	157	129
CB	161	136
CC	166	133

Table 8.
Microhardness of KNN ceramics as a function of calcination conditions and mixing duration.

conditions is presented in **Table 8**. The KNN ceramics obtained by 2 h mixing recorded higher hardness values than their counterparts obtained by 3 h mixing, in agreement with the density evolution in **Table 7**. On the other hand, the obtained values show a different evolution of microhardness with the calcination condition if the mixtures were obtained for varying mixing time, that is, for the case of smaller particles in 2 h mixture, the best conditions to improve the microhardness appear as CC, where higher calcination temperature is combined with a faster heating rate and 2 h dwell. In the case of larger particles in 3 h mixture, the similar values obtained for calcinations CB and CC indicate tailoring could be achieved by considering the proper combination of dwelling, calcination temperature and heating rate, where dwelling exhibits a marked influence.

4. Conclusion

A novel approach based on mechanochemical activation of the mixture of downsized precursor powders is proposed for the conventional solid-state reaction synthesis method of KNN ceramics. The mechanochemical activation and downsizing of the precursors have been performed by ball milling with alumina bodies. The approach was tested in varying conditions of calcination, including temperature, heating rate and dwelling. The potential of the proposed approach for the control of the powder reactivity toward the suitable alkaline content, K/Na ratio, presence of secondary phase and grain homogeneity in the final KNN ceramic was demonstrated by the decreased reaction temperature in agreement with TGA results and particle size, as well as XRD ones. The results indicate that the application of ball milling with alumina bodies for 2 h achieves small enough particles to enable proper reaction by calcination at a temperature as low as 800° where the heat rate is $10°min^{-1}$ and the dwelling is 2 h and achieve the best K content. Such results could be exploited in the fabrication of KNN with suitable piezoelectric properties for a large range of applications.

Acknowledgements

This publication is part of the grant PID2021-128548OB-C21 and CNS2023-144190 "NextGenerationEU"/PRTR funded by MCIN/AEI/10.13039/501100011033, and CIGRIS/2022/077 project funded by Generalitat Valenciana.

Conflict of interest

The authors declare no conflict of interest.

Author details

Alina Pruna*, Ashley Bonilla, Rut Benavente, Maria D. Salvador-Moya, David Busquets-Mataix and Amparo Borrell
Institute for Materials Technology (IUTM), Universitat Politècnica de València, Valencia, Spain

*Address all correspondence to: apruna@itm.upv.es

References

[1] Panda PK, Sahoo B. PZT to lead free piezo ceramics: A review. Ferroelectrics. 2015;**474**(1):128-143. Available from: https://www.tandfonline.com/doi/abs/10.1080/00150193.2015.997146

[2] Shao T, Du H, Ma H, Qu S, Wang J, Wang J, et al. Potassium–sodium niobate based lead-free ceramics: Novel electrical energy storage materials. Journal of Materials Chemistry A. 2017;**5**(2):554-563. Available from: https://pubs.rsc.org/en/content/articlehtml/2017/ta/c6ta07803f

[3] Chauhan A, Vaish R, Bowen C. Piezoelectric material selection for ultrasonic transducer and actuator applications. Proceedings of the Institution of Mechanical Engineers, Part L: Journal of Materials: Design and Applications. 2013;**229**(1):3-12. Available from: https://journals.sagepub.com/doi/10.1177/1464420713493591

[4] Saito Y, Takao H, Tani T, Nonoyama T, Takatori K, Homma T, et al. Lead-free piezoceramics. Nature. 2004;**432**(7013):84-87. Available from: https://pubmed.ncbi.nlm.nih.gov/15516921/

[5] Wang K, Li JF. Domain engineering of lead-free Li-modified (K,Na)NbO3 polycrystals with highly enhanced piezoelectricity. Advanced Functional Materials. 2010;**20**(12):1924-1929. Available from: https://onlinelibrary.wiley.com/doi/full/10.1002/adfm.201000284

[6] Li JF, Wang K, Zhu FY, Cheng LQ, Yao FZ. (K, Na)NbO3-based lead-free piezoceramics: Fundamental aspects, processing technologies, and remaining challenges. Journal of the American Ceramic Society. 2013;**96**(12):3677-3696. Available from: https://onlinelibrary.wiley.com/doi/full/10.1111/jace.12715

[7] Rubio-Marcos F, Romero JJ, Martín-Gonzalez MS, Fernández JF. Effect of stoichiometry and milling processes in the synthesis and the piezoelectric properties of modified KNN nanoparticles by solid state reaction. Journal of the European Ceramic Society. 2010;**30**(13):2763-2771

[8] Zhang BP, Li JF, Wang K, Zhang H. Compositional dependence of piezoelectric properties in NaxK1–xNbO3 lead-free ceramics prepared by spark plasma sintering. Journal of the American Ceramic Society. 2006;**89**(5):1605-1609. Available from: https://onlinelibrary.wiley.com/doi/full/10.1111/j.1551-2916.2006.00960.x

[9] Singh R, Patro PK, Kulkarni AR, Harendranath CS. Synthesis of nano-crystalline potassium sodium niobate ceramic using mechanochemical activation. Ceramics International. 2014;**40**(7):10641-10647

[10] Jiten C, Rawat M, Bhattacharya A, Singh KC. (Na0.5K0.5)NbO3 nanocrystalline powders produced by high energy ball milling and corresponding ceramics. Materials Research Bulletin. 2017;**90**:162-169

[11] Zhang Y, Zhai J, Xue S. Effect of three step sintering on piezoelectric properties of KNN-based lead free ceramics. Chemical Physics Letters. 2020;**758**:137906

[12] Matsubara M, Yamaguchi T, Sakamoto W, Kikuta K, Yogo T, Hirano SI. Processing and piezoelectric properties of lead-free (K,Na) (Nb,Ta) O3 ceramics.

Journal of the American Ceramic Society. 2005;**88**(5):1190-1196. Available from: https://onlinelibrary.wiley.com/doi/full/10.1111/j.1551-2916.2005.00229.x

[13] Rubio-Marcos F, Ochoa P, Fernandez JF. Sintering and properties of lead-free (K,Na,Li)(Nb,Ta,Sb)O3 ceramics. Journal of the European Ceramic Society. 2007;**27**(13-15):4125-4129

[14] Zhen Y, Li JF. Normal sintering of (K,Na)NbO3-based ceramics: Influence of sintering temperature on densification, microstructure, and electrical properties. Journal of the American Ceramic Society. 2006;**89**(12):3669-3675. Available from: https://onlinelibrary.wiley.com/doi/full/10.1111/j.1551-2916.2006.01313.x

[15] Ahn CW, Park CS, Choi CH, Nahm S, Yoo MJ, Lee HG, et al. Sintering behavior of lead-free (K,Na)NbO3-basedpiezoelectric ceramics. Journal of the American Ceramic Society. 2009;**92**(9):2033-2038. Available from: https://onlinelibrary.wiley.com/doi/full/10.1111/j.1551-2916.2009.03167.x

[16] Wang K, Li JF. (K, Na)NbO3-based lead-free piezoceramics: Phase transition, sintering and property enhancement. Journal of Advanced Ceramics. 2012;**1**(1):24-37. Available from: https://link.springer.com/article/10.1007/s40145-012-0003-3

[17] Vats G, Vaish R. Selection of optimal sintering temperature of K0.5Na0.5NbO3 ceramics for electromechanical applications. Journal of Asian Ceramic Societies. 2014;**2**(1):5-10

[18] Wu J, Xiao D, Wu W, Zhu J, Wang J. Effect of dwell time during sintering on piezoelectric properties of (Ba0.85Ca0.15)(Ti0.90Zr0.10)O3 lead-free ceramics. Journal of Alloys and Compounds. 2011;**509**(41):L359-L361

[19] Kambale KR, Shroff S, Butee SP, Singh R, Kulkarni AR. Effect of addition of V2O5 on the densification, dielectric and ferroelectric behavior of lead free potassium sodium niobate ceramics. Ferroelectrics. 2017;**518**(1):94-102. Available from: https://www.tandfonline.com/doi/abs/10.1080/00150193.2017.1360600

[20] Beltrami R, Mercadelli E, Baldisserri C, Galassi C, Braghin F, Lecis N. Synthesis of KNN powders: Scaling effect of the milling step. Powder Technology. 2020;**375**:101-108

[21] Maeda T, Takiguchi N, Ishikawa M, Hemsel T, Morita T. (K,Na)NbO3 lead-free piezoelectric ceramics synthesized from hydrothermal powders. Materials Letters. 2010;**64**(2):125-128

[22] Lide DR, editor. CRC Handbook of Chemistry and Physics. 84th ed (National Institute of Standards and Technology). Boca Raton: CRC Press LLC; 2003. 2616 pp. $139.95. ISBN 0-8493-0484-9. J Am Chem Soc. 2003;126(5):1586-1586. Available from: https://pubs.acs.org/doi/full/10.1021/ja0336372

[23] Feizpour M, Ebadzadeh T, Jenko D. Synthesis and characterization of lead-free piezoelectric (K0.50Na0.50)NbO3 powder produced at lower calcination temperatures: A comparative study with a calcination temperature of 850°C. Journal of the European Ceramic Society. 2016;**36**(7):1595-1603

[24] Matsubara M, Yamaguchi T, Kikuta K, Hirano SI. Sintering and piezoelectric properties of potassium sodium niobate ceramics with newly developed sintering aid. Japanese Journal of Applied Physics, Part 1: Regular

Papers and Short Notes and Review Papers. 2005;**44**(1A):258-263. Available from: https://iopscience.iop.org/article/10.1143/JJAP.44.258

[25] Yang Z, Chang Y, Liu B, Wei L. Effects of composition on phase structure, microstructure and electrical properties of (K0.5Na0.5)NbO3–LiSbO3 ceramics. Materials Science and Engineering A. 2006;**432**(1-2):292-298

[26] Malic B, Jenko D, Bernard J, Cilensek J, Kosec M. Synthesis and sintering of (K,Na)NbO3 based ceramics. Materials Research Society Symposium—Proceedings. 2003;**755**(1):83-88. Available from: https://link.springer.com/article/10.1557/PROC-755-DD4.4

[27] Yang H, Lin Y, Zhu J, Wang F. An efficient approach for direct synthesis of K0.5Na0.5NbO3 powders. Powder Technology. 2009;**196**(2):233-236

[28] Vendrell X, García JE, Bril X, Ochoa DA, Mestres L, Dezanneau G. Improving the functional properties of (K0.5Na0.5)NbO3 piezoceramics by acceptor doping. Journal of the European Ceramic Society. 2015;**35**(1):125-130

[29] Cheng LQ, Wang K, Yu Q, Li JF. Structure and composition characterization of lead-free (K, Na) NbO3 piezoelectric nanorods synthesized by the molten-salt reaction. Journal of Materials Chemistry C. 2014;**2**(8):1519-1524. Available from: https://pubs.rsc.org/en/content/articlehtml/2014/tc/c3tc32148g

[30] Politova ED, Kaleva GM, Mosunov AV, Sadovskaya NV, Kiselev DA, Ilina TS, et al. Microstructure and properties of lead-free perovskite ceramics on the base of KNN perovskite. Diffusion Foundations. 2020;**27**:90-98

[31] Zhang H, Yang S, Yang S, Kong D, Zhang BP, Zhang Y. Reliability enhancement in nickel-particle-dispersed alkaline niobate piezoelectric composites and actuators. Journal of the European Ceramic Society. 2011;**31**(5):795-800

[32] Zhang HJ, Li JX, Chu WY, Su YJ, Qiao LJ. Effect of humidity and hydrogen on the promotion of indentation crack growth in lead-free ferroelectric ceramics. Materials Science and Engineering B. 2010;**167**(3):147-152

Chapter 11

Technological Advances in Ceramic Membranes for Water Treatment

Fabiola Colmenero Fonseca, Amparo Borrell, María Dolores Salvador Moya, Rut Benavente and Juan Francisco Palomino Bernal

Abstract

A study on the technological innovations in the development of ceramic membranes for water treatment is provided. These membranes have proven to be effective in removing contaminants in the water, such as heavy metals and pathogens. It discusses the different materials based on oxides such as Al_2O_3, TiO_2, SiO_2, and ZrO_2, used in manufacturing, as well as fabrication techniques and surface modifications that have improved their properties. In addition, the environmental and economic impact on water treatment plants is addressed. From a financial point of view, implementing membranes can be an investment, both in acquiring the membranes and in the infrastructure. Membrane maintenance and operating costs are often high, which can affect the economic viability of the treatment plant. The results suggest ceramic membranes are a viable and efficient option for treating polluted water.

Keywords: ceramic membranes, water treatment, technological advances, pollutants, environmental impact

1. Introduction

Water treatment is a global challenge that requires innovative and efficient solutions to remove contaminants and ensure the availability of clean water. Ceramic membranes have emerged as a promising technology in this field due to their ability to efficiently remove a wide range of contaminants, including heavy metals and pathogens [1, 2]. These advances in ceramic membranes are driven by the development of new materials and manufacturing techniques that improve their properties and performance [3].

Ceramic membranes are primarily manufactured from oxides such as Al_2O_3, TiO_2, SiO_2, and ZrO_2, each providing unique properties that benefit the filtration process [4]. These membranes stand out for their chemical and mechanical resistance, high selectivity, and ability to operate in extreme temperature and pH conditions [5].

IntechOpen

Manufacturing techniques, such as thermal sintering, make it possible to obtain membranes with specific porous structures, essential for the performance of the filtration process [6].

In the context of sintering technology, the development of cold sintering processes represents a significant innovation. These processes allow for the densification of materials at much lower temperatures (below 300°C), which can significantly reduce the costs and problems associated with high-temperature sintering, such as the formation of blocking layers due to side reactions [7]. Cold sintering is based on the dissolution of the sharp edges of the particles by a solvent and the subsequent filling of the voids between the particles through the evaporation of the solvent and the precipitation of the dissolved species [8]. This innovative mechanism has proven effective in densifying a wide range of inorganic and organic-inorganic compounds, and its application in solid electrolytes and solid-state batteries has shown promising results [9].

To maximize the performance of ceramic membranes in water treatment, it is essential to understand and optimize their physicochemical and structural properties. Advances in characterization techniques have made it possible to analyze the microstructure of membranes in detail, facilitating the design of membranes with better filtration properties [10]. In addition, surface modifications, such as functionalization with specific chemical agents, can improve the selectivity and efficiency of membranes for the removal of specific contaminants [11].

The environmental and economic impact of the implementation of ceramic membranes in water treatment plants is an aspect to consider, while the acquisition and infrastructure necessary for the implementation of these membranes can represent a significant investment, maintenance, and operating costs tend to be high, which can affect the long-term economic viability [12]. However, environmental benefits, such as reduced pollution and improved water quality, justify continued exploration and investment in this technology.

The research of this chapter focuses on the application of cold sintering processes for the manufacture of ceramic membranes, to improve their electrochemical and structural properties. It is proposed that low-temperature densification by cold sintering can improve the ionic conductivity and structural stability of membranes, which is critical for their performance in water treatment applications [13]. In addition, the use of additives with high solubility and ionic conductivity to improve the binding between grain boundaries will be investigated, which could expand the applicability of these techniques in various target materials [14].

Therefore, the development of ceramic membranes for water treatment, in combination with innovative cold sintering techniques, represents a significant advance in water purification technology. This approach not only improves pollutant removal efficiency but also offers more sustainable and economically viable solutions to meet global water management challenges [15].

Cold sintering is an innovative process that allows the densification of ceramic materials at significantly low temperatures, using the combination of pressure and a solvent that facilitates the dissolution and recrystallization of particles at grain boundaries. This method is especially relevant in the context of technological development for the production of ceramic membranes, due to its ability to reduce energy and operating costs while minimizing the defects associated with sintering at high temperatures. The low sintering temperature not only improves the efficiency and performance of the membranes but also contributes to greater structural stability and ionic conductivity, which are essential for its application in the treatment of contaminated water.

The use of cold sintering in the manufacture of ceramic membranes represents a significant advance in terms of technological innovation. This process offers a sustainable and economically viable alternative to conventional sintering methods, enabling the production of membranes with optimized characteristics for the removal of contaminants in water. The ability of this method to operate at low temperatures and with lower energy requirements makes it a promising option for the future of water treatment technologies, thus promoting the adoption of more efficient and environmentally responsible solutions in the industry.

2. Filtration mechanisms and working principles of ceramic membranes

Ceramic membranes use a porous matrix to separate contaminants from water using physical and chemical mechanisms. According to [1, 16], these membranes operate primarily by pore size, where fine pores allow water to pass through while retaining larger particles and contaminants. This approach is supported by studies that highlight the importance of controlled porosity and pore size distribution to optimize filtration efficiency [1].

In the study of membranes, it is essential to distinguish between charged and neutral membranes, as their characteristics influence the performance of the filtration process. Distinguishing between charged and neutral membranes is essential in the field of separation and filtration, as each type of membrane has unique characteristics that directly affect process performance. Charged membranes, which can have positive or negative charges, interact differently with the suspended particles and solutes in the solution. This interaction can improve the selectivity of the filtration process, allowing for more efficient separation of contaminants, especially those with opposite charges.

On the other hand, neutral membranes lack surface charge, which can result in a different behavior against suspended matter and solutes. This type of membrane tends to have a lower affinity for certain particles, which can influence the rate of fouling (membrane clogging) and, consequently, the efficiency of the filtration process.

In addition, the surface charge of membranes affects the phenomenon of electrolysis and ion transport, which can result in variations in permeate flow and in the quality of the final product. In summary, the choice between charged and neutral membranes should be based on the specific needs of the filtration process, considering feed characteristics and separation goals to maximize system efficiency and effectiveness.

The asymmetrical configuration of the membranes plays a fundamental role in the optimization of the flow, allowing greater efficiency in the separation of solutes. In addition, flow density is directly related to the type of tangential flow applied, which can significantly affect system performance. It is also essential to consider the characteristics of the feed water, since its composition impacts the permeate flow and, therefore, the quality of the final product. Finally, the surface charge properties of the membrane are decisive in the regulation of permeate flow, as they influence the interaction with the solutes present in the water, which can modify the efficiency of the filtration process.

These interactions become especially relevant in systems where the purity of the permeate is sought to be maximized. For example, in wastewater treatment applications, the asymmetrical configuration can facilitate the removal of contaminants, while proper tangential flow helps minimize fouling, thus extending the life of the membrane. Variability in feedwater quality, which can include different levels of turbidity, concentration of dissolved solids, and presence of chemical compounds, requires membranes to be designed with specific characteristics to accommodate these conditions. Consequently, a thorough understanding of these factors allows for the design of more efficient and sustainable filtration systems, optimizing both performance and operating costs in the long term.

On the other hand, Avasarala and Bose [3] point out that in addition to pore size, surface chemical interactions play an important role in the filtration process. These interactions can include electrostatic adsorption, acid-base interactions, and hydrogen bonding, depending on the chemical composition of the ceramic membrane and the contaminants present in the water [3].

From a more analytical perspective, the research of Avasarala and Bose [3] highlights the importance of surface modification to improve the selectivity and efficiency of ceramic membranes in water treatment applications. This modification may include techniques such as nanoparticle coating or chemical functionalization to increase affinity for specific contaminants. In addition, understanding the mechanisms of mass transport within the porous matrix is critical. According to He et al. [1], porous structure of ceramic membranes not only affects the filtration efficiency but also the mechanical strength and long-term durability under different operating conditions.

In conclusion, while [1, 17] emphasize the importance of pore size and porous structure in ceramic membrane filtration mechanisms, Avasarala and Bose [3] broaden the focus by highlighting surface chemical interactions and surface modification as key strategies to improve membrane efficiency and selectivity in water treatment.

3. Recent advances in ceramic materials for water treatment membranes

Following recent advances in the field of ceramic materials for water treatment membranes (see **Table 1**), significant progress has been observed in improving the selectivity and efficiency of these membranes. Recent research in **Table 1** highlights

Aspect	Chemical function	Advances in engineering	Significant contributions to the global market
Development of selectivity and efficiency	Advances in materials engineering have enabled the development of ceramic membranes with significant improvements in filtration selectivity and efficiency [11].	Innovations in manufacturing methods have optimized porous structure and pore size distribution to improve contaminant separation [18].	Contributing to more effective applications in industrial wastewater treatment and brackish water desalination [19].
Chemical surface functionalization	Chemical functionalization of ceramic surfaces has been shown to increase the ability of membranes to separate specific contaminants [18].	Implementation of coatings and surface treatments that improve selectivity and reduce unwanted adsorption, increasing the durability of membranes [11].	Improving water quality by efficiently removing specific contaminants and reducing environmental impact in water treatment processes [19].

Source: Own elaboration.

Table 1.
Research analysis matrix on ceramic materials for water treatment membranes (2020–2024).

DOI: http://dx.doi.org/10.5772/intechopen.1008828

that "advances in materials engineering have allowed the development of ceramic membranes with improved filtration and durability properties" [3]. In addition, it has been shown that "chemical functionalization of ceramic surfaces can increase the ability of membranes to separate specific contaminants" [18]. These technological developments have paved the way for more effective applications in industrial wastewater treatment and brackish water desalination, thus contributing to improved water quality and environmental sustainability [19].

This matrix provides a framework for analyzing how these investigations have contributed from a chemical and engineering point of view, as well as their potential impact on the global water treatment technology market.

4. Designing and optimizing ceramic membranes to improve efficiency and durability

The methodology for cold sintering of ceramic membranes (see **Figure 1**) involves the preparation of a homogeneous suspension of ceramic powders in a solvent with soluble and conductive additives, followed by their compaction in a mold under high pressure to form a dense green body, and finally, a low-temperature heat treatment (<300°C) with continuous pressure to densify the material by dissolving particles and filling gaps [15]. This process reduces energy and operating costs by eliminating the need for high temperatures and expensive equipment, improving efficiency through high densification and reduced defects, and minimizing environmental impact [15].

The problem to be solved is the need for efficient and economical sintering processes to manufacture ceramic membranes for water treatment; cold sintering offers a promising solution by facilitating the production of membranes with high ionic conductivity and structural stability, as demonstrated in previous studies [15].

Table 2 A detailed and structured overview of the process of designing and optimizing ceramic membranes using cold sintering is provided. This innovative method

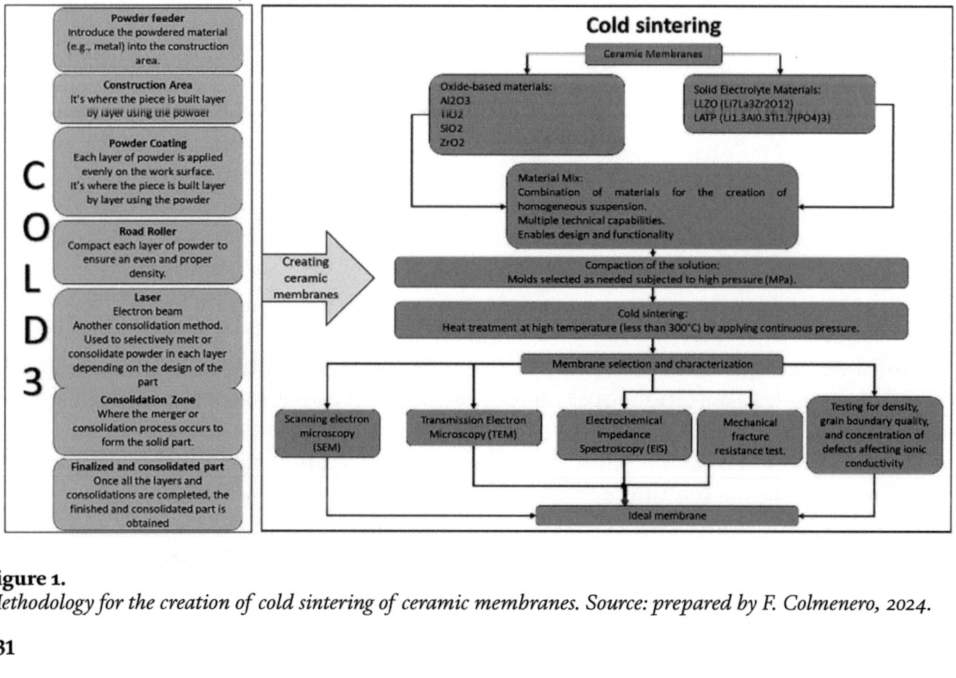

Figure 1.
Methodology for the creation of cold sintering of ceramic membranes. Source: prepared by F. Colmenero, 2024.

Aspect	Description
Objective	Reduce operating and energy costs, improve efficiency through high densification, and minimize environmental impact.
Problem to solve	Need for efficient and economical sintering processes to manufacture ceramic membranes for water treatment.
Material selection and preparation	Oxides: Al_2O_3, TiO_2, SiO_2, ZrO_2. • Electrolitos sólidos: LLZO ($Li_7La_3Zr_2O_{12}$), LATP ($Li_{1.3}Al_{0.3}Ti_{1.7}(PO_4)_3$).
Innovation	Expansion of technical capabilities and new possibilities in the design and functionality of ceramic membranes [8, 20].
Applications	Fundamental in high-performance electrochemical devices, solid-state batteries, gas sensors, and chemical analysis.
Cold sintering process	Suspension preparation: Dispersion of ceramic powders in solvents with additives. • Compaction: Molded under high pressure to form a compact green body. • -Sintering: Low-temperature heat treatment with continuous pressure to densify the material without high temperatures.
Membrane characterization	Evaluation of microstructure, grain limits, ionic conductivity, and mechanical properties for specific applications.

Source: The table was prepared based on the research and detailed analysis carried out by F. Colmenero, 2024, regarding Refs. [15].

Table 2.
Design and optimization process of ceramic membranes by cold sintering.

not only seeks to reduce operating and energy costs by eliminating the need for high temperatures, but also improves efficiency through high densification and minimizes the environmental impact. The strategic selection of materials, such as oxides and solid electrolytes of high ionic conductivity, opens up new possibilities in the design and functionality of these membranes, which finds applications in advanced electrochemical devices, solid-state batteries, and precision sensors. Detailed characterization of membranes, including microstructure, ionic conductivity, and mechanical properties, is critical to optimizing their performance in specific applications such as water treatment and energy storage.

This diagram in table form provides a concise and structured overview of the process, objectives, problems to be solved, materials used, innovations, applications, and key characteristics of the cold sintering process for ceramic membranes.

4.1 Selection and preparation of materials

For the manufacture of ceramic membranes by cold sintering, materials based on oxides such as Al_2O_3, TiO_2, SiO_2, and ZrO_2 will be selected. In addition, it works with solid electrolyte materials such as LLZO ($Li_7La_3Zr_2O_{12}$) and LATP ($Li_{1.3}Al_{0.3}Ti_{1.7}(PO_4)_3$), known for their high ionic conductivities and possible applications in solid-state batteries [15].

The combination of solid oxides and electrolytes not only expands the technical capabilities of ceramic membranes, but also opens up new possibilities in terms of design and functionality. These materials are very important in the manufacture of advanced membranes due to their ability to facilitate ion conduction through ceramic layers, which are critical for applications in high-performance electrochemical devices.

In the field of energy, ceramic membranes based on these materials can be used in solid-state batteries, promoting greater safety and energy density compared to conventional batteries. Its high ionic conductivity allows for more efficient charging and discharging, thus contributing to improved battery durability and cycle life. In addition, in sensor technology, these membranes can play an important role in selective gas detection and chemical component analysis, taking advantage of their chemical stability and resistance to high temperatures. This makes them ideal candidates for applications in environmental monitoring, medical diagnostics, and industrial process control, where accuracy and reliability are critical.

The strategic combination of solid oxides and electrolytes opens new frontiers for the manufacture of advanced ceramic membranes, driving significant advances in several key technology sectors.

4.2 Cold sintering process

The cold sintering process involves the following steps:

- Preparation of the ceramic suspension: Commercial ceramic powders are dispersed in a suitable solvent, usually water or a mixture of organic solvents, to form a homogeneous suspension. Additives with high solubility and ionic conductivity can be added to improve the bond between grain boundaries and increase the density and ionic conductivity of the final material [15].

- Suspension compaction: The suspension is compacted in a mold under high pressure (up to several hundred MPa) to form a green body with high initial density.

- Cold sintering: The compacted green body is heat-treated at low temperature (less than 300°C), while continuous pressure is applied. During this process, the solvent facilitates the dissolution of the sharp edges of the particles, and the subsequent evaporation of the solvent and precipitation of the dissolved species fill the gaps between the particles, achieving high densification [15].

- Characterization of the membranes: To evaluate the quality and performance of cold-sintered ceramic membranes, the following characterizations will be carried out:

  a. Microstructure and grain boundaries: Techniques such as scanning electron microscopy (SEM) and transmission electron microscopy (TEM) will be used to study the microstructure and grain boundaries of membranes.

  b. The chemistry, composition, and structure of grain boundaries will be analyzed to understand their relationship to enhanced ionic conductivity [15].

  c. Ionic conductivity: Ionic conductivity shall be measured by electrochemical impedance spectroscopy (EIS).

  d. How density, grain boundary quality, and defect concentration affect the ionic conductivity of membranes will be evaluated [15].

e. Mechanical properties: Hardness and fracture resistance tests will be carried out to determine the mechanical properties of the sintered membranes.

f. These properties are critical for durability and performance in water treatment and energy storage applications [15].

5. Advanced applications of ceramic membranes in wastewater treatment

To improve the scalability of cold sintering, several key technological innovations are required. First, it is important to maintain stable temperatures and humidity levels within the sintering chambers to prevent defects and ensure the quality of the material [1]. In addition, the sintering mechanisms at the atomic and microstructural levels must be deepened to achieve high densification at low temperatures [3]. Precise control of parameters such as pressure and sintering time is essential to prevent the formation of defects such as blocking layers and pores [1]. Bringing cold sintering to large-scale industrial production requires the design of suitable furnaces and feeding systems, facing challenges of temperature uniformity and low-temperature material handling [3].

In addition, optimization of specific formulations is necessary to enable efficient densification without compromising desired properties [1]. Integrating cold sintering into industrial applications such as water treatment and energy storage involves optimizing ionic conductivity and structural stability on a large scale, which demands

Problems and variables	Explanation	Applications	References
Pore size	Pore size controls how efficiently the membrane can separate particles of different sizes.	Tratamiento de agua potable y residual	[1, 3]
Pore size distribution	Pore size distribution affects the uniformity and efficiency of filtration across the membrane.	Water purification for industrial use	[1]
Surface chemical interactions	Chemical interactions such as adsorption, acid-base interactions, and hydrogen bonding can influence the membrane's ability to retain specific contaminants.	Filtration of contaminated wastewater	[3]
Surface modification	Modifying the membrane surface with coatings or chemical functionalization can improve filtration selectivity and efficiency.	Industrial effluent treatment	[3]
Mass transport mechanisms	Understanding how contaminants move through pores and the porous matrix is critical to optimizing separation efficiency.	Treatment of water contaminated with heavy metals	[1]
Mechanical strength and durability	The strength and durability of the ceramic membrane under different operating conditions determine its long-term service life and performance.	Seawater desalination	[1]

Source: Prepared by F. Colmenero, 2024. Based on existing membrane studies.

Table 3.
Matrix of problems, variables, and applications in filtration mechanisms and principles of operation of ceramic membranes in water treatment.

innovations in design and characterization [3]. Finally, the development of certification standards and procedures will be of vital importance to guarantee the quality and safety of products manufactured using this method [1, 3].

The categories are analyzed, addressing various key aspects of the operation and application of ceramic membranes in water treatment, while highlighting the importance of each identified variable and problem in enhancing membrane technology (**Table 3**).

6. Challenges and future perspectives in ceramic membrane research and development

The main challenges to the industrial scalability of cold sintering include the design of furnaces and feeding systems suitable to ensure temperature uniformity and efficient low-temperature material handling, as well as the precise control of parameters such as temperature, humidity, and atmosphere using advanced monitoring and control systems. In addition, the composition of materials must be optimized to enable efficient densification at low temperatures and ensure effective integration into applications such as water treatment and energy storage, addressing challenges related to ionic conductivity and large-scale structural stability. Finally, the development of certification standards and procedures will be essential to ensure the quality and safety of products manufactured through this adapted process of [1, 3].

The main technological challenges in the implementation of cold sintering include:

- Temperature and humidity control: Maintaining stable temperatures and humidity levels within sintering chambers is important to avoid defects and ensure material quality. Temperature and humidity control systems must be accurate and reliable.

- Densification at low temperatures: Achieving high densification of materials at much lower temperatures than conventional sintering is a key technical challenge. This requires a deep understanding of the mechanisms of sintering at the atomic and microstructural levels.

- Defect minimization: Low-temperature sintering must prevent the formation of blocking layers, pores, and other defects common in high-temperature sintering. This involves precise control of the process parameters.

- Industrial scalability: Bringing cold sintering from laboratory scale to large-scale industrial production presents engineering challenges related to the design of furnaces, powder feeding systems, and automated processes.

- Integration with applications: Integrating cold sintering efficiently into applications such as water treatment and energy storage requires optimizing the ionic conductivity and structural stability of materials.

- In summary, the main technical challenges include the precise control of processing conditions, the understanding of the fundamental mechanisms of low-temperature sintering, and the scalability of the process for industrial implementation.

Cold sintering offers several significant benefits:

- Cost reduction: By operating at much lower temperatures than conventional sintering, energy and operating costs are greatly reduced. In addition, cold sintering can eliminate the need for expensive high-temperature equipment, reducing capital investment [5].
- Improved efficiency: The ability to achieve high densification at low temperatures can improve the ionic conductivity and structural stability of membranes, critical to their performance in water treatment and energy storage applications [14].
- Minimization of defects: Low-temperature sintering can prevent the formation of blocking layers and other defects associated with high-temperature sintering, improving the overall quality of the material [8].
- Environmental sustainability: By reducing energy consumption and minimizing the generation of toxic byproducts, cold sintering is a more sustainable and environmentally friendly technology [16].

6.1 Case studies on the successful implementation of ceramic membranes in water treatment projects

In Europe, ceramic membranes have been successfully implemented in several water treatment projects, focusing mainly on improving drinking water quality and wastewater management (**Figure 2**).

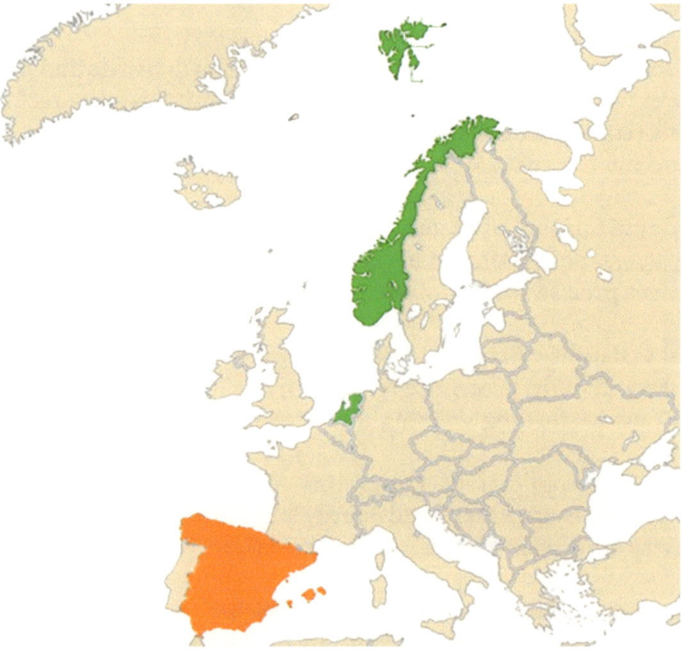

Figure 2.
Successful implementation projects of ceramic membranes in water treatment projects. Source: Own elaboration.

1. DEMOWARE project (the Netherlands):

    - Description: The DEMOWARE project, carried out in the Netherlands, focused on the implementation of ceramic membranes for advanced municipal wastewater treatment. Ceramic membrane technology was used for the efficient removal of persistent pollutants such as microplastics and emerging pollutants.

    - Results: Ceramic membranes were shown to be highly effective for contaminant removal, with high abrasion resistance and chemical stability, making them suitable for long-term wastewater treatment applications [21].

2. Oslo Water Treatment Plant (Norway):

    - Description: The Oslo water treatment plant implemented ceramic membranes for drinking water treatment, standing out for its ability to provide an effective barrier against pathogens and organic contaminants. This project focused on improving the safety of the drinking water supply through the effective elimination of microorganisms and organic compounds.

    - Results: The ceramic membranes proved to be robust and capable of operating with high efficiency even under fluctuating water quality conditions, ensuring high-quality drinking water standards in a sustainable manner [22].

3. España

    - In Spain, the application of ceramic membranes in water treatment projects has also shown significant progress, especially in the context of improving the sustainability and efficiency of water treatment systems.

- Wastewater treatment plant in Barcelona:

    - Description: The wastewater treatment plant in Barcelona implemented ceramic membranes for the advanced treatment of industrial effluents. These membranes were used for water recovery and the removal of specific contaminants present in industrial wastewater.

    - Results: Ceramic membrane technology allowed a significant reduction in the pollutant load of treated water, complying with the strictest environmental regulations and facilitating the safe reuse of water in industrial processes [22].

4. LIFE REMEMBrANE project (Valencia):

    - Description: The LIFE REMEMBrANE project in Valencia focused on the implementation of ceramic membranes for desalination and brackish water treatment. This technology was applied to improve the energy efficiency and sustainability of desalination processes, reducing energy consumption and minimizing waste generation.

- Results: The ceramic membranes developed within the framework of the LIFE REMEMBrANE project demonstrated high selectivity and durability, providing a cost-effective and environmentally friendly solution for the production of desalinated water in coastal regions [23].

The implementation of ceramic membranes in water treatment projects in Europe and Spain has shown promising results in terms of filtration efficiency, durability, and sustainability. These case studies highlight the ability of ceramic membranes to improve the quality of treated water, effectively manage wastewater, and promote the reuse of water resources, thereby contributing to environmental sustainability and water supply security.

6.2 Key references to recent research and technological advances in the field of ceramic membranes

The LIFE REMEMBrANE project in Valencia has made a significant contribution to several key aspects related to sustainability and energy efficiency in the treatment of brackish water using ceramic membranes.

Below are details of how this project has positively impacted and how we could extrapolate these benefits to cold sintering:

Contributions to the LIFE REMEMBrANE Project in Valencia:

1. Energy efficiency and sustainability:

   - Impact: The use of ceramic membranes in the LIFE REMEMBrANE project has significantly improved the energy efficiency of desalination processes. These membranes allow for greater water recovery and a considerable reduction in energy consumption compared to conventional desalination methods.

   - Benefits: This has led to a decrease in greenhouse gas emissions associated with desalinated water production, thus contributing to mitigating climate change and promoting more sustainable practices in water management.

2. Waste reduction and environmental impact:

   - Impact: Developed ceramic membranes are more durable and have a longer lifespan compared to other membrane technologies. This minimizes the generation of waste associated with frequent membrane replacement and reduces the environmental footprint of the desalination process.

   - Benefits: The reduction of solid and liquid waste contributes to the preservation of natural resources and the protection of sensitive aquatic ecosystems in the coastal areas where these projects are implemented.

Potential application in cold sintering:

To extrapolate these benefits to the field of cold sintering, we could consider the following points:

1. Energy efficiency:

    - Cold sintering seeks to reduce energy consumption compared to conventional methods of sintering, such as high-temperature sintering. As in the LIFE REMEMBrANE project, where energy efficiency was improved by using ceramic membranes, in cold sintering advanced ceramic materials could be explored that allow efficient densification at lower temperatures, thus reducing the overall energy consumption of the process.

2. Sustainability and waste reduction:

    - Cold sintering could benefit from ceramic materials that are not only energy-efficient but also generate less waste during the manufacturing process. This could be achieved by developing ceramic formulations that optimize densification and mechanical strength without requiring high temperatures or contaminating additives.

The LIFE REMEMBrANE project in Valencia has set an important precedent in the use of ceramic membranes to improve sustainability and efficiency in brackish water desalination. By extrapolating these principles to the field of cold sintering, we could move toward more sustainable and energy-efficient ceramic material manufacturing processes, benefiting both the environment and the manufacturing industry.

From cold sintering in Valencia, several strategies could be implemented to improve sustainability and efficiency in water treatment, benefiting not only the local region but also areas with severe water stress such as Guanajuato and Jalisco in Mexico. Here are some ideas and how they might be applied:

1. Development of advanced ceramic membranes:

    - Objective: To research and develop advanced ceramic membranes using cold sintering techniques. These membranes could be applied in filtration systems for water treatment, improving quality and reducing the pollutant load.

    - Benefits: The ceramic membranes developed could offer high selectivity and durability, reducing the need for maintenance and increasing the efficiency of drinking water and wastewater treatment in Valencia.

2. Application in water reuse projects:

    - Objective: To use the ceramic membranes developed for the reuse of treated wastewater in agricultural, industrial, or urban applications.

    - Benefits: Reduction of freshwater consumption, protection of local water resources, and improvement in sustainable water management in Valencia.

3. Optimization of industrial processes:

    - Objective: To implement cold sintering technologies for the manufacture of ceramic components used in water treatment plants and desalination systems.

- Benefits: Improvement in the energy efficiency of industrial processes related to water treatment, reduction of operating costs, and minimization of environmental impacts.

According to a study by the World Resources Institute (WRI) that utilized various climate models and socioeconomic scenarios, future water stress—a measure of competition for and depletion of surface water—was assessed and ranked in 167 countries for the years 2020, 2030, and 2040. The study revealed that 33 countries are projected to experience extremely high levels of water stress by 2040 (see **Figure 3**). Notably, Chile, Estonia, Namibia, and Botswana may see a particularly sharp rise in water stress during this period. As a result, businesses, agriculture, and communities in these nations could become increasingly susceptible to water scarcity compared to current conditions.

In Latin America, and particularly in western Mexico, water stress is a growing challenge due to multiple factors, including increased demand for water due to population and economic growth, contamination of surface and groundwater sources, and the effects of climate change that alter precipitation patterns. Regions such as Guanajuato and Jalisco face chronic water shortages, which exacerbate the vulnerability of communities to significant socioeconomic and environmental impacts. In this context, the application of advanced ceramic membranes through technologies such as cold sintering becomes essential. Not only do these membranes offer an efficient and sustainable solution for the treatment of contaminated and brackish water, but they can also contribute to improving the quality of life by ensuring access to safe and reliable drinking water for affected populations.

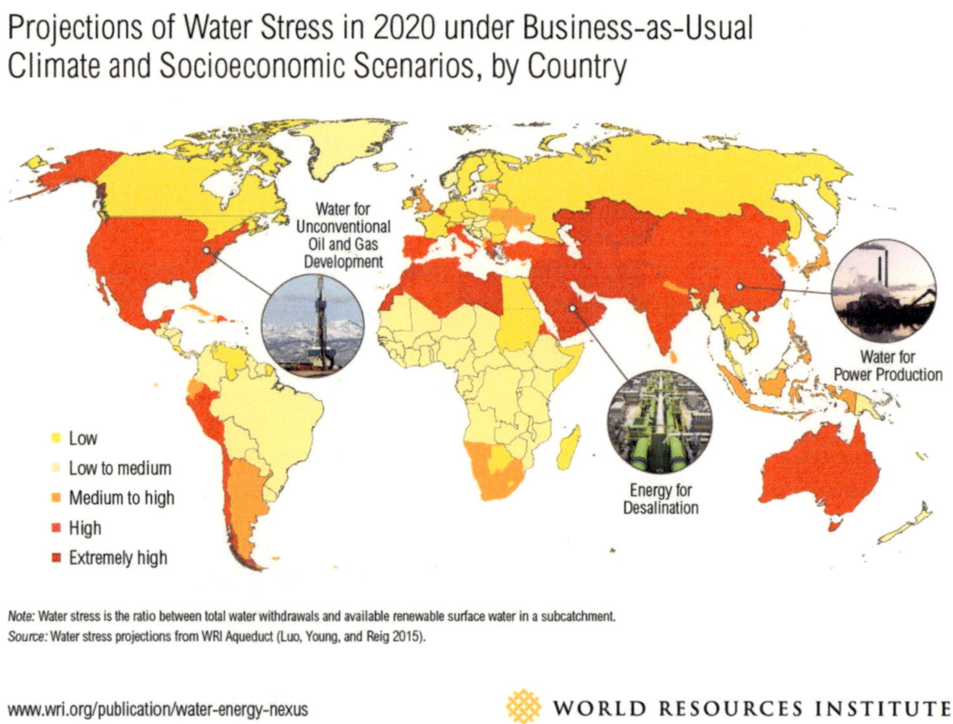

Figure 3.
Water stress by country toward the year 2040. Source: World Resources Institute. www.wri.org.

DOI: http://dx.doi.org/10.5772/intechopen.1008828

Water stress is measured on a scale of 1–5, where 5 means the use of 100% of the available water.

A study by the World Resources Institute (See **Figure 4**) reveals that approximately one-third of the world's population lives in countries at high risk of water scarcity shortly, with Mexico standing out in this category. The country is at a high level of water stress, consuming between 40 and 80% of its available water resources annually and ranking 24th out of 164 nations assessed globally. However, detailed analysis indicates that certain regions, especially in Northern and Central Mexico, face extremely high-water stress, using between 80 and 100% of available water. This situation underscores the urgency of taking preventive and strategic measures to improve water management and mitigate associated risks, such as increasing agricultural efficiency and investing in infrastructure for water treatment and reuse.

The states in Mexico most at risk of water crises include Baja California Sur, followed closely by Guanajuato, Mexico City, Aguascalientes, the State of Mexico, Querétaro, Hidalgo, Chihuahua, Zacatecas, Sonora, Sinaloa, Nuevo León, Morelos, Jalisco, and Tamaulipas. Conversely, several states, particularly those in the Gulf

Figure 4.
Water stress in Latin America. Source: https://cemeri.org/mapas/m-estres-hidrico-america-latina-lt.

region, are experiencing a medium to high risk of drought. These include Tabasco, Veracruz, Tamaulipas, Tlaxcala, Yucatán, Colima, and Puebla (**Figure 5**).

The application of cold sintering in west-central Mexico represents an innovation to improve the quality of life by addressing challenges related to water access and management. Through the development of advanced ceramic membranes, cold sintering technology offers efficient and sustainable solutions for the treatment of drinking and wastewater. According to Ref. [24], ceramic membranes manufactured by cold sintering are highly selective and durable, capable of effectively filtering contaminants and improving the purity of the treated water. This not only ensures access to safe drinking water but also reduces the health risks associated with drinking contaminated water.

In addition, the application of cold sintering in brackish water desalination projects can be especially beneficial in arid and semi-arid regions of west-central Mexico, where access to fresh water is limited. According to Ref. [25], cold sintering allows the development of ceramic membranes that are energy efficient and can operate at lower temperatures, thus reducing the overall energy consumption of the desalination process. This technology not only improves the availability of drinking water but also promotes the conservation of water resources and resilience to adverse weather conditions.

Finally, the implementation of cold sintering not only offers advanced technological solutions but also drives economic and social development by promoting technological autonomy and local training in the manufacture and maintenance of ceramic membranes. According to Ref. [26], local capacity building strengthens community infrastructure and fosters sustainable water management practices, thereby improving the quality of life of affected populations. Cold sintering emerges as an ultimate invention that not only addresses critical water challenges but also empowers communities to achieve more equitable and sustainable development.

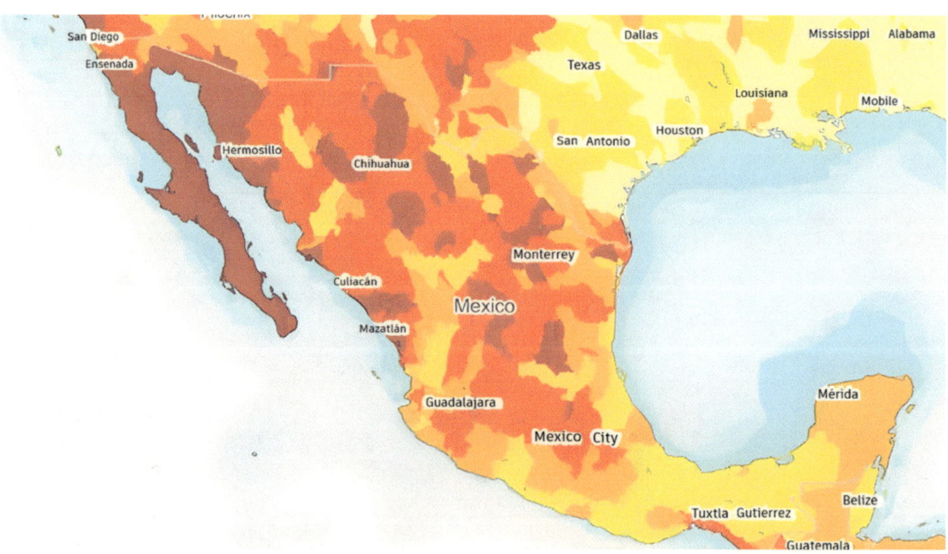

Figure 5.
Map of water stress in Mexico. Darker tones identify regions with greater stress. Fuente: https://mexiconewsdaily.com/news/day-zero-for-water/.

Benefits for Guanajuato and Jalisco, Mexico:

1. Technology and knowledge transfer:

  - Valencia could collaborate with institutions in Guanajuato and Jalisco to transfer cold sintering technologies and knowledge in the development of ceramic membranes. This could help establish local capacities to improve water treatment in water-stressed areas.

2. Implementation of sustainable solutions:

  - The application of cold sintering technologies in Guanajuato and Jalisco could contribute to implementing sustainable solutions for wastewater treatment and brackish water desalination. This could improve the availability of safe and potable water for local communities.

3. Reducing dependence on external water resources:

  - The adoption of advanced technologies such as cold sintering would allow Guanajuato and Jalisco to reduce their dependence on external sources of water, promoting water self-sufficiency and resilience to climate change and water scarcity.

Cold sintering in Valencia can play a very important role in improving sustainability and efficiency in water use, both locally and globally. By collaborating with regions such as Guanajuato and Jalisco in Mexico, innovative solutions could be implemented that significantly contribute to addressing the challenges of water stress and improving the quality of life of local populations.

7. Conclusions

The comprehensive examination of ceramic membranes' filtration mechanisms and operational principles in water treatment unveils critical variables and diverse applications that are crucial to addressing global water challenges. Central to this discourse is the problem matrix that highlights key factors—such as pore size and distribution, surface chemical interactions, modifications, mass transport mechanisms, and mechanical strength—that fundamentally influence the efficiency and selectivity of these membranes across various applications, including drinking water purification, desalination, and the treatment of industrial effluents.

The ongoing research and development of ceramic membranes, particularly those fabricated through the innovative cold sintering process, present both formidable challenges and significant prospects. Notably, overcoming technological barriers such as the precise control of temperature and humidity during manufacturing, achieving effective densification at low temperatures, minimizing structural defects, and scaling production for industrial applications is essential. Furthermore, the integration of these membranes into advanced water treatment systems requires a concerted effort to tailor their properties to meet specific environmental and operational demands.

Case studies from Europe, such as the DEMOWARE initiative in the Netherlands and the LIFE REMEMBrANE project in Valencia, exemplify the advancements in filtration efficiency, durability, and reduced environmental impacts achievable with ceramic membrane technologies. These projects serve as inspiring benchmarks, particularly for regions like Guanajuato and Jalisco in Mexico, which are grappling with severe water stress. By demonstrating the feasibility of these technologies in real-world applications, they pave the way for broader adoption and implementation.

The transition toward cold sintering not only signifies a promising advancement in the sustainability and efficiency of global water treatment efforts but also embodies a proactive approach to resilience against the burgeoning challenges posed by climate change and dwindling water resources. By enabling the production of ceramic membranes at lower, energy-efficient temperatures, cold sintering substantially lowers energy consumption relative to traditional sintering methods. This reduction not only decreases operational costs but also enhances the economic viability of water treatment technologies, particularly for resource-constrained communities.

Moreover, cold sintering aligns with environmental sustainability goals by diminishing the carbon footprint and mitigating the toxic waste generated during the manufacturing of ceramic membranes. The utilization of materials that demand less energy for production and yield fewer harmful by-products underscores a commitment to greener practices in water management.

In terms of water quality, ceramic membranes produced through cold sintering exhibit remarkable efficiency and longevity, effectively filtering out a range of contaminants, including microorganisms, heavy metals, and various chemical compounds. This enhancement in water quality is crucial for reducing health risks associated with contaminated drinking water and improving overall public health outcomes.

Furthermore, the application of advanced ceramic membranes created *via* cold sintering facilitates the safe reuse of treated wastewater for agricultural, industrial, and urban applications. This not only conserves precious water resources but also alleviates pressure on freshwater sources, particularly in drought-prone regions.

In conclusion, cold sintering transcends a mere technological innovation in ceramic membrane production; it offers transformative, practical, and sustainable solutions that can significantly enhance access to safe drinking water while simultaneously safeguarding the environment. This innovative approach holds the potential to effectively tackle pressing global water challenges, thereby contributing to a more sustainable and resilient future for communities worldwide. As we advance toward implementing these technologies, the collaborative efforts of researchers, policymakers, and industry stakeholders will be paramount in realizing the full potential of cold sintering in addressing the water crises faced globally.

The research on the filtration mechanisms and working principles of ceramic membranes using cold sintering not only highlights their potential as an innovative solution for water treatment but also underlines their relevance in the context of global challenges related to access to drinking water and environmental sustainability. The results obtained from case studies in Europe, combined with the ability of these membranes to operate efficiently at lower temperatures and with lower energy consumption, strengthen the viability of their implementation in vulnerable regions, such as Guanajuato and Jalisco in Mexico. By offering practical solutions that improve water quality and promote the safe reuse of water resources, this research not only contributes to water science and technology, but also directly addresses the urgent needs of communities affected by water scarcity, pointing toward a more sustainable and resilient future.

Acknowledgements

This publication is part of the R&D projects PID2021-128548OB-C21 and CNS2023-144190, funded by MCIN/AEI/10.13039/501100011033 "ERDF A way to do Europe" and by the European Union "NextGenerationEU"/PRTR, respectively.

Conflict of interest

The authors declare no conflict of interest.

Author details

Fabiola Colmenero Fonseca*, Amparo Borrell, María Dolores Salvador Moya, Rut Benavente and Juan Francisco Palomino Bernal
Institute of Materials Technology, Universitat Politècnica de València, Valencia, Spain

*Address all correspondence to: fcolfon@upvnet.upv.es

References

[1] He Z, Liu Z, Gu Q, Zhang L, Wang J. Ceramic-based membranes for water and wastewater treatment. Colloids and Surfaces A: Physicochemical and Engineering Aspects. 2019;**578**:123513

[2] Kim DY, Chun SH, Jung Y, Mohamed DFMS, Kim HS, Kang DY, et al. Phthalate plasticizers in children's products and estimation of exposure: Importance of migration rate. International Journal of Environmental Research and Public Health. 2020;**17**:8582. DOI: 10.3390/IJERPH172 28582

[3] Avasarala S, Bose S. 2D nanochannels and huge specific surface area offer unique ways for water remediation and adsorption: Assessing the strengths of hexagonal boron nitride in separation technology. Functional Composite Materials. 2023;**4**(1):5

[4] Pabby AK. Handbook of Membrane Separations [Internet]. CRC Press; 2008 [cited 2025 Jan 13]. Available from: https://www.taylorfrancis.com/books/mono/10.1201/9781420009484/handbook-membrane-separations-syed-rizvi-ana-maria-sastre-requena-anil-kumar-pabby

[5] Li C, Sun W, Lu Z, Ao X, Yang C, Li S. Systematic evaluation of TiO_2-GO-modified ceramic membranes for water treatment: Retention properties and fouling mechanisms. Chemical Engineering Journal. 2019;**378**:122138

[6] Gitis V, Rothenberg G. Ceramic Membranes: New Opportunities and Practical Applications [Internet]. John Wiley & Sons; 2016 [cited 2025 Jan 16]. Available from: https://www.researchgate.net/publication/316934666_Ceramic_membranes_New_opportunities_and_practical_applications

[7] Hubadillah SK, Jamalludin MR, Dzarfan Othman MH, Iwamoto Y. Recent progress on low-cost ceramic membrane for water and wastewater treatment. Ceramics International. 2022;**48**(17):24157-24191

[8] Guo J, Floyd R, Lowum S, Maria JP, Herisson de Beauvoir T, Seo JH, et al. Cold sintering: Progress, challenges, and future opportunities. Annual Review of Materials Research. 2019;**49**(1):275-295

[9] Guo J, Baker AL, Guo H, Lanagan M, Randall CA. Cold sintering process: A new era for ceramic packaging and microwave device development. Journal of the American Ceramic Society. 2017;**100**(2):669-677

[10] Guo J, Berbano SS, Guo H, Baker AL, Lanagan MT, Randall CA. Cold sintering process of composites: Bridging the processing temperature gap of ceramic and polymer materials. Advanced Functional Materials. 2016;**26**(39):7115-7121

[11] Arumugham T, Kaleekkal NJ, Gopal S, Nambikkattu J, Rambabu K, Aboulella AM, et al. Recent developments in porous ceramic membranes for wastewater treatment and desalination: A review. Journal of Environmental Management. 2021;**293**:112925

[12] Miller DJ, Dreyer DR, Bielawski CW, Paul DR, Freeman BD. Surface modification of water purification membranes. Angewandte Chemie International Edition. 2017;**56**(17):4662-4711

[13] Echevarría C, Pastur M, Valderrama C, Cortina JL, Vega A, Mesa C, et al. Techno-economic

assessment of decentralized polishing schemes for municipal water reclamation and reuse in the industrial sector in coastal semiarid regions: The case of Barcelona (Spain). Science of the Total Environment. 2022;**815**:152842

[14] Grasso S, Biesuz M, Zoli L, Taveri G, Duff AI, Ke D, et al. A review of cold sintering processes. Advances in Applied Ceramics. 2020;**119**(3):115-143

[15] Ferrer-Nicomedes S, Mormeneo-Segarra A, Vicente-Agut N, Barba-Juan A. Introducing an ionic conductive matrix to the cold-sintered Li1.3Al0.3Ti1.7(PO4)3-based composite solid electrolyte to enhance the electrical properties. Journal of Power Sources. 2023;**581**:233494

[16] Hood ZD, Zhu Y, Miara LJ, Chang WS, Simons P, Rupp JLM. A sinter-free future for solid-state battery designs. Energy and Environmental Science. 2022;**15**(7):2927-2936

[17] Polak D, Zielińska I, Szwast M, Kogut I, Małolepszy A. Modification of ceramic membranes with carbon compounds for pharmaceutical substances removal from water in a filtration—Adsorption system. Membranes (Basel). 2021;**11**(7):481

[18] Liangdy A, Lee WJ, Bao Y, Oh WD, Lim TT. Hybrid process of persulfate-based advanced oxidation with MeOx-functionalized catalytic ceramic membrane for synergistic removal of micropollutants: Recent developments, new insights, and prospects. Chemical Engineering Journal. 2023;**466**:143280

[19] Ayaz M, Namazi MA, Din MA, ud, Ershath MIM, Mansour A, Aggoune el HM. Sustainable seawater desalination: Current status, environmental implications, and future expectations. Desalination. 2022;**540**:116022

[20] Saeed W, Liu Z, Yan R, Li Y, Xu H, Tian Y, et al. Nanostructured compliant interconnections for advanced micro-electronic packaging. Materials and Design. Jul 2024. pp. 7-15

[21] da Silva Biron D, dos Santos V, Zeni M. Ceramic Membranes Applied in Separation Processes [Internet]. Cham: Springer International Publishing; 2018. Available from: https://link.springer.com/book/10.1007/978-3-319-58604-5 [Accessed: January 13, 2025]

[22] Issaoui M, Jellali S, Zorpas AA, Dutournie P. Membrane technology for sustainable water resources management: Challenges and future projections. Sustainable Chemistry and Pharmacy. 2022;**25**:100590

[23] European Union. Life Cycle Assessment of REMEB Ceramic Membranes [Internet]. 2018. Available from: https://ec.europa.eu/research/participants/documents/downloadPublic?documentIds=080166e5ba592ecf&appId=PPGMS [Accessed: January 12, 2025]

[24] Han Y, Zou D, Kang Y, Zhong Z, Xing W. One-step sintering process for high-performance SiC membranes for efficient filtration of dust-laden gas. Journal of Membrane Science. 2024;**692**:122265

[25] Tai ZS, Abd Aziz MH, Othman MHD, Mohamed Dzahir MIH, Hashim NA, Koo KN, et al. Ceramic membrane distillation for desalination. Separation and Purification Reviews. 1 Oct 2020;**49**(4):317-356

[26] Guo L, Wu Y, Huang F, Jing P, Huang Y. An approach to complex transboundary water management in Central Asia: Evolutionary cooperation in transboundary basins under the water-energy-food-ecosystem nexus. Journal of Environmental Management. Feb 2024;**351**:119940

Chapter 12

Cold Sintering: A New Sintering Technology for Advanced Ceramic Materials

Chunchun Li, Guobin Zhu, Xiaowei Zhu and Siyu Xiong

Abstract

Ceramic sintering is the process of solidifying ceramic powder into a dense bulk material through the migration of matter, which is the necessary path for the body to transform into a high-strength, dense ceramic body. Low-temperature sintering technology introduces electric fields, solvents, pressure, etc., to change the thermodynamic and kinetic conditions of sintering, which has been widely studied by global scholars since the twentieth century. Currently, there are common sintering processes such as cold sintering, dielectric barrier discharge plasma sintering, hot isostatic pressing sintering, flash sintering, and microwave sintering. Among them, cold sintering has gained widespread attention due to its advantages of simple equipment, convenient operation, and low sintering temperature. Copyright belongs to the author. Commercial reprint requires authorization from the author, non-commercial reprint please indicate the source.

Keywords: ceramics, sintering, cold sintering, low-temperature sintering, dielectric materials

1. Introduction

Ceramic materials have long played a crucial role in modern technology due to their exceptional physical and chemical properties. From everyday porcelain to electronic components and aerospace parts in high-tech fields, the application of ceramic materials spans a wide range, highlighting their significance. However, conventional sintering processes often necessitate high temperatures, which not only consume significant energy but can also adversely impact the material's properties.

In recent years, cold sintering technology has gained significant attention within the materials science community as an innovative method for sintering. Compared to traditional processes, this technology enables the densification of ceramic materials at lower temperatures, resulting in a substantial reduction in energy consumption and production costs while maintaining or even enhancing material properties (**Figure 1**) [1]. The emergence of cold sintering technology not only offers new possibilities for ceramic material production but also paves the way for advancements in materials science.

In this chapter, we initially examine the early concepts and preliminary experiments of cold sintering technology, followed by a focused exploration of significant

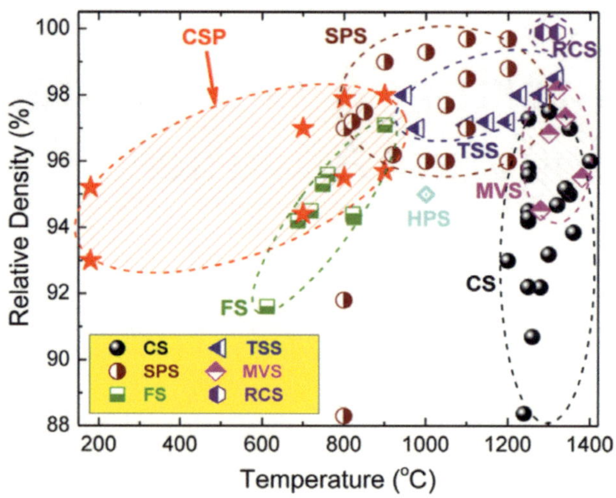

Figure 1.
The relative density of materials produced by the cold sintering process at low temperatures in comparison with other sintering techniques [1].

technological breakthroughs that have emerged during its development. These advancements have not only propelled the evolution of cold sintering technology but also contributed to a substantial enhancement in its current level of maturity. Through an in-depth analysis of these advancements, we can gain a better understanding of the progressive steps taken in the advancement of cold sintering technology and evaluate its present status within the realm of advanced ceramic materials sintering.

Additionally, this chapter introduces the physical and chemical principles underlying cold sintering technology, compares the mechanisms of traditional sintering and cold sintering, and discusses the procedural steps and operational methods involved in cold sintering. Through these discussions, we aim to elucidate the key parameters and influencing factors in the cold sintering process, providing readers with a comprehensive framework for comprehending this technology.

Finally, this chapter provides a comprehensive overview of the fundamental advantages of cold sintering technology and its application value across diverse fields, underscoring the criticality of ongoing research and development efforts while offering practical recommendations for both industry and academia. Through this chapter, readers will acquire a lucid comprehension of cold sintering technology's transformative potential as well as a foundational understanding of its pivotal role in advancing the realm of ceramic materials.

2. The development of cold sintering

The early concepts of cold sintering technology originated from a reflection and challenge to traditional high-temperature sintering methods, with its earliest traces dating back to the 1980s. As concerns over energy consumption and the preservation of material properties grew, scientists began exploring the possibility of densifying ceramic materials at lower temperatures. Early experiments, through the introduction of specific chemical additives and solvents, successfully densified materials and enhanced their properties at temperatures far below those of conventional sintering.

For instance, in 1986, the Yamasaki team from Kochi University in Japan [2] was the first to combine hydrothermal methods with uniaxial pressure, successfully sintering ceramics at 350°C, a process they termed hydrothermal hot pressing (HHP). However, due to complexities in equipment, limitations in product size and shape, and lower densification, this technology did not receive widespread research attention. In 2014, Jantunen et al. [3] successfully prepared well-densified Li_2MoO_4 ceramics by moistening the powder with deionized water and pressing it at 130 MPa, followed by sintering at 120°C. In 2016, the Randall team [4, 5] from Pennsylvania State University, inspired by natural crystallization phenomena, utilized transient solvents such as aqueous solutions under uniaxial pressures ranging from 100 to 500 MPa and within short durations, successfully sintering various ceramic materials like NaCl, Li_2MoO_4, V_2O_5, and $BaTiO_3$ at temperatures ranging from room temperature to 300°C, with densification levels exceeding 90% [6, 7]. This technique, named the Cold Sintering Process (CSP) for the first time, has the potential to shorten sintering times and reduce energy consumption, significantly improving sintering efficiency. Although these preliminary experiments achieved limited success at this time, they laid the foundation for subsequent cold sintering technologies and revealed the feasibility of achieving material densification at lower temperatures.

Subsequently, research teams from Europe and Asia have also successfully implemented low-temperature densification of over 70 ceramic materials [8–11], including $CaCO_3$, Al_2SiO_5-NaCl, Al_2O_3-NaCl, and $Na_{0.5}Bi_{0.5}MoO_4$-Li_2MoO_4, using this cold sintering method. As research continued, cold sintering technology made significant progress and breakthroughs in various fields. Researchers successfully utilized cold sintering to prepare ceramic-polymer composites, addressing the issue of co-sintering ceramics with polymers at high temperatures, thus opening new avenues for the development of novel composite materials. Additionally, cold sintering achieved breakthroughs in multi-material integration, particularly in the production of solid-state batteries, where it enabled the intimate combination of different materials, paving the way for the production of solid-state batteries with superior performance. Microwave dielectric materials, such as Li_2MoO_4, prepared using cold sintering, exhibited excellent microwave dielectric properties, maintaining high densification and good microwave dielectric performance even after sintering at low temperatures. Cold sintering technology began to be applied in industrial production, especially in the field of power metallurgy, where its advantages in reducing energy consumption and shortening production cycles made it highly commercially viable. In environmental governance, cold sintering technology was explored for the solidification treatment of heavy metals in fly ash, offering advantages such as volume reduction, weight reduction, and low energy consumption, providing a new solution for fly ash treatment.

In terms of technological breakthroughs, a significant milestone was the development of equipment and methods capable of precisely controlling the sintering atmosphere. By sintering in specific atmospheres, the porosity and grain size of materials could be effectively controlled, resulting in high-performance ceramic materials with specific functions. For example, sintering in a reducing atmosphere could produce solid oxide fuel cell (SOFC) electrolyte layers with excellent electrochemical performance [12]. On the other hand, researchers conducted in-depth studies on the reaction steps and microscopic mechanisms of cold sintering, establishing the fundamental principles and densification mechanisms of cold sintering technology, and providing a theoretical basis for further optimization and application of the technology. Internationally, research institutions such as the University of Oulu in Finland

and Pennsylvania State University in the United States have made significant progress in the study of cold sintering technology, bringing new research directions to the field of materials science. These advancements not only propelled the development of cold sintering technology but also demonstrated its broad prospects and commercial potential in theoretical research, material preparation, and industrial applications, indicating that cold sintering technology is expected to play an increasingly important role in future materials science and industrial production.

In summary, cold sintering technology has evolved from a conceptual validation stage to a relatively mature sintering method, showing significant potential in both laboratory settings and future industrial production. Compared to traditional sintering technologies, cold sintering offers clear advantages in energy efficiency, cost, and material performance. Currently, this technology has been applied in various fields such as electronics, energy, and biomedicine. For example, in the production of high-performance ceramic substrates, solid oxide fuel cell electrolyte layers, and biocompatible implant materials. The development trajectory of cold sintering technology reflects its growing maturity and expanding application scope, indicating that it will provide more efficient, economical, and environmentally friendly solutions for the production of advanced ceramic materials. Cold sintering technology offers a broad new approach to the low-temperature sintering of ceramics. The following sections will provide a detailed introduction to the densification principles and process methods of cold sintering technology.

3. The principle of cold sintering technology

Due to the relatively short history of cold sintering technology, research on its densification mechanisms has not yet fully matured. Currently, researchers generally believe that the densification mechanisms of cold sintering mainly include two types: one is the non-equilibrium dissolution-precipitation process caused by the intermediate liquid phase, the other is the plastic deformation process caused by pressure [13, 14]. Based on the theory of liquid phase sintering and microstructure observations on a broad number of systems, a schematic illustrating the three stages of the dissolution-precipitation process is shown in **Figure 2** [4]. The first stage is the dissolution-rearrangement process, where the intermediate liquid phase uniformly wets the ceramic powder, forming a liquid film on the surface of the ceramic particles, which increases the slipperiness between the ceramic particles. Under the action of the intermediate liquid phase, the dissolution of the sharp edges of the ceramic

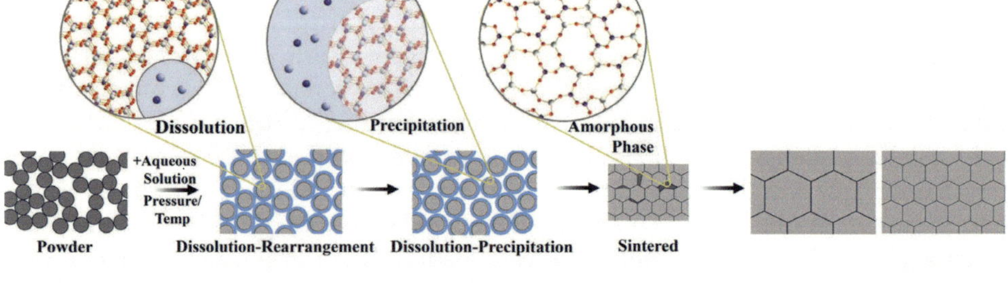

Figure 2.
Schematic diagram of the densification mechanism of CSP [4].

particles reduces the interfacial area, facilitating the rearrangement in the subsequent sintering stage. Under appropriate pressure and temperature conditions, the liquid phase redistributes and diffuses into the pores between the particles [4].

The second stage is the dissolution-precipitation process, which triggers a significant chemical driving force, bringing the solid and liquid phases into equilibrium. The intermediate liquid phase evaporates, causing the solution to reach saturation under certain temperature and pressure conditions. Under the action of capillary forces and external pressure, atomic clusters or particles far from high chemical potential contact zones migrate to areas of lower chemical potential and precipitate. This process eliminates pores by reducing the material's surface-free energy, thereby achieving densification of the ceramic [8, 15].

The third stage is the grain growth stage. The crystalline phase or amorphous phase will be formed in the final sintering process. If the crystal phase is formed, the ceramic particles will grow obviously. If there is an amorphous phase, the amorphous phase will limit the grain boundary diffusion and migration, thereby inhibiting the further growth of ceramic particles. In addition to the dissolution-precipitation mechanism, plastic deformation through viscous flow or dislocation movement is also considered one of the densification mechanisms of cold sintering, which helps to achieve densification under high pressure [16, 17].

3.1 Comparison in the mechanism of traditional sintering and cold sintering

Traditional sintering and cold sintering are two different material densification processes, and they have significant differences in mechanism. Ceramic materials have a high melting point, so the sintering temperature of ceramics is higher in the traditional sintering technology, and the sintering temperature is usually more than 1000°C [18–21]. At high temperatures, the atoms or molecules inside the material move through a diffusion mechanism, thereby reducing porosity and increasing the compactness of the material. Diffusion can be volume diffusion, grain boundary diffusion, or surface diffusion. In some cases, especially in the sintering with liquid phase participation, the plastic flow of the material can promote particle rearrangement and densification. As the sintering time increases, small particles are gradually absorbed by large particles, increasing particle size and thereby reducing pores [22]. The traditional high-temperature sintering technology has some disadvantages. It not only consumes a lot of energy but also volatilizes unstable additives (such as Bi, Pb, Li, Na, K, and other elements) at high sintering temperature, resulting in the imbalance of stoichiometric ratio, resulting in the change of crystal structure and affecting the performance [23, 24]. It also makes some ceramic materials and metal electrodes co-fire, and other aspects face challenges. When different materials are co-sintered, high temperatures will also cause unexpected chemical reactions, delamination, cracking, and other problems in the product [25, 26].

The cold sintering technology proposed in recent years can reduce the sintering temperature to below 400°C [7]. Using the transient solvent in the form of the liquid phase and uniaxial pressure, the rapid densification of ceramic materials can be achieved through the dissolution-precipitation process of ceramic particles. During the cold sintering process, the applied pressure can activate the interface between the particles, promote the migration of atoms or molecules, and achieve densification even at a lower temperature [6, 27]. Compared with conventional sintering, cold sintering usually does not involve the formation of liquid phases, so it is suitable for materials that decompose or deteriorate at high temperatures. The cold sintering

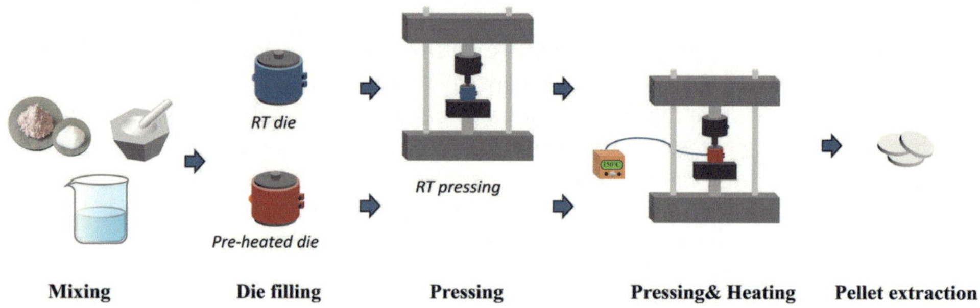

Figure 3.
Scheme of the cold sintering route.

technology has the characteristics of low sintering temperature and short sintering time and has received extensive attention since its development [28].

3.2 Steps and operation methods of the cold sintering process

A standardized and unique cold sintering process has not been defined yet, especially regarding how the temperature and pressure shall be applied. Here, a description of the approach adopted in previous works is introduced. The solvent is added to the ceramic powder after mixing for a few minutes to promote close contact between the liquid and the solid. The ceramic slurry is then introduced into the mold (usually cylindrical) and pressed by a hydraulic or mechanical press. The device is also equipped with a heating system, usually a heating sleeve wrapped around the mold, or two heating plates placed above and below the mold. **Figure 3** resumes schematically the CSP process [14, 29, 30].

3.3 Key parameters and influencing factors in the cold sintering process

The ceramic material system is huge, and the factors affecting the densification of different material systems are different. It is generally affected by factors such as particle size, solubility of particles in the intermediate liquid phase, type of intermediate liquid phase, temperature, and pressure. **Figure 4** shows the key sintering parameters for cold sintering to prepare different material systems [29, 31].

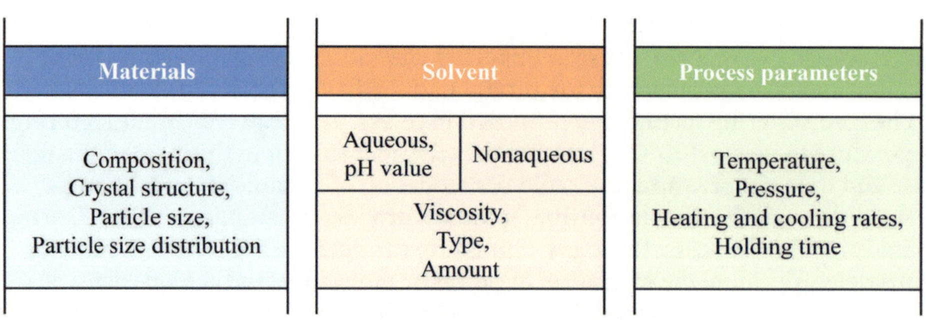

Figure 4.
The key parameters that the cold sintering process can take for different materials [29].

4. Material systems in cold sintering

Among the various ceramic materials prepared using CSP technology, dielectric ceramics have been the most extensively studied, followed by semiconductors, solid-state batteries, and thermoelectric materials. The following sections will provide a comprehensive yet succinct overview of the current significant research progress in these materials.

4.1 Dielectric ceramics

4.1.1 Microwave dielectric ceramics

In recent years, numerous microwave dielectric ceramic material systems have achieved promising preparation outcomes using CSP technology, including molybdates, tungstates, borates, chlorides, and fluorides, among others.

Molybdates and their composites are among the earliest and most extensively studied materials within the realm of cold CSP. In 2014, Kähäri et al. [3] successfully produced Li_2MoO_4 ceramics with densities ranging from 87 to 93%. Post-heat treatment led to dielectric properties characterized by a relative permittivity (ε_r) ranging from 4.6 to 5.2 and a quality factor ($Q \times f$) of 10,200–18,500 GHz. In 2016, Guo et al. [4] broadened the scope of research to include $Na_2Mo_2O_7$ and $K_2Mo_2O_7$, showcasing the versatility of CSP across various molybdate materials. Moreover, the application of CSP in composite materials has demonstrated effective control over the frequency-temperature coefficient (τ_f) of the composites [32].

The CSP has made significant strides in the development of tungstate and borate ceramics. In 2019, Hong et al. [33] employed CSP to prepare HBO_2-II and H_3BO_3 ceramics, achieving high densities of 94.5% and 97.7%, respectively, and excellent microwave dielectric properties (HBO_2-II: $\varepsilon_r \sim 4.21$, $Q \times f \sim 47{,}500$ GHz; H_3BO_3: $\varepsilon_r \sim 2.84$, $Q \times f \sim 146{,}000$ GHz), setting a new record for $Q \times f$ values. In 2020, Ding et al. [34] demonstrated an efficient method for dry pressing H_3BO_3 ceramics at room temperature. That same year, Hao et al. [35] investigated the impact of Na_2WO_4 grain size on microwave dielectric properties using CSP, resulting in high density (> 92%) and superior dielectric performance ($\varepsilon_r \sim 5.8$, $Q \times f \sim 22{,}000$ GHz). Further research indicated that performance could be further enhanced through optimized sintering conditions and heat treatment ($\varepsilon_r \sim 5.7$, $Q \times f \sim 70{,}000$ GHz) [36]. In 2021, Chen et al. [37] successfully prepared Na_2WO_4-$x$$Ni_{0.2}Cu_{0.2}Zn_{0.6}Fe_2O_4$ composites with densities of up to 87% and outstanding properties. These achievements underscore the potential of CSP in the fabrication of tungstate and borate materials.

Beginning in 2016, CSP has also achieved notable advancements in the preparation of chloride and fluoride microwave dielectric ceramics. In 2016, Guo et al. [4] pioneered the use of CSP to synthesize NaCl ceramics with high density, although the performance metrics were not reported at that time. By 2017, Induja et al. [9] showcased the microwave properties of NaCl ceramics ($\varepsilon_r \sim 5.2$, $Q \times f \sim 12{,}000$ GHz) and the performance of Al_2SiO_5–NaCl composites ($\varepsilon_r \sim 4.52$, $Q \times f \sim 22{,}350$ GHz). In 2018, Li et al. [38] explored the impact of varying amounts of deionized water on the microwave properties of NaCl ceramics, finding that cold sintering under specific pressure conditions could yield outstanding performance ($\varepsilon_r \sim 5.55$, $Q \times f \sim 49{,}600$ GHz). In 2020 and 2021, Liu et al. [39, 40] significantly enhanced the density and $Q \times f$ values of LiF ceramics through CSP and subsequent heat treatment (up to $\varepsilon_r \sim 8.45$, $Q \times f \sim 134{,}050$ GHz). Madhuri et al. [41], Zhou et al. [42], and Jin et al. [43] also achieved comparable results with $NaCa_2Mg_2V_3O_{12}$–xNaCl, LiF–$CaTiO_3$, and BaF_2 ceramics, respectively, underscoring the potential of CSP to improve the performance of fluoride ceramics.

4.1.2 Ferroelectric materials

The CSP technology has shown significant potential in the field of ferroelectric materials, successfully preparing a variety of materials including KH_2PO_4, $NaNO_2$, and BaTiO3 [5]. Studies have found that even under low-temperature and high-pressure conditions, the performance of KH_2PO_4 and $NaNO_2$ can match that of conventionally sintered materials [44]. After annealing at 900°C, $BaTiO_3$'s performance matches that of materials sintered at high temperatures in **Figure 5** [6]. CSP streamlines the fabrication process and improves the properties of the material, thus paving new paths for the application of ferroelectric materials.

Ceramics prepared by Guo et al. [6] and Ma et al. [45] using cold sintering technology exhibit submicron grain sizes, high relative density, high relative permittivity, and low dielectric losses. These ceramics also feature high breakdown strength, energy storage density, and energy storage efficiency. In addition, specific fluxes such as $Ba(OH)_2/TiO_2$ aqueous suspensions and NaOH-KOH molten hydroxides make it possible to densify $BaTiO_3$ at even lower temperatures, achieving single-step densification at 150°C [6].

4.1.3 Piezoelectric ceramic materials

In 2019, Huang et al. [27] demonstrated that cold sintering at 180°C followed by heat treatment at 900°C can produce $Na_{0.5}Bi_{0.5}TiO_3$ piezoelectric ceramics with comparable performance to those made by traditional high-temperature sintering methods. Besides, Ma et al. [46] successfully employed a cold sintering-assisted sintering

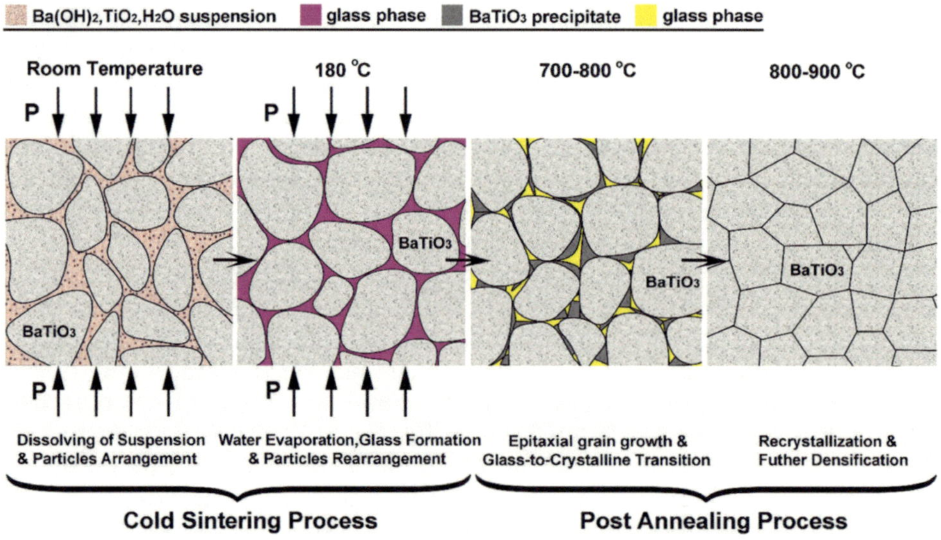

Figure 5.
Schematic illustration of the primary stages during cold sintering and post-annealing processes in the case of $BaTiO_3$ nanocrystalline ceramics preparation [6].

method to fabricate dense $K_{0.5}Na_{0.5}NbO_3$ (KNN) lead-free piezoelectric ceramics with a density exceeding 98%. The cold sintering process, which involved the formation of a potassium-rich second phase at grain boundaries, enhanced the ceramic's properties and reduced the required post-sintering temperature. Furthermore, doping with 10% $LiBiO_3$ resulted in transparent KNN-0.1LB ceramics with 74% transparency [47], further demonstrating the versatility of cold sintering in producing high-performance lead-free piezoelectric and transparent ceramics.

4.2 Semiconductor materials: ZnO ceramics

The CSP has revolutionized the processing of ZnO-based semiconductor materials, effectively addressing the issues of excessive grain growth and performance degradation caused by traditional high-temperature sintering. By utilizing transient liquid phases such as acetic acid solution, CSP achieves high densification (up to 98%) and excellent electrical conductivity at temperatures below 300°C [48], which is significantly lower than the 1400°C required for conventional sintering. Research has shown that additional methods such as adding deionized water [49], optimizing sintering pressure and temperature, and post-sintering heat treatments can further enhance the performance of ZnO materials. For instance, post-sintering treatments can increase electrical conductivity from 0.0005 to 16.4 S cm^{-1} [50], while doping with aluminum oxide can optimize the performance of AZO ceramics, achieving a densification of 99.23% and an extremely low resistivity [51]. These advancements indicate that the cold sintering technology holds great potential for enhancing the electrical properties and controlling the microstructure of ZnO materials, opening new avenues for the development and application of semiconductor materials.

4.3 Solid-state battery materials

Solid-state batteries, utilizing solid electrolytes and electrodes, provide improved safety and energy density but necessitate high-temperature sintering above 1000°C to minimize grain boundary resistance and enhance conductivity. Excessive sintering temperatures may result in the evaporation of volatile elements and the formation of secondary phases, which can impair the ion transport properties of the solid electrolyte [12]. Additionally, prolonged sintering durations can limit the advancement of sodium-ion solid electrolyte materials.

4.3.1 Solid electrolyte

The Seth [1] team was the first to use CSP to prepare the $Li_{1.5}Al_{0.5}Ge_{1.5}(PO_4)_3$ solid electrolyte, achieving 80–88% densification by holding at 120°C for 20 min. However, the amorphous phase resulting from cold sintering lowered the room-temperature ionic conductivity. Post-heat treatment at 650°C led to recrystallization of the amorphous phase, increasing the ionic conductivity to 5.4×10^{-5} S cm^{-1}. By incorporating PVDF-HFP polymer, the room-temperature ionic conductivity was further improved to 1.0×10^{-4} S cm^{-1} in **Figure 6** [1], demonstrating the potential of cold sintering technology in the preparation of solid electrolytes [52].

Wang et al. [53] utilized cold sintering technology to prepare the garnet-type $Li_{6.1}Al_{0.3}La_3Zr_2O_{12}$ solid electrolyte, achieving 87.7% densification of LLZO electrolyte at 350°C through the "dissolution-precipitation" mechanism of CSP. However, the precipitated β-Li_5AlO_4 impurity significantly affected its conductivity to only

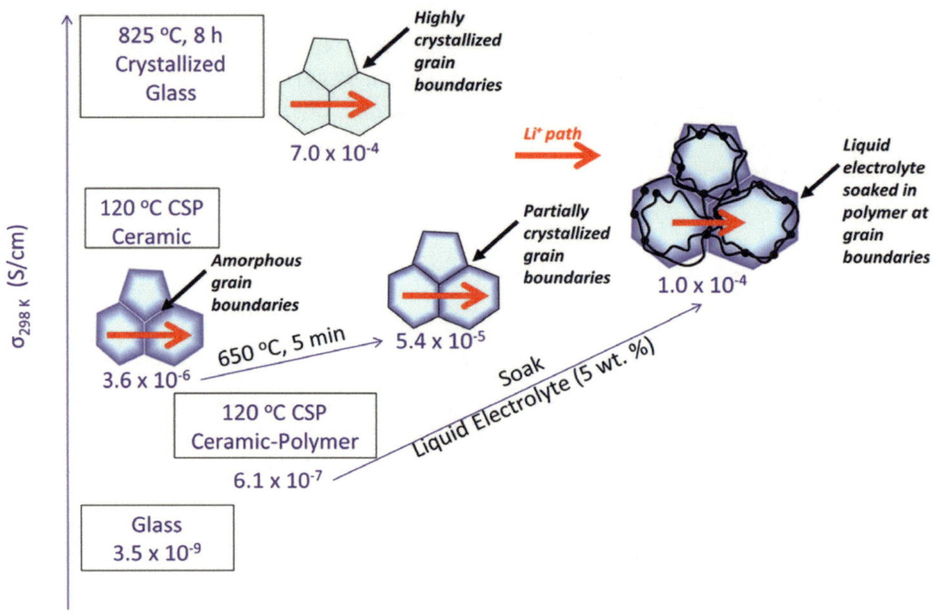

Figure 6.
Summarizes the processing conditions for achieving the conductivity at 25°C and the proposed lithium-ion conduction mechanisms [1].

3.38×10^{-9} S cm^{-1}. Subsequently, Seo et al. [54] prepared an LLZO-SM composite electrolyte with conductivity up to 4×10^{-4} S cm^{-1} by doping Mg and Sr. elements, under a sintering temperature of 120°C and a pressure of 400 MPa.

Leng et al. [11] prepared a dense Mg-doped $Na_{3.256}Mg_{0.128}Zr_{1.872}Si_2PO_{12}$ electrolyte material with 83% densification using 30 wt% deionized water as a transient liquid phase in cold sintering at 140°C. After post-heat treatment at 800–1100°C, the conductivity increased to 1.36 mS cm^{-1}, significantly higher than that of the 1000°C sintered sample without liquid phase (0.1 mS cm^{-1}). Subsequently, Leng [55] further increased the conductivity to 1.5 mS cm^{-1} by adding 1.1 wt% Bi_2O_3 to the Mg-doped NASICON. Silva used different liquid phases such as water, acetic acid, nitric acid, and strong base to prepare NASICON solid electrolytes via cold sintering. Studies showed that using a strong base as the liquid phase achieved the highest densification (95%), but the conductivity of the cold sintered samples was two orders of magnitude lower than that of conventionally sintered samples. This was attributed to the smaller grain size in cold sintered samples, which increased the total grain boundary resistance and thus reduced the conductivity [56]. In addition, Nakaya et al. [57] successfully prepared CsH_2PO_4 proton electrolyte material with 98% densification using cold sintering technology at 200°C, achieving an ionic conductivity of 2.30×10^{-4} S cm^{-1}. Thabet et al. [58] prepared $BaCe_{0.8}Zr_{0.1}Y_{0.1}O_3$ proton electrolyte material with 83% densification at 180°C, which was improved to 94% densification after post-treatment at 700°C, reaching a total conductivity of 4×10^{-2} S cm^{-1}.

4.3.2 Electrode material

Seo et al. [59] successfully addressed the challenge of low capacity in $LiFePO_4$ cathodes by creating a composite material with improved density and performance. By mixing 80% $LiFePO_4$ with 10% conductive carbon and 10 wt% PVDF and using cold sintering at 240°C with LiOH solution as an intermediate liquid phase, they achieved a relative density of 89% for the $LiFePO_4$-PTFE cathode composite material. This material exhibited a volume capacity of 340 mAh/cm^3 at a charge-discharge rate of 0.03–0.1C, surpassing the performance of other high-performance $LiFePO_4$ cathodes [60]. Furthermore, the composite material retained 87% of its initial capacity after 40 cycles at 0.2C, indicating excellent capacity retention. V_2O_5, a semiconductor, acts as a cathode in lithium-ion batteries when combined with Li^+ ions [61]. To enhance its conductivity and stability, it is mixed with conductive polymers. Cold sintering technology produced a V_2O_5/poly(3,4-ethylenedioxythiophene): poly(styrenesulfonate) composite material with a densification of 90.2% and a sheet conductivity of 4.8×10^{-4} S cm^{-1} [61, 62].

Seo et al. [63] fabricated a binder-free $Li_4Ti_5O_{12}$ (LTO) based anode using the cold sintering process. Initially, a composite anode was formed by tape casting with a binder, followed by heat treatment to remove the binder. The binder-free composite was then humidified to provide a transient liquid phase and subsequently cold sintered at 120°C under a uniaxial pressure of 500 MPa, directly deposited on a current collector. The resulting LTO/CNF composite anode had a density of 2.82 g/cm^3 (87% relative density), and its volumetric capacity density was found to be approximately 380 mAh/cm^3.

4.4 Thermoelectric materials

Santos [64] conducted a comparative study on the performance differences of thermoelectric ceramic materials $Ca_3Co_4O_9$ prepared by conventional sintering and cold sintering. Compared to the 58% density of conventionally sintered samples, the cold sintered samples exhibited a significant increase in density to 78.2%, along with an electrical conductivity of 6459 S/m.

4.5 Other materials

4.5.1 Ceramic-polymer composites

The introduction of CSP has enabled the one-step sintering of ceramic-polymer composites, providing a straightforward and effective method for the preparation of traditionally incompatible materials [65, 66]. Notable advancements include Guo's successful creation of Li_2MoO_4-PTEE microwave dielectric material and $Li_{1.5}Al_{0.5}Ge_{1.5}(PO_4)_3$/(PVDF-HFP) composite electrolyte, demonstrating significant improvements in material densification and properties [52].

4.5.2 Metastable ceramic materials

The CSP technology has shown potential in processing metastable ceramic materials, which are prone to decomposition at high temperatures [67]. Research by Bang et al. [68] and Yang et al. [69] has demonstrated the successful preparation of SnO and ZrW_2O_8 materials with high relative densities and good thermoelectric properties, highlighting the technology's ability to control microstructure and enhance performance [70].

4.5.3 Nuclear and spent waste treatment

The CSP technology has been applied to the treatment of radioactive elements in nuclear and spent waste, offering a new environmental treatment method. Hassan and colleagues have immobilized radioactive iodine in apatite and sodalite, achieving high densification and retention rates, and low leaching rates, indicating the technology's effectiveness in immobilizing radioactive materials [71–74].

In addition to these materials, cold sintering technology has also been successfully applied in the preparation of various ceramic materials, including nano-TiO_2 [75, 76], $CaSO_4$ [77], CeO_2 [78], Y_2O_3 [79], ZrO_2 [80, 81], and $InGaZnO_4$ [82] targets.

5. Application fields of cold sintering

Cold sintering technology has demonstrated significant potential and adaptability in the field of ceramics. By utilizing lower temperatures and higher pressures, this technique effectively enhances the densification process of materials, significantly improving the efficiency of material densification and the precision of microstructural design. This method notably reduces processing time, energy consumption, and production costs, making it particularly suitable for applications requiring precise control over the microstructure and properties of materials [29].

To date, nearly 100 different materials have successfully been densified through cold sintering technology, with the majority being dielectric ceramics. Consequently, the primary application areas of cold sintering include piezoelectric media, microwave dielectrics, semiconductors, capacitors, patch antennas, and solid-state batteries. This technology not only provides a new method for material preparation but also exhibits significant potential and broad application prospects in practical applications [83–85].

5.1 Advancing research fields

As an innovative method for preparing ceramics and ceramic matrix composites, cold sintering technology has shown its vast potential and application prospects. The technique enables the rapid production of dense materials by applying external pressure, under limited temperature conditions, and adding a liquid phase. Compared to traditional sintering methods, this process significantly lowers the processing temperature, shortens production time, simplifies chemical treatment steps, and greatly reduces production costs. The unique aspect of cold sintering is its ability to fabricate hybrid composites in a single step, such as embedding polymer fillers within ceramic matrices. The primary densification mechanism is pressure-solution creep, and the addition of a liquid phase facilitates a smoother path for powder consolidation and atomic diffusion [28, 44]. Therefore, cold sintering technology not only enhances material performance but also offers new solutions for the fields of materials science and engineering.

5.2 Future application potential and challenges

In the future, cold sintering technology holds great potential, especially in the automotive and biomedical fields. It is expected to be used for producing ceramic composite brake pads and ceramic-polymer composites for orthopedic implants and drug delivery systems. By reducing energy consumption and production costs, cold sintering technology is anticipated to play a key role in promoting the development of a green economy and providing high-performance materials for the development of new electronic devices [29].

6. Current challenges and future outlook

6.1 The main challenges facing

However, the promotion and application of cold sintering technology face several challenges. First, the acquisition and maintenance costs of high-pressure equipment are high, increasing the initial investment burden. Additionally, to ensure the high quality and performance of the materials produced, strict control over process conditions, including temperature, pressure, and time, is necessary. Moreover, the long-term stability and material compatibility of cold sintering technology are also issues that need to be addressed. Overcoming these challenges will help cold sintering technology to be applied in a wider range of fields and provide innovative solutions to the fields of materials science and engineering [86–88].

To overcome these challenges, future research should focus on the following aspects.

6.1.1 Optimizing process parameters to enhance material performance

This involves precise control over key process parameters such as temperature, pressure, and time to further refine the cold sintering process, thereby improving the performance and quality of materials. A thorough understanding of the relationship between processing parameters, microstructure, and material properties is essential.

6.1.2 Exploring new applications in emerging fields

Cold sintering technology holds great potential in new areas such as the automotive industry and biomedicine. For instance, the development of ceramic composite brake pads and ceramic-polymer composites can reduce energy consumption and production costs while providing high-performance material solutions for these sectors.

6.1.3 Contributing to the green economy

Cold sintering technology is poised to become a pivotal force in promoting the development of a green economy. It can provide high-performance materials for the manufacture of new electronic devices, thereby facilitating the sustainable development of the electronics industry.

6.2 Exploring future research directions and technological advancements

The further development of cold sintering technology requires close collaboration among multiple disciplines, including materials science, chemical engineering, and mechanical engineering, to drive technological innovation and application breakthroughs. Future research directions should focus on the following four core themes [14, 89–91].

6.2.1 Optimizing process parameters and expanding application areas

By fine-tuning process parameters such as temperature, pressure, and time, the efficiency and material performance of cold sintering technology can be further enhanced. At the same time, the potential for applying cold sintering technology in more fields, such as aerospace and automotive manufacturing, should be explored.

6.2.2 Deepening the understanding of the cold sintering mechanism

In-depth research into the physical and chemical mechanisms of the cold sintering process, including the mechanism of liquid-phase-assisted sintering and the impact of pressure on material densification, is necessary to achieve more precise control and prediction of the cold sintering process.

6.2.3 Interdisciplinary research collaboration

Encouraging interdisciplinary collaboration, combining knowledge and technology from materials science, chemical engineering, mechanical engineering, and other fields, to jointly address current challenges faced by cold sintering technology, such as material compatibility and process stability.

6.2.4 Cross-disciplinary integration of experimental research and applications

Combining experimental research with applications in different fields, such as electronic ceramics, energy materials, and bioceramics, to explore the cross-application of cold sintering technology. This can lead to the development of new composite materials with specific functions, meeting the demand for high-performance materials across various industries.

Through the in-depth exploration of these research directions, cold sintering technology is expected to achieve broader applications in the fields of materials science and engineering, providing innovative solutions to related industries.

Author details

Chunchun Li[1,2]*, Guobin Zhu[2], Xiaowei Zhu[2] and Siyu Xiong[2]

1 School of Electronic and Information Engineering, Xi'an Jiaotong University, Xi'an, Shaanxi, China

2 Guangxi University Key Laboratory of Non-ferrous Metal Oxide Electronic Functional Materials and Devices, College of Material Science and Engineering, Guilin University of Technology, Guilin, China

*Address all correspondence to: lichunchun2003@126.com

References

[1] Berbano SS, Jing G, Hanzheng G. Cold sintering process of $Li_{1.5}Al_{0.5}Ge_{1.5}(PO_4)_3$ solid electrolyte. Journal of the American Ceramic Society. 2017;**100**:2123-2135. DOI: 10.1111/jace.14727

[2] Yamasaki N, Yanagisawa K, Nishioka M. A hydrothermal hot-pressing method: Apparatus and application. Journal of Materials Science Letters. 1986;**5**:355-356. DOI: 10.1007/BF01748104

[3] Hanna K, Merja T, Jari J. Dielectric properties of lithium molybdate ceramic fabricated at room temperature. Journal of the American Ceramic Society. 2014;**97**:3378-3379. DOI: 10.1111/jace.13277

[4] Jing G, Hanzheng G, Baker AL. Cold sintering: A paradigm shift for processing and integration of ceramics. Angewandte Chemie International Edition. 2016;**55**:11457-11461. DOI: 10.1002/anie.201605443

[5] Hanzheng G, Amanda B, Jing G. Cold sintering process: A novel technique for low-temperature ceramic processing of ferroelectrics. Journal of the American Ceramic Society. 2016;**99**:3489-3507. DOI: 10.1111/jace.14554

[6] Hanzheng G, Jing G, Amanda B. Hydrothermal-assisted cold sintering process: A new guidance for low-temperature ceramic sintering. ACS Applied Materials & Interfaces. 2016;**8**:20909-20915. DOI: 10.1021/acsami.6b07481

[7] Hanzheng G, Amanda B, Jing G. Protocol for ultralow-temperature ceramic sintering: An integration of nanotechnology and the cold sintering process. ACS Nano. 2016;**10**:10606-10614. DOI: 10.1021/acsnano.6b03800

[8] Florian B, Studart AR. Geologically-inspired strong bulk ceramics made with water at room temperature. Nature Communications. 2017;**8**:14655. DOI: 10.1038/ncomms14655

[9] Induja IJ, Sebastian MT. Microwave dielectric properties of mineral sillimanite obtained by conventional and cold sintering process. Journal of the European Ceramic Society. 2017;**37**:2143-2147. DOI: 10.1016/j.jeurceramsoc.2017.01.007

[10] Induja IJ, Sebastian MT. Microwave dielectric properties of cold sintered Al_2O_3-NaCl composite. Materials Letters. 2018;**211**:55-57. DOI: 10.1016/j.matlet.2017.09.083

[11] Haoyang L, Jiajia H, Jiuyuan N. Cold sintering and ionic conductivities of $Na_{3.256}Mg_{0.128}Zr_{1.872}Si_2PO_{12}$ solid electrolytes. Journal of Power Sources. 2018;**391**:170-179. DOI: 10.1016/j.jpowsour.2018.04.067

[12] Yulong L, Qian S, Dawei W. Development of the cold sintering process and its application in solid-state lithium batteries. Journal of Power Sources. 2018;**393**:193-203. DOI: 10.1016/j.jpowsour.2018.05.015

[13] Changli F, Xiaomeng L, Jing G. Recent progress of dielectric materials prepared via cold sintering process (in Chinese). Journal of Shaanxi Normal University (Natural Science Edition). 2021;**49**:30-42. DOI: 10.15983/j.cnki.jsnu.2021.02.011

[14] Anna G, Sglavo VM. The cold sintering process: A review on processing

features, densification mechanisms and perspectives. Journal of the European Ceramic Society. 2021;**41**:1-17. DOI: 10.1016/j.jeurceramsoc.2021.09.024

[15] Bin HW, Lei L, Meng C. Plastic deformation and effects of water in room-temperature cold sintering of NaCl microwave dielectric ceramics. Journal of the American Ceramic Society. 2018;**101**:4038-4043. DOI: 10.1111/jace.15572

[16] Matthias H, Florian B, Cristina R-A. Cold densification and sintering of nanovaterite by pressing with water. Journal of the European Ceramic Society. 2020;**40**:893-900. DOI: 10.1016/j.jeurceramsoc.2019.10.034

[17] Shenglin K, Hongxia G, Jinbin W. Influence of surface coating on the microstructures and dielectric properties of $BaTiO_3$ ceramic via a cold sintering process. RSC Advances. 2020;**10**:30870-30879. DOI: 10.1039/d0ra03849k

[18] Mingwei W, Haifeng Z, Jinfang W. Review of ceramic materials prepared by cold sintering technology (in Chinese). China Ceramics. 2021;**57**:1-10. DOI: 10.16521/j.cnki.issn.1001-9642.2021.03.001

[19] Peiteado M, De La Rubia MA, Velasco MJ. Bi_2O_3 vaporization from ZnO-based varistors. Journal of the European Ceramic Society. 2005;**25**:1675-1680. DOI: 10.1016/j.jeurceramsoc.2004.06.006

[20] Jingjing F, Youran Z, Mingsheng M. Current status and development trend of cold sintering process (in Chinese). Journal of Inorganic Materials. 2023;**38**:125-136. DOI: 10.15541/jim20220338

[21] Smith BL, Schäffer TE, Mario V. Molecular mechanistic origin of the toughness of natural adhesives, fibres and composites. Nature. 1999;**399**:761-763. DOI: 10.1038/21607

[22] Arnaud N, Sengul MY, Hwi BS. Comparing hydrothermal sintering and cold sintering process: Mechanisms, microstructure, kinetics and chemistry. Journal of the European Ceramic Society. 2020;**40**:1312-1324. DOI: 10.1016/j.jeurceramsoc.2019.11.049

[23] Kousuke N, Kenji S, Naoto T. Liquid-phase sintering of highly Na^+ ion conducting $Na_3Zr_2Si_2PO_{12}$ ceramics using Na_3BO_3 additive. Journal of the American Ceramic Society. 2017;**101**:1255-1265. DOI: 10.1111/jace.15288

[24] Maliha S, Juliewatty MJ, Arifin AZ. Structural, piezoelectric, and dielectric properties of PZT-based ceramics without excess lead oxide. Journal of the Australian Ceramic Society. 2019;**56**:371-377. DOI: 10.1007/s41779-019-00337-3

[25] Matej S, Urban T, Hana U. Investigating the feasibility of preparing metal-ceramic multi-layered composites using only the aerosol-deposition technique. Materials. 2021;**14**:4548. DOI: 10.3390/ma14164548

[26] Runzu J, Juan L. Research progress of cold sintering technology of ceramics (in Chinese). Journal of Guiyang University (Natural Science Edition). 2021;**16**:60-66. DOI: 10.16856/j.cnki.52-1142/n.2021.04.014

[27] HuanQi H, Jian T, Jun L. Preparation of $Na_{0.5}Bi_{0.5}TiO_3$ ceramics by hydrothermal-assisted cold sintering. Ceramics International. 2019;**45**:6753-6758. DOI: 10.1016/j.ceramint.2018.12.166

[28] Tong Y, Jiang C, Li L. Current understanding and applications of the cold sintering process. Frontiers

of Chemical Science and Engineering. 2019;**13**:654-664. DOI: 10.1007/s11705-019-1832-1

[29] Jing G, Richard F, Sarah L. Cold sintering: Progress, challenges, and future opportunities. Annual Review of Materials Research. 2019;**49**:275-295. DOI: 10.1146/annurev-matsci-070218-010041

[30] Jing G, Baker AL, Hanzheng G. Cold sintering process: A new era for ceramic packaging and microwave device development. Journal of the American Ceramic Society. 2017;**100**:669-677. DOI: 10.1111/jace.14603

[31] Arnaud N, Sengul MY, Takao S. Roadmap for densification in cold sintering: Chemical pathways. Open Ceramics. 2020;**2**:100019. DOI: 10.1016/j.oceram.2020.100019

[32] Jing G, Baker AL, Hanzheng G. Cold sintering process: A new era for ceramic packaging and microwave device development. Journal of the American Ceramic Society. 2016;**100**:669-677. DOI: 10.1111/jace.14603

[33] Bin HW, Lei L, Han Y. Cold sintering and microwave dielectric properties of dense HBO_2-II ceramics. Journal of the American Ceramic Society. 2019;**102**:5934-5940. DOI: 10.1111/jace.16450

[34] Bin HW, Lei L, Han Y. Room-temperature-densified H_3BO_3 microwave dielectric ceramics with ultra-low permittivity and ultra-high Qf value. Journal of Materiomics. 2020;**6**:233-239. DOI: 10.1016/j.jmat.2020.02.006

[35] Jianyu H, Jing G, Chuansheng M. Cold sintering of Na_2WO_4 ceramics using a Na_2WO_4-$2H_2O$ chemistry. Journal of the European Ceramic Society. 2021;**41**:6029-6034. DOI: 10.1016/j.jeurceramsoc.2021.05.019

[36] Jianyu H, Jing G, Changli F. The effects of cold sintering parameters on the densification of Na_2WO_4 ceramics using $Na_2WO_4{\cdot}2H_2O$ dry powders. Journal of the American Ceramic Society. 2022;**105**:5058-5068. DOI: 10.1111/jace.18470

[37] Naichao C, Bin X, Xinwei X. Cold-sintered Na_2WO_4-$Ni_{0.2}Cu_{0.2}Zn_{0.6}Fe_2O_4$ ceramics with matched permittivity and permeability for miniaturized antenna. Journal of the American Ceramic Society. 2021;**104**:2125-2133. DOI: 10.1111/jace.17649

[38] Lei L, Bin HW, Shuang Y. Effects of water content during cold sintering process of NaCl ceramics. Journal of Alloys and Compounds. 2019;**787**:352-357. DOI: 10.1016/j.jallcom.2019.02.112

[39] Bing L, Lei L, Xin SK. Enhancement of densification and microwave dielectric properties in LiF ceramics via a cold sintering and post-annealing process. Journal of the European Ceramic Society. 2021;**41**:1726-1729. DOI: 10.1016/j.jeurceramsoc.2020.09.073

[40] Bing L, Ke S, Qiang JY. High quality factor cold sintered LiF ceramics for microstrip patch antenna applications. Journal of the European Ceramic Society. 2021;**41**:4835-4840. DOI: 10.1016/j.jeurceramsoc.2021.03.052

[41] Rakhi M, Santha NI, Subodh G. Insights into the microstructure and dielectric properties of cold sintered $NaCa_2Mg_2V_3O_{12}$ based composites. Frontiers in Materials. 2021;**8**:665033. DOI: 10.3389/fmats.2021.665033

[42] Fei ZM, Xuan ZY, Ke S. A novel strategy to assemble core-shell-structured LiF-$CaTiO_3$ microwave dielectric ceramics via cold sintering and post-annealing treatment. Journal of Materials Science: Materials in

Electronics. 2021;**32**:28447-28453. DOI: 10.1007/s10854-021-07225-2

[43] Hao J, D, Bing L, Xin SK. Boosting densification and microwave dielectric properties in cold sintered BaF_2 ceramics for 5.8 GHz WLAN applications. Journal of Alloys and Compounds. 2021;**886**:161141. DOI: 10.1016/j.jallcom.2021.161141

[44] Salvatore G, Mattia B, Luca Z. A review of cold sintering processes. Advances in Applied Ceramics. 2020;**119**:115-143. DOI: 10.1080/17436753.2019.1706825

[45] Jiaping M, Xiaoming C, Wanqing O. Microstructure, dielectric, and energy storage properties of $BaTiO_3$ ceramics prepared via cold sintering. Ceramics International. 2018;**44**:4436-4441. DOI: 10.1016/j.ceramint.2017.12.044

[46] Jianzhang M, Hanying L, Huajing W. Composition, microstructure and electrical properties of $K_{0.5}Na_{0.5}NbO_3$ ceramics fabricated by cold sintering assisted sintering. Journal of the European Ceramic Society. 2019;**39**:986-993. DOI: 10.1016/j.jeurceramsoc.2018.11.044

[47] Lin C, Wang H, Ma J. Effect of dwell time on cold sintering assisted sintering based highly transparent $0.9K_{0.5}Na_{0.5}NbO_3$-$0.1LiBiO_3$ ceramics. Journal of Alloys and Compounds. 2020;**826**:154249. DOI: 10.1016/j.jallcom.2020.154249

[48] Shuichi F, Jing G, Hanzheng G. Demonstration of the cold sintering process study for the densification and grain growth of ZnO ceramics. Journal of the American Ceramic Society. 2016;**100**:546-553. DOI: 10.1111/jace.14617

[49] Gonzalez-Julian J, Neuhaus K, Bernemann M. Unveiling the mechanisms of cold sintering of ZnO at 250 °C by varying applied stress and characterizing grain boundaries by Kelvin probe force microscopy. Acta Materialia. 2018;**144**:116-128. DOI: 10.1016/j.actamat.2017.10.055

[50] Yang J, Nengneng L, Shaohai W. Remarkably improved electrical conductivity of ZnO ceramics by cold sintering and post-heat-treatment. Ceramics International. 2018;**44**:20570-20574. DOI: 10.1016/j.ceramint.2018.07.192

[51] Hong XY, Jiang XP, Zhu GS. The preparation of high-density aluminum-doped zinc oxide ceramics by cold sintering process. Journal of Alloys and Compounds. 2020;**832**:153241. DOI: 10.1016/j.jallcom.2019.153241

[52] Jing G, Berbano SS, Hanzheng G. Cold sintering process of composites: Bridging the processing temperature gap of ceramic and polymer materials. Advanced Functional Materials. 2016;**26**:7115-7121. DOI: 10.1002/adfm.201602489

[53] Xinchao W, Jinzhu W, Fuzhen L. Influence of cold sintering process on the structure and properties of garnet-type solid electrolytes. Ceramics International. 2020;**46**:18544-18550. DOI: 10.1016/j.ceramint.2020.04.160

[54] Joo-Hwan S, Hiroto N, Yuki T. Broad temperature dependence, high conductivity, and structure-property relations of cold sintering of LLZO-based composite electrolytes. Journal of the European Ceramic Society. 2020;**40**:6241-6248. DOI: 10.1016/j.jeurceramsoc.2020.06.050

[55] Haoyang L, Jiuyuan N, Jian L. Combining cold sintering and Bi_2O_3-activated liquid-phase sintering to fabricate high-conductivity Mg-doped NASICON at reduced temperatures.

Journal of Materiomics. 2019;**5**:237-246. DOI: 10.1016/j.jmat.2019.02.005

[56] Da Silva P, Gustavo J, Martin B, Laptev AM. Sintering of a sodium-based NASICON electrolyte: A comparative study between cold, field assisted and conventional sintering methods. Journal of the European Ceramic Society. 2019;**39**:2697-2702. DOI: 10.1016/j.jeurceramsoc.2019.03.023

[57] Hiroto N, Masato I, Herisson DBT. Applying cold sintering process to a proton electrolyte material: CsH_2PO_4. Journal of the European Ceramic Society. 2019;**39**:396-401. DOI: 10.1016/j.jeurceramsoc.2018.09.001

[58] Thabet K, Quarez E, Joubert O. Application of the cold sintering process to the electrolyte material $BaCe_{0.8}Zr_{0.1}Y_{0.1}O_{3-\delta}$. Journal of the European Ceramic Society. 2020;**40**:3445-3452. DOI: 10.1016/j.jeurceramsoc.2020.03.043

[59] Joo-Hwan S, Jing G, Hanzheng G. Cold sintering of a Li-ion cathode: $LiFePO_4$-composite with high volumetric capacity. Ceramics International. 2017;**43**:15370-15374. DOI: 10.1016/j.ceramint.2017.08.077

[60] Joo-Hwan S, Kris V, Jing G. Cold sintering approach to fabrication of high rate performance binderless $LiFePO_4$ cathode with high volumetric capacity. Scripta Materialia. 2018;**146**:267-271. DOI: 10.1016/j.scriptamat.2017.12.005

[61] Baba H, DS, Jing G, Joo-Hwan S. Microstructures and electrical properties of V_2O_5 and carbon-nanofiber composites fabricated by cold sintering process. Japanese Journal of Applied Physics. 2018;**57**:025702. DOI: 10.7567/jjap.57.025702

[62] Jing G, Hanzheng G, Baba HDS. Semiconducting properties of cold sintered V_2O_5 ceramics and co-sintered V_2O_5-PEDOT:PSS composites. Journal of the European Ceramic Society. 2017;**37**:1529-1534. DOI: 10.1016/j.jeurceramsoc.2016.11.021

[63] Joo-Hwan S, Kris V, Ramakrishnan R. Cold sintering process for fabrication of a high volumetric capacity $Li_4Ti_5O_{12}$ anode. Materials Science and Engineering: B. 2019;**250**:114435. DOI: 10.1016/j.mseb.2019.114435

[64] Moraes DSA, Daniel T, Virginia GM. Cold sintering and thermoelectric properties of $Ca_3Co_4O_9$ ceramics. Ceramics International. 2020;**46**:14064-14070. DOI: 10.1016/j.ceramint.2020.02.206

[65] Xuetong Z, Jing G, Ke W. Introducing a ZnO-PTFE (polymer) nanocomposite varistor via the cold sintering process. Advanced Engineering Materials. 2018;**20**:1700902. DOI: 10.1002/adem.201700902

[66] Arnaud N, Kosuke T, Ke W. Sintering mechanisms and dielectric properties of cold sintered (1-x) SiO_2-x PTFE composites. Journal of the European Ceramic Society. 2019;**39**:4743-4751. DOI: 10.1016/j.jeurceramsoc.2019.07.048

[67] Campo CM, Rodríguez JE, Ramírez AE. Thermal behaviour of romarchite phase SnO in different atmospheres: A hypothesis about the phase transformation. Heliyon. 2016;**2**:e00112. DOI: 10.1016/j.heliyon.2016.e00112

[68] Hwi BS, Thomas HDB, Randall Clive A. Densification of thermodynamically unstable tin monoxide using cold sintering process. Journal of the European Ceramic Society. 2019;**39**:1230-1236. DOI: 10.1016/j.jeurceramsoc.2018.11.026

[69] Cheng Y, Jinping L, Dongliang Y. ZrW_2O_8 with negative thermal expansion fabricated at ultralow temperature: An energy-efficient strategy for metastable material fabrication. ACS Sustainable Chemistry & Engineering. 2019;7:14747-14755. DOI: 10.1021/acssuschemeng.9b02682

[70] Cheng Y, Jinping L, Haofan S. Effects of the liquid phase content on the microstructure and properties of the ZrW_2O_8 ceramics with negative thermal expansion fabricated by the cold sintering process. Journal of the European Ceramic Society. 2020;**40**:6079-6086. DOI: 10.1016/j.jeurceramsoc.2020.05.045

[71] Muhmood ul H, Jin RH. Cold sintering and durability of iodate-substituted calcium hydroxyapatite (IO-HAp) for the immobilization of radioiodine. Journal of Nuclear Materials. 2019;**514**:84-89. DOI: 10.1016/j.jnucmat.2018.11.024

[72] Muhmood ul H, Suriya V, Jin RH. Non-volatile immobilization of iodine by the cold-sintering of iodosodalite. Journal of Hazardous Materials. 2020;**386**:121646. DOI: 10.1016/j.jhazmat.2019.121646

[73] Suriya V, Muhmood ul H, Jin RH. Adsorption and immobilization of radioactive ionic-corrosion-products using magnetic hydroxyapatite and cold-sintering for nuclear waste management applications. Journal of Nuclear Materials. 2019;**514**:40-49. DOI: 10.1016/j.jnucmat.2018.11.026

[74] Muhmood UH, Sajid I, Jong-Il Y. Immobilization of radioactive corrosion products by cold sintering of pure hydroxyapatite. Journal of Hazardous Materials. 2019;**374**:228-237. DOI: 10.1016/j.jhazmat.2019.04.038

[75] Valentina M, Francesca S, Riccardo B. Nano-to-macroporous TiO_2 (anatase) by cold sintering process. Journal of the European Ceramic Society. 2019;**39**:2453-2462. DOI: 10.1016/j.jeurceramsoc.2019.02.047

[76] Falk GS, Sergio YGG, Dachamir H. Low-energy microwave synthesis and cold sintering of nanograined TiO_2-Nb_2O_5. Materials Letters. 2020;**278**:128418. DOI: 10.1016/j.matlet.2020.128418

[77] Lei L, Han Y, Bin HW. Dense gypsum ceramics prepared by room-temperature cold sintering with greatly improved mechanical properties. Journal of the European Ceramic Society. 2020;**40**:4689-4693. DOI: 10.1016/j.jeurceramsoc.2020.06.003

[78] Thitirat C, Usa S, Taras K. Enhancing the densification of ceria ceramic at low temperature via the cold sintering assisted two-step sintering process. Ceramics International. 2018;**44**:S54-S57. DOI: 10.1016/j.ceramint.2018.08.253

[79] Doyun L, Jung-A L, Young-Woo H. The effect of yttrium nitrate addition on the densification behaviour of Y_2O_3 ceramics during the cold sintering process. Journal of the European Ceramic Society. 2020;**40**:3208-3214. DOI: 10.1016/j.jeurceramsoc.2020.02.025

[80] Hanzheng G, Bayer Thorsten JM, Jing G. Current progress and perspectives of applying cold sintering process to ZrO_2-based ceramics. Scripta Materialia. 2017;**136**:141-148. DOI: 10.1016/j.scriptamat.2017.02.004

[81] Hanzheng G, Bayer Thorsten JM, Jing G. Cold sintering process for 8 mol%Y_2O_3-stabilized ZrO_2 ceramics. Journal of the European Ceramic Society. 2017;**37**:2303-2308. DOI: 10.1016/j.jeurceramsoc.2017.01.011

[82] Jiang-An L, Chen-Hui L, Jing-Jing S. Preparation of high-density $InGaZnO_4$ target by the assistance of cold sintering. Materials Science in Semiconductor Processing. 2018;**84**:17-23. DOI: 10.1016/j.mssp.2018.04.030

[83] Jon-Paul M, Xiaoyu K, Floyd RD. Cold sintering: Current status and prospects. Journal of Materials Research. 2017;**32**:3205-3218. DOI: 10.1557/jmr.2017.262

[84] Jing G, Xuetong Z, Thomas HDB. Recent progress in applications of the cold sintering process for ceramic-polymer composites. Advanced Functional Materials. 2018;**28**:1801724. DOI: 10.1002/adfm.201801724

[85] Yuchen L, Xuetong Z, Shenglin K. Recent progress of cold sintering process on functional ceramic materials. Journal of Materials Science: Materials in Electronics. 2023;**34**:2105. DOI: 10.1007/s10854-023-11460-0

[86] Cekdar V, Levent K. Cold sintering of ceramics and glasses: A review. Current Opinion in Solid State and Materials Science. 2020;**24**:100807. DOI: 10.1016/j.cossms.2020.100807

[87] Dawei W, Linhao L, Juan J. Cold sintering of microwave dielectric ceramics and devices. Journal of Materials Research. 2021;**36**:333-349. DOI: 10.1557/s43578-020-00029-w

[88] Dong Z, Yiqun F. State-of-the-art developments in fabricating ceramic membranes with low energy consumption. Ceramics International. 2021;**47**:14966-14987. DOI: 10.1016/j.ceramint.2021.02.195

[89] Honghui W, Fangping Z, Huimin Q. Polymer-/ceramic-based dielectric composites for energy storage and conversion. Energy & Environmental Materials. 2022;**5**:486-514. DOI: 10.1002/eem2.12237

[90] Ningjie G, Guisheng Z, Huarui X. Preparation of CaF_2 transparent ceramics by cold sintering. Ceramics International. 2022;**48**:34184-34189. DOI: 10.1016/j.ceramint.2022.08.339

[91] Radhakrishnan J, Subramani S, Ocaña JL. Cold sintering behaviors of barium titanates: Recent progress and impact on microstructure, densification and dielectric-ferroelectric response. Coordination Chemistry Reviews. 2024;**502**:215621. DOI: 10.1016/j.ccr.2023.215621

Chapter 13

Thermoelectric Ceramics: Multidimensional Renewable Materials

Vaishali Poddar

Abstract

The scientific world has urged us to improve the sustainability of our electricity base because now we need electricity for just about everything. The use of conventional electrical energy resources to accommodate the growing electricity demand is quite alarming. We are not only creating imbalance in nature but also increasing the levels of greenhouse gases and polluting emissions (e.g., power generation: carbon footprint; refrigeration: mercury, CFC, freon emission). Renewable energy sources (based on solar energy, wind energy, hydro energy, etc.) stand in contrast to conventional sources with respect to their diversity, abundance, and applicability on and around the planet, but above all, no greenhouse gas emissions throughout their life cycle. Fascinatingly, thermoelectric (TE) ceramic is a class of semiconducting material with huge scope in the future as a renewable energy source. The peculiarity about thermoelectric ceramic material is that the same material can be used for producing power (Seebeck effect) and for removing or adding heat (Peltier effect). This chapter highlights the peculiar characteristics of thermoelectric ceramics and their prominent applications in the current era of energy crisis.

Keywords: thermoelectric, semiconducting ceramics, Seebeck effect, Peltier effect, photovoltaic

1. Introduction

Today we need electricity for just about everything. To accommodate the growing demand for electricity, the increase in the use of conventional and exhaustible electrical energy resources is quite alarming. This scenario is creating imbalance in nature by increasing the levels of greenhouse gases and polluting emissions (e.g.,power generation: carbon footprint; refrigeration: mercury, chlorofluorocarbon (CFC), freon emission) in our environment. And for this reason, the scientific world has urged us now to improve the sustainability of our electricity base.

Renewable energy sources (based on solar energy, wind energy, hydro energy, etc.) stand in contrast to the conventional sources with respect to their diversity, abundance, and applicability on and around the planet, but above all, no greenhouse gas emissions throughout their life cycle [1]. Fascinatingly, thermoelectric (TE) is a unique class of materials with huge scope in the future as a renewable energy source. The peculiarity about thermoelectric material is that the material can be used for producing power (Seebeck effect) as well as for removing or adding heat (Peltier effect) [2]. Of course, depending on the output requirement, i.e., power or heat, a slight modification in the circuit needs to be made. Therefore, thermoelectric may be used for both heating and cooling, thereby making it highly suitable for precise temperature control applications, e.g., thermocouple. Thermoelectricity is shown by various material families such as tellurides, chalcogenides, clathrates, skutterudites, half-Heusler alloys, silicides, oxides, and polymer composites [3–4]. These materials have a large range of possible options with respect to composition and structure. Most of these materials are ceramic in nature because of large effective masses and under certain limitations based on two criteria: (1) bond characteristics; and (2) temperature. The bonding found in thermoelectric ceramics is a combination of ionic and covalent types with the dominance of ionic characteristics [5]. The operating temperature should be such that the charge carrier concentration is low and the corresponding activation energy is in the range of less than 0.08 eV. Thermoelectric ceramic material has been recently emerging, and there are very few reports on their performance. Thermoelectricity is also exhibited in all metals, but to a very low extent. Metals, although they have high carrier concentration and easy processing, promoting, thermoelectricity, but they have disadvantages such as sensitivity to impurities and oxidation, which degrade their thermoelectric behavior.

1.1 Thermodynamics of thermoelectric ceramics

According to the second law of thermodynamics, the energy input to a system is never completely converted to usable energy output. The difference between the two constitutes the unutilized heat, and it is, though in variable quantities, available in almost every system. Thermoelectric materials, used in power generation, scavenge the unutilized heat from the source and convert it to a usable electrical form of energy [6]. The unutilized heat may be of high temperature, medium temperature, or low temperature grade. That means heat is dumped, for example, from industrial processes like hot metallurgical processing (high temperature: greater than 600°C), from exhaust gases released by the combustion units (medium temperature: 250–600°C) or from hot surfaces of the indirectly involved processing units (low temperature: less than 250°C) into the environment, causing thermal pollution [7]. Surprisingly, around 66% of the unutilized heat in industries is available in the low-temperature range of around 66–230°C. Thermoelectric can work as a renewable energy source in low-temperature regimes or as a supplement to conventional energy sources in high-temperature regimes. In the medium temperature range, thermoelectric can work in either way depending on the system or the purpose for which it is used [8]. Therefore, thermoelectric comes under the category of sustainable energy-efficient technology with the additional advantages of simple configuration, no moving parts, no vibration, no noise, no pollution, and high stability [9]. However, the challenge in using a thermoelectric system is that for a particular temperature, it should be inclusive of all forms of heat source, be able to run continuously, have a long life, be low cost, and be light in weight [9, 10].

1.2 Ceramic as thermoelectric generator

A ceramic is suitable as a thermoelectric generator (TEG) when the supply of thermal energy produces many numbers of charge carriers due to the Edison effect [11]. Also, these generated charge carriers should move from the hot end of material to the cold end (that is, across the thermal gradient) due to the Seebeck effect with sufficient mobility. Under the effect of a thermal gradient, dc voltage develops across the length of the thermoelectric ceramic and a net current flows through the circuit by means of the charge carriers. At high temperatures, the charge carriers are moving, vibrating, and rotating with more energy compared to those at low temperatures. However, the transition of charge carriers from the hot end to the cold end will undergo scattering a gazillion times due to the atoms present in their path. This scattering is remarkable in ceramic due to its high effective mass and a cloud of polarization with a certain spatial diameter. The ceramic structure does not have enough space left for the movement of charge carriers without any interaction. During transition, charge carriers interact with atoms, which results in a change in direction of the charge carrier and/or generation of a photon or phonon. Thermoelectric ceramic transfers the bulk of the heat from the hot end to the cold end in the form of phonons rather than photons. However, most of the phonons are consumed in the generation and transfer of charge carriers from the hot end to the cold end of the thermoelectric. Thus, in thermoelectric, unlike metals, the number of phonons reaching the cold end is as low as possible.

1.3 Thermoelectric cooler

For a thermoelectric cooler (TEC), sometimes called a Peltier cooler, the material should respond to a low-voltage DC power input. The response is in the form of the transfer of heat from one side of the TEC to the other and is directly dependent upon the DC voltage. Depending on the direction of current, the side will get heated or cooled. With a change in the polarity of current, the hot side will now get cooled down and the cold side will get heated up. In other words, a thermoelectric cooler works as a semiconductor-based small heat pump. Consequently, it can be used for heating or cooling, although in practice the main application is cooling. However, there is a limit to the flow of current through a thermoelectric cooler because this current is prone to the generation of heat, which adds to the overall heat dissipation [2].

A change in DC voltage is reflected as a change in temperature, which can be amplified with the use of a proper TE material pair (p-type TE to n-type TE combination) in the circuit. One such ceramic material pair for this cooling application can be $Ca_3Co_4O_9$ (p-type) and $Ca_{0.95}Sm_{0.05}MnO_3$ (n-type) [12]. The controlled DC input to the thermoelectric cooler allows keeping control over the electron–hole pair combination rate in TEC. This merit is highly suitable for use of TEC in precise temperature control applications like aerospace (extremely demanding conditions), imaging technology (constant viscosity of ink), and temperature measuring units. Thermoelectric coolers also provide refrigeration and temperature control in electronic packages and medical instruments.

1.4 The figure-of-merit

Thermoelectric ceramic materials are a member of the advanced materials family, wherein tailoring of constituent elements, following a systematic synthesis route, and engineering of material structure is of crucial importance. The performance of TE is characterized by Figure-of-Merit expressed as Z with unit K^{-1}. To make it dimensionless, Figure-of-Merit, Z is normally expressed as ZT, where T is the average temperature of hot side and cold side of the TE. The maximum efficiency of the energy conversion process in TEG and TEC at a given point in the TE material is determined by its ZT value and is given in Eq. (1).

$$ZT = \frac{S^2 \sigma T}{k} \quad (1)$$

where S, σ, k, and T are the Seebeck coefficients, (μV K^{-1}), electrical conductivity (Ω^{-1} m^{-1}), thermal conductivity (W m^{-1} K^{-1}), and absolute working temperature (K), respectively [13]. From Eq. (1), it is observed that the TE material should behave like a glass to have low thermal conductivity (k), a metal to have high electrical conductivity (σ), and a semiconductor to have a high Seebeck coefficient (S) all at the same time [3]. In other words, a contradictory combination of material behavior is required in TE materials.

As shown in **Figure 1**, each of the performance parameters for the thermoelectric depends on the temperature of operation, T. To increase the ZT value, the Seebeck coefficient, S, and electrical conductivity, σ, should be high and the thermal conductivity, k, should be low. Though interrelated through classical physics, due to the contradictory requirements by each of these factors, improvement in one property leads to a trade-off in which the other properties get lowered. For example, if the electrical conductivity is increased by a high carrier concentration in the thermoelectric material, the Seebeck coefficient and thermal conductivity are decreased. It is to be noted that k is the total thermal conductivity inclusive of the lattice and electronic thermal conductivity components of the TE material [9]. It can be concluded that with the increase in temperature, both the numerator and denominator of Eq. (1) should show an increasing trend. But there is a limit to it. Thermoelectric ceramic is also subjected to a bipolar effect at high temperatures. This effect is predominant in doped TEs. The number of minority charge carriers increases and competes with the majority charge carriers. Now these carriers have opposite signs, and therefore they increase the Seebeck coefficient and electrical conductivity [14]. With minority charge carrier contribution, the Seebeck coefficient increases to a maximum and then decreases. This inter-relation of temperature, Seebeck coefficient, and band gap has been rightly put forth by Goldsmid-Sharp as band gap [15], as given by Eq. (2).

$$E_g = 2e|S|_{max} T_{max} \quad (2)$$

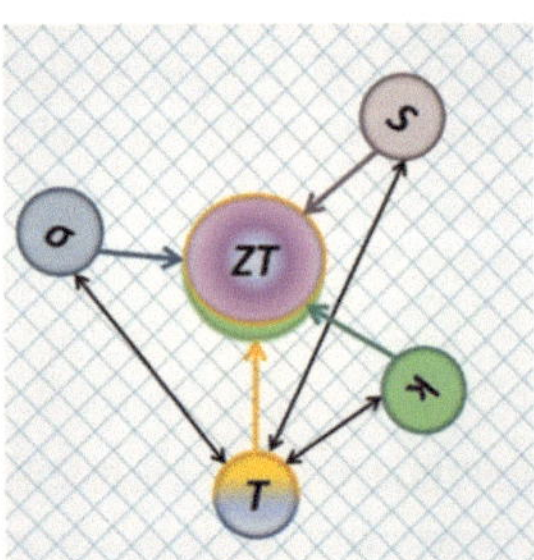

Figure 1.
Dependency schematic of Figure-of-Merit, ZT, Seebeck coefficient, S, electrical conductivity, σ, thermal conductivity, k, and absolute temperature of the performance evaluation, T of the TE material. The parameter indicated by the head of the arrow is strongly influenced by the parameter present at the other end.

The ZT values of 1 are considered good, but a value of 3–4 makes the TE material compete with other mechanical devices available for the same purpose. Synthesis of higher ZT (ZT > 1) TE material is a challenge for their proper use in refrigeration and electrical power generation applications. For materials having similar thermal conductivities, the term power factor ($PF = S^2\sigma$) is used to define the performance of TE [9]. To increase the Seebeck coefficient and electrical conductivity, valence band engineering (convergence) and conduction band engineering for TE materials are focused, respectively.

2. Thermoelectric effect

Thermoelectric effect is the phenomenon by which a temperature difference across a material sets up electric potential in it. It is exhibited by all metals, but TE materials are identified as a separate class of materials. This is because they are good at converting the temperature difference into voltage [9].

Thermoelectric effects result from the interaction of charge flow and heat flow across the length of a TE. Thermoelectric phenomena are known more specifically as the Seebeck effect (converting temperature to current), Peltier effect (converting current to temperature), and Thomson effect (conductor heating/cooling). All materials exhibit thermoelectric effects, but the name 'thermoelectric materials' is used to describe the materials that are good at converting heat to electricity and vice versa. It's interesting to note that these are all old effects, but they are making a significant contribution to the recent green technologies.

2.1 Seebeck effect

A discussion of thermoelectric materials and devices starts with one of the most fundamental phenomena, the Seebeck effect. In 1821, Seebeck observed that if two dissimilar metals were joined together and the junctions were held at different temperatures (T and T + ΔT), then a voltage difference (ΔV) proportional to the temperature difference (ΔT) developed [8]. The ratio of the voltage developed to the temperature difference (ΔV/ΔT) is related to the intrinsic property of the material termed the Seebeck coefficient (S or α) or thermopower as described in Eq. (3). This developed voltage is used to drive a current through the device or a load resistance [6, 9].

$$S = \frac{\Delta V}{\Delta T} \tag{3}$$

This effect can be understood by establishing a temperature gradient across a material. Energetic electrons flow from higher potential to a lower potential until an electric field is established to impede further flow of electrons. The Seebeck coefficient has the units of volts per degree Kelvin and is usually expressed in $\mu V\ K^{-1}$. The type of charge carrier determines the sign of the Seebeck: $S < 0$ for electrons and $S > 0$ for holes in a semiconductor. S is very low for metals (only a few microvolts per degree Kelvin, e.g., Al: $-0.2\ \mu V\ K^{-1}$ at 100°C) and is much higher for semiconductors (typically a few hundred microvolts per degree Kelvin, e.g., Ge: $-210\ \mu V\ K^{-1}$ at 700°C) [16].

2.2 Peltier effect

A few years after the discovery of the Seebeck effect, Peltier observed that if an electrical current is passed through the junction of two dissimilar metals, heat is either absorbed or rejected at the junction, depending on the direction of the current. This phenomenon is known as the Peltier effect. This effect is attributed to the difference in Fermi energies of two materials. The Seebeck effect and the Peltier effect are related to each other through thermodynamics as per the following Kelvin Law mentioned as Eq. (4).

$$\Pi = ST \tag{4}$$

Π is called the Peltier coefficient. The Peltier coefficient is simply the Seebeck coefficient times the absolute temperature [2].

The rate at which Peltier heat is liberated or rejected at the junction ($\dot{Q}_p$) is given by the following relation represented in Eq. (5).

$$\dot{Q}_p = SIT \tag{5}$$

Here I is the current through the junction, and T is the temperature in degree Kelvin. Therefore, now using a low-temperature TE device, water can be frozen with the current in one direction and boiled with the current in the opposite direction [16].

2.3 Thomson effect

The Thomson effect was discovered in 1854 by the British physicist William Thomson (Lord Kelvin). In the Thomson effect, heat is absorbed or produced when current flows in a material with a temperature gradient. In many materials, the Seebeck coefficient is not constant with temperature, and so a spatial gradient in temperature can result in a gradient in the Seebeck coefficient (dS/dT). If a current is driven through this gradient, then a continuous version of the Peltier effect occurs. According to Thomson, the heat (H) is proportional to both the electric current (I) and the temperature gradient (ΔT), and the proportionality constant, known as the Thomson coefficient, K, is related by thermodynamics to the Seebeck coefficient [9]. This can be presented as Eq. (6) and (7).

$$H = K(I \times \Delta T) \tag{6}$$

$$K = T\frac{dS}{dT} \tag{7}$$

Though Seebeck and Peltier effects are reversible, it is rather more difficult to demonstrate experimentally the Peltier effect than the Seebeck effect because of the unintentional involvement of Joule heating. Thus, in practice, thermoelectric energy conversion is mostly expressed in terms of the Seebeck coefficient rather than the Peltier coefficient. Eq. (4) suggests that the Seebeck and Peltier coefficients are interdependent and that, if required, the Peltier coefficient can be evaluated from it.

3. Electrical conductivity in thermoelectric ceramics

When a temperature gradient is applied across the TE material, the charge carrier mobility is affected, as shown in **Figure 2**. The kinetic energy of the charge carriers at the

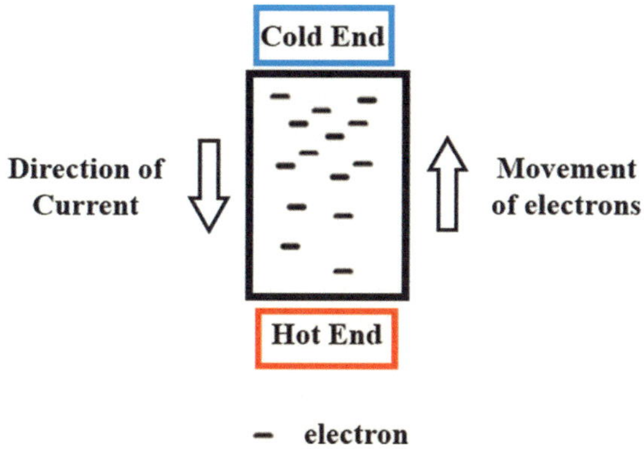

Figure 2.
Electron gradient across an n-type TE material because of temperature gradient.

hot end is greater than the charge carriers at the cold end. So, the number of charge carriers diffusing from the hot end (high energy) to the cold end (low energy) is higher compared to the charge carriers diffusing from the cold end to the hot end. This creates a charge gradient across the TE material corresponding to the temperature gradient.

The most common methods for determining the electrical conductivity of TE material are the Hall Effect and the four-point probe method. The Hall Effect is measured by applying magnetic fields perpendicular to the surface of the sample and electrical currents along the edges of the sample while measuring the voltage across the sample. The applied magnetic field and current create a force that pushes mobile charges to one side of the material, creating an electric field [17]. Four-probe apparatus is one of the standard and most widely used apparatus for the measurement of resistivity of semiconductors. In this method, it is considered that the conductivity of material is uniform in the area of measurement [18].

If there is a minority carrier injection into the semiconductor by the current-carrying electrodes, most of the carriers recombine near electrodes so that their effect on conductivity is negligible. The surface on which the probes rest is flat with no surface leakage. The four probes used for resistivity measurement contact the material surface at points that lie in a straight line. The diameter of the contact between metallic probes and the semiconductor should be small compared to the distance between the probes. The boundary between the current-carrying electrodes and the bulk material is hemispherical and small in diameter. It is assumed that the size of the metal tip is infinitesimal and the sample thickness is greater than the distance between the probes; then the resistivity is determined according to Eq. (8).

$$\rho_0 = \frac{V}{I} \times 2\pi s \tag{8}$$

where V is the potential difference between the inner probes in volts, I is current through the outer pair of probes in amperes and s is spacing between the probes in meters.

4. Thermal conductivity in thermoelectric ceramics

Thermal conductivity is the sum of electronic (k_{el}) and lattice (k_{latt}) thermal conductivity. In a material, the transport of heat across its length takes place through the movement of charge carriers (electrons and/or holes) and thermal carriers (phonons). In metals and degenerate semiconductors, the transport of heat is majorly through charge carriers, and it is well explained by Wiedeman-Franz-Lorentz Law, $k_{el} = L\sigma T$, which states that the electronic thermal conductivity is directly proportional to electrical conductivity, and the constant of proportionality is the Lorentz number, 2.45×10^{-8} W·Ω·K^{-2}. In this law, it is considered that heat and electrical transport take place through charge carriers only [19]. However, for non-degenerate semiconductors like thermoelectric ceramics, a decrease in L takes place with an increase in temperature. This decrease in L is correlated with an increase in the value of the Seebeck coefficient (S). Using these L values, k_{el} was determined with respect to temperature for all types of thermoelectric ceramics. In practice, the electronic contribution to the thermal conductivity is negligible, i.e., k_{el} is very small as compared to k_{latt} ($k_{el} < < k_{latt}$), and hence k_{el} is neglected [19].

The concept of lattice thermal conductivity, k_{latt}, was developed earlier by Born and Debye in 1913. Lattice thermal conductivity is contributed by the phonons with different wavelengths [10]. It is approximated by the classical kinetic theory, which treats phonons as gas particles, and following all the assumptions of the theory, k_{latt} is given as Eq. (9).

$$k_{latt} = \tfrac{1}{3} C_v l v_s \quad (9)$$

where C_v is the specific heat at constant volume, l is the mean free phonon path, and ν_s is the average velocity of sound. At very low temperatures (below 40 K), when lattice vibrations are negligible, Debye law is followed, which states that C_v is proportional to T^3. Generated phonons are fewer in number and have long wavelengths. Above this temperature, the number of generated phonons increases, and their wavelength reduces. Also, they start interacting with each other (phonon–phonon scattering) as well as with the phase structure [20].

As the amount of heat absorbed by the TE material varies across the length extending from hot side to cold side, the phonons generated can exhibit a wide range of wavelengths. The defects or the particle size should match the mean free path of these phonons so as to scatter them. This requires proper engineering of the TE ceramic structure to ensure maximum heat-carrying phonons are utilized, which will reduce thermal conductivity [21]. This is one of the main design characteristics to be considered while developing the thermoelectric ceramic material of high ZT.

Ioffe et al. studied the physical problems of thermoelectricity and presented their study on the mobility of charge carriers and factors affecting the same. The mobility of charge carriers is limited by their scattering because of thermal vibrations and lattice defects. If the Debye temperature is less, then the amplitude is large, which will reduce the mobility [10].

The dependency of Z on temperature and mobility can be expressed roughly by the relation given in Eq. (10).

$$Z \sim \frac{\mu}{k_{latt}} T^{3/2} \quad (10)$$

where Z depends on the type of impurity introduced in the system. The metallic impurity atoms act as donors irrespective of their valency, and non-metals act as acceptors [10].

If there is a temperature gradient in a solid body, the number of phonons moving from the hot side to the cold side is greater than the number moving in the opposite direction. As a result of collisions, the phonons can drag electrons along in their own direction of motion, and at the cold side of the specimen a negative charge will appear (similarly a positive charge will appear at the hot side) until the potential difference produced balances the effect of charge flow. This is the dominant phenomenon happening in the high-temperature thermoelectric ceramics, so that their electrical conductivity increases at higher temperatures.

5. Thermoelectric material systems

In the early 1960s, a requirement for autonomous sources of electrical power arose from the exploration of space, the exploitation of the Earth's resources, and advances in medical physics. But TEs have long been too inefficient to be cost effective in most of the applications. A. F. Ioffe initiated wide research in thermoelectric semiconductors. Along with coworkers, he demonstrated that the thermoelectric properties in bulk materials can be enhanced by tuning or doping techniques to vary the lattice thermal conductivity without affecting the electrical conductivity. The first commercial TE generator used an oil burning lamp as the heat source and ZnSb ceramic and constantan as TE materials. This TE generator developed in the twentieth century powered a radio. Similar work was carried out, and a focus was made on finding small-band gap semiconductors made from heavy elements [10, 22].

However, a revival of interest in thermoelectric began in the mid-1990s when theoretical predictions suggested that thermoelectric efficiency could be greatly enhanced through nanostructure engineering, which led to experimental efforts to demonstrate the proof-of-principle and high-efficiency materials [9]. In 1995, Slack et al. developed a TE material based on the "phonon glass and electron crystal (PGEC)" concept. The voids in the TE crystal structure are filled with heavy element atoms, which act as rattlers, increasing the number of phonon scattering centers, hence reducing the lattice thermal conductivity significantly [20]. The constituent atoms had small differences in the electronegativity and high charge mobility. The main characteristic of these TE materials is that they can carry the electric current as a crystal but behave as amorphous materials with respect to the lattice thermal conductivity. Such development was verified on TE materials like skutterudites, clathrates, ceramics, half-Heusler alloys, and β-Zn_4Sb_3 phases.

From **Figure 3**, it can be seen that in the low-temperature range, TE is used as solar TE—a renewable energy source—or it can be used along with photovoltaic (PV) to improve the PV efficiency. Also, Peltier coolers are included in this category to carry out refrigeration. In high-temperature applications, TE is employed for waste heat recovery. There are many kinds of TE materials, from semiconductors to ceramics, from oxides to organics, from bulk to superlattice, and from nanoparticles to nanowires. Each TE material is suitable for a particular temperature range. Higher the *ZT*, more is the suitability for that particular temperature. **Figure 3** illustrates the mostly preferred TE material to be used at a particular temperature. For good efficiency, materials with high electrical conductivity, low thermal conductivity, and a high Seebeck coefficient are needed. The widespread commercial integration of these materials has been impeded for several reasons, the most crucial being the need for materials with high *ZT* that also have a long life, are low cost, and are light in weight. Some of the TE materials under study are tellurides, chalcogenides, clathrates, skutterudites, half-Heusler alloys, silicides, oxides, polymer composites, etc. Apart from the search for high *ZT* values, research on new TE materials now focuses on eco-

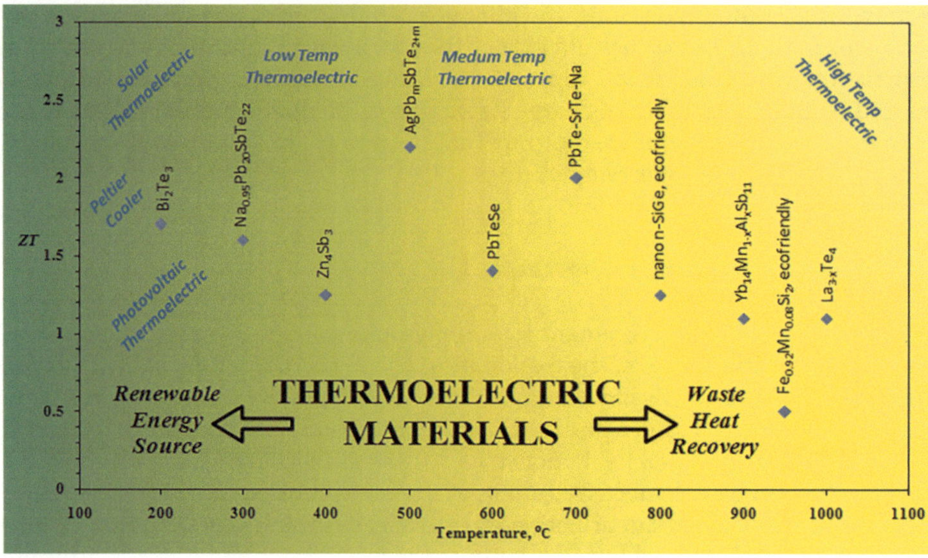

Figure 3.
Highest ZT thermoelectric materials to be used at various temperatures for various applications [12, 16, 23, 24].

friendly lead and/or tellurium-free materials with good mechanical and thermal stability [25]. Among thermoelectric materials with high efficiency and lacking the non-desired tellurium or lead, skutterudites (e.g., $Ba_{0.08}La_{0.05}Yb_{0.04}Co_4Sb_{12}$) and silicides (e.g., Mg_2Si, $MnSi_{1.73}$, $Fe_{0.92}Mn_{0.08}Si_2$) stand out with a promising future [26]. The research is also heading toward finding all-round user-friendly TE material with respect to cost, material availability, and scale-up viability.

6. Synthesis of bulk thermoelectric ceramics

Various techniques for synthesis of TE bulk materials are (1) melting-casting method (2) Rapid Solidification Process (RSP) (3) Self-propagating high-temperature synthesis (SHS) process (4) generic powder metallurgy (PM) and Mechanical Alloying (MA) method. TE produced by conventional ingot technology generally has a rather coarse eutectic structure, which cannot be engaged in thermoelectricity [21]. In many cases, this first method, based on the classical melting process, does not yield a homogeneous TE bulk product directly because of the various transformation reactions (e.g., eutectic, peritectic, eutectoid, peritectoid, etc. depending on the equilibrium phase diagram) [27]. In the case of the SHS process, very low density is achieved, which reduces the phase formation tendency and the electrical conductivity of the TE material [21]. Rapid solidification followed by hot pressing is found to be an effective way to produce fine grain structures and improve the phase transformation speed and thermoelectricity properties. However, the fine grain sizes in these alloys grow rapidly up to about a few μm after hot pressing and impairing the TE properties [28]. PM and MA are found to be very good candidates for the demanding TE ceramic technology, but a lot of work has yet to be done to get rid of problems like inhomogeneity in phase formation, dispersion of dopants, etc. [29, 30]. To sum up, many techniques have been employed for the synthesis of TE material, but there is no such specific route that can lead us to the synthesis of this material on an industry level.

7. Electrical contacts for TE circuit

Establishment of proper electrical contact is essential to improve the TE power efficiency. The contact region is the only way to assist in the transfer of charge carriers from the TE material across the load resistance. Discontinuity in the contact region will reduce the TE performance [6]. The contacts used may be silver, aluminum or copper based depending upon the cost, shape and size of TE legs, temperature of application and the operating conditions. The interface between contact and TE material becomes crucial because it affects the electrical properties (transport) of the TE devices. This interface is called the junction. An ideal junction is the one where there are no defects formed at the interface. However, in practice most of the contacts have defects at the interface, which add to the contact resistance between the contact and the TE leg. The contact resistance depends on the material's surface roughness, interstitial conditions, contact pressure, and temperature. The first three process parameters affect the contact area, whereas the fourth parameter, i.e., temperature, has a strong influence over the resistivity of the contact material [31]. For low-temperature TE, mostly Al is used for connections. In the case of intricate designs, physical vapor deposition of Al is carried out to create the connections. Cu is the preferred connector over Al in most of the applications, e.g., Bi_2Te_3 TE circuit. This is because of the better combination of electrical and thermal conductivities of Cu than Al. In high-temperature applications, Ag is used, as in the case of β-$FeSi_2$ TE circuit [32, 33]. Here, the silver foil with 99.9% Ag was used for circuit formation with a resistivity at room temperature of about 1.59 μΩ-cm. The thermal conductivity of Ag varies with respect to temperature following the relation, $y = -2.0714x^2 - 0.4714x + 432.4$, $R^2 = 0.9962$ [33]. The hot side and the cold side of the TE need to be completely and uniformly layered by the silver paste so as to reduce the contact resistance. For this, conductive silver paste (particle size 10 nm) was used. The applied surfaces were then exposed to UV light for 45 min for curing. The silver foil of thickness 20 μm was properly cut to the size of the hot side and cold side. It was placed over the silver paste, and a pressure of around 4 MPa was applied, and then the TE leg was kept in the muffle furnace at 800°C for 1 h to ensure a perfect contact.

8. Physical and mathematical model for TE

Evaluation of TE ceramic circuits with respect to the electrical performances can be made through either physical mode or mathematical mode. This study is required to find out the usable and worthy temperature regime and the energy efficiency of a particular TE ceramic. The discussion in this section, related to the thermoelectric power generator (TEG) and refrigeration circuits, is generic and can be applied to any thermoelectric systems with minor changes. It includes the Seebeck effect, the Peltier effect, a parameterization of the Joule heat, as well as all thermal and electrical resistances [2]. Geometry optimization and investigations of the influence of process parameters and load resistance, which affect the output power, are also discussed. The study is carried out stepwise in three stages: (1) TE single leg, (2) TE unicouple, and (3) TE module, and is applied to iron disilicide TE to cite an example.

8.1 TE single leg

The first stage includes the study of a single TE leg made up of p-type material. The p-type TE leg is connected electrically using silver film or silver wire as an element of

TE generator (TEG) setup as shown in **Figure 4(a)**. The two joints are held at different temperatures to have thermal gradient [35]. A similar study is applied to the n-type leg.

The voltages generated by p-type and n-type legs are respectively evaluated using Eq. (3) and are given as:

$$\Delta V_p = S_p(T_H - \mathrm{T}_C) \tag{11}$$

$$\Delta V_n = S_n(T_H - T_C) \tag{12}$$

where ΔV_p and ΔV_n are voltages generated, and S_p and S_n are Seebeck coefficients of p-type and n-type leg respectively. The sign of S is positive if the emf tends to drive an electric current from the hot side to the cold side.

The rate of heat flow through the TE leg is:

$$\dot{Q}_p = \frac{k_p A(T_H - T_C)}{l} \tag{13}$$

$$\dot{Q}_n = \frac{k_n A(T_H - T_C)}{l} \tag{14}$$

where k_p and k_n are thermal conductivities of p-type and n-type legs, respectively, and A and l are cross-sectional area and length of each leg, respectively. The internal resistances are described as $r_{inp} = r_p + 2r_c + 2r_{Ag}$ and $r_{inn} = r_n + 2r_c + 2r_{Ag}$ for p-type and n-type circuits, respectively, as shown in **Figure 4(b)**. They include TE resistance r_p or r_n; contact resistance at each side, r_c and resistance of the contact material, here Ag, r_{Ag}. In TE, the hot side and cold side are having different temperatures, and r_{Ag} here, have to be calculated at the average temperature. The same circuit is closed in power generation mode using a variable load resistance R_L (for refrigeration mode, dc voltage difference is applied instead of R_L) as shown in **Figure 4(c)** and

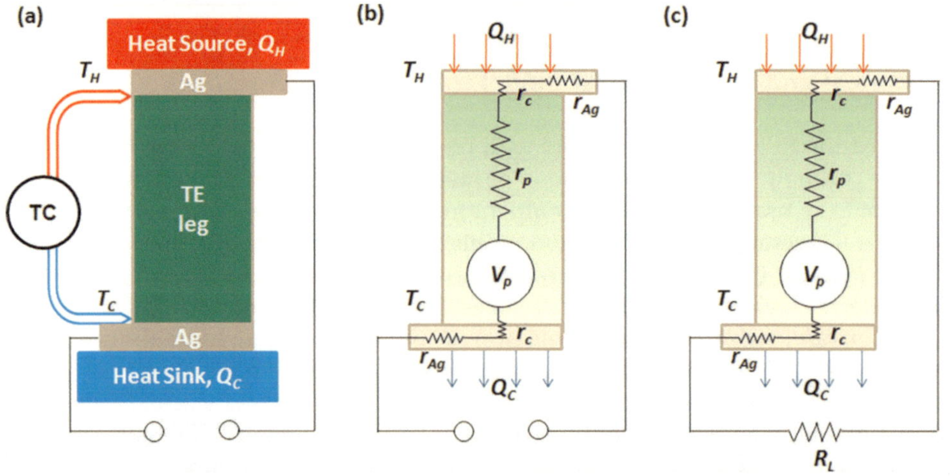

Figure 4.
Single thermoelectric leg made of p-type material under steady-state heat transfer conditions. The p-type leg is connected electrically using a silver connector. The two joints are held at different temperatures to have thermal gradient. A similar study can be applied to n-type leg. (a) Schematic showing physical setup of thermoelectric generator (TEG); (b) Thermal circuit representation of (a) under no load condition; (c) Thermal circuit in power generation mode [34].

correspondingly an electric current I_p flows in the circuit [36]. The current I_p *and* I_n flowing through the p-type and n-type circuits, respectively, is given as:

$$I_p = \frac{\Delta V_p}{r_{inp} + R_L} \tag{15}$$

$$I_n = \frac{\Delta V_n}{r_{inn} + R_L} \tag{16}$$

where r_{inp} and r_{inn} are the internal resistances for p-type and n-type legs, respectively, and all these factors are taken care of during generator circuit construction by following proper assembly procedure as mentioned in Section 6.3.

The energy conversion efficiency, η, of the TE leg is defined as the ratio of the power output (used or stored) to the total heat energy input. To evaluate the physical condition, let the circuit be divided into two halves, the upper hot side and the lower cold side. The power output is to the cold side and is represented as P_L. The heat input, or in other words, the power input, is to the hot side and is represented as $\dot{Q_{Hp}}$.

Now let us determine the power output, P_L, across load resistance R_L as:

$$P_L = I^2 R_L \tag{17}$$

Thus, from Eqs. (10), (14), and (16),

$$P_{Lp} = \frac{S_p{}^2 (T_H - T_C)^2}{\left(r_{inp} + R_L\right)^2} R_L \tag{18}$$

Similarly, from Eqs. (11), (15), and (16),

$$P_{Ln} = \frac{S_n{}^2 (T_H - T_C)^2}{\left(r_{inn} + R_L\right)^2} R_L \tag{19}$$

where P_{Lp} and P_{Ln} are the power outputs for p-type and n-type legs, respectively.

Power output and energy conversion efficiency are the primary parameters to characterize TE performance, and they should be as maximized as possible, and thus the condition for the same needs to be evaluated. For this purpose, the Maximum Power Transfer theorem is used for TE. From **Figure 4(c)**, it can be predicted that the maximum power output depends on the value of the load resistance. To find the maximum power, differentiate Eq. (18) or Eq. (19) with respect to resistance R_L and equate it to zero.

$$\frac{dP_{Lp}}{dR_L} = \frac{\left[S_p{}^2 (T_H - T_C)^2\right]\left[\left(r_{inp} + R_L\right)^2 - 2R_L\left(r_{inp} + R_L\right)\right]}{\left(r_{inp} + R_L\right)^4} = 0 \rightarrow R_L = r_{inp} \tag{20}$$

The condition for maximum power output, according to the Maximum Power Transfer theorem, is that the load resistance, R_L, should be equal to the internal resistance, r_{inp} or r_{inn}. The power and efficiency calculations are carried out considering $r_{in} = R_L$, which is also termed the "matched condition" [37]. Applying this condition, the Eqs. (18) and (19) are now become, respectively:

$$P_{Lp} = \frac{{S_p}^2(T_H - T_C)^2}{4r_{inp}} \tag{21}$$

$$P_{Ln} = \frac{{S_n}^2(T_H - T_C)^2}{4r_{inn}} \tag{22}$$

The heat absorbed at the hot side or the power input consists of the heat getting absorbed due to flow of current (Peltier effect), rate of heat flow across the length (Seebeck effect), and heat transport boundary condition (Joules effect). An assumption made over here is that the convection and radiation at the overall surfaces of the legs are negligible. Each of these factors is related to $\dot{Q_{Hp}}$, as mentioned in Eqs. (23)–(24).

$$\dot{Q_{Hp}} = S_p I T_H + \frac{k_p A(T_H - T_C)}{l} - \frac{1}{2} I^2 r_{inp} \tag{23}$$

$$\dot{Q_{Hn}} = S_n I T_H + \frac{k_n A(T_H - T_C)}{l} - \frac{1}{2} I^2 r_{inn} \tag{24}$$

The energy conversion efficiency of the p-type and n-type TE legs is respectively given as:

$$\eta_p = \frac{P_{Lp}}{\dot{Q}_{Hp}} \tag{25}$$

$$\eta_n = \frac{P_{Ln}}{\dot{Q}_{Hn}} \tag{26}$$

Rearranging the terms in Eqs. (21), (23) and (25) can be rewritten as;

$$\eta_p = \frac{\frac{T_h - T_c}{4}}{\left(\frac{T_h}{2} + \frac{k_p \rho_p}{S_p^2} - \frac{T_h - T_c}{8}\right)} \tag{27}$$

Similarly, for n-type leg Eq. (26) can be rewritten as;

$$\eta_n = \frac{\frac{T_h - T_c}{4}}{\left(\frac{T_h}{2} + \frac{k_n \rho_n}{S_n^2} - \frac{T_h - T_c}{8}\right)} \tag{28}$$

where $\rho_p = \frac{r_{inp} A}{l}$ and $\rho_n = \frac{r_{inn} A}{l}$ are the resistivity of p-type and n-type legs, respectively. Direct energy conversion relies on the physical transport properties ρ, k, and S of the TE material, and their energy conversion efficiency can be written in terms of the Figure-of-Merit, Z.

The Eqs. (26–27), can be rewritten in terms of Z as;

$$\eta_p = \frac{2Z_p(T_h - T_c)}{8 + Z_p(3T_h + T_c)} \tag{29}$$

$$\eta_n = \frac{2Z_n(T_h - T_c)}{8 + Z_n(3T_h + T_c)} \tag{30}$$

where Z_p and Z_n are Figure-of-Merit for p-type and n-type legs, respectively. It can be seen from Eqs. (29–30) that the temperature difference across the TE leg and the TE Figure-of-Merit (Z) has a major impact on the energy conversion efficiency. The temperature difference between the hot side and the cold side is to be kept as high as possible to have maximum energy conversion efficiency [38]. This is strongly influenced by the thermoelectric generator (TEG) design and thus can be controlled by engineering the design parameters of a TEG. Figure-of-Merit, Z is the material property and is an important consideration while classifying any material as TE.

As shown in **Figure 5**, the direction of the current decides which TE side will get heated up or cooled down. If the mechanical setup of TE is made to be used for heating, then by just changing the direction of current, the same setup can be used for cooling. Normally, the Peltier or TE coolers are used till 200°C. The important criterion to evaluate the performance of a TE cooler is the Coefficient of Performance (COP). The COP is the ratio of heat absorbed at the cold side (Q_C) to the input power (Q_{iP}) of the TE leg.

$$COP = \frac{Q_C}{Q_{iP}} \tag{31}$$

The heat dissipated by the heat sink, Q_H, is the sum of heat absorbed at the cold side and the input power.

$$Q_H = Q_C + Q_{iP} \tag{32}$$

From Eq. (1), the condition for maximum COP is that Q_{iP} is to be kept at minimal, i.e. minimum amount of heat to be dissipated by the heat sink. In other words, COP depends on the operating current, I, or the temperature difference ($\Delta T = T_h - T_c$) [2].

The current I flowing through the cold side of TE absorbs Peltier heat at a rate of $Q_{\dot{iP}}$. Using Eqs. (4) and (5),

$$Q_{\dot{iP}} = \Pi I \tag{33}$$

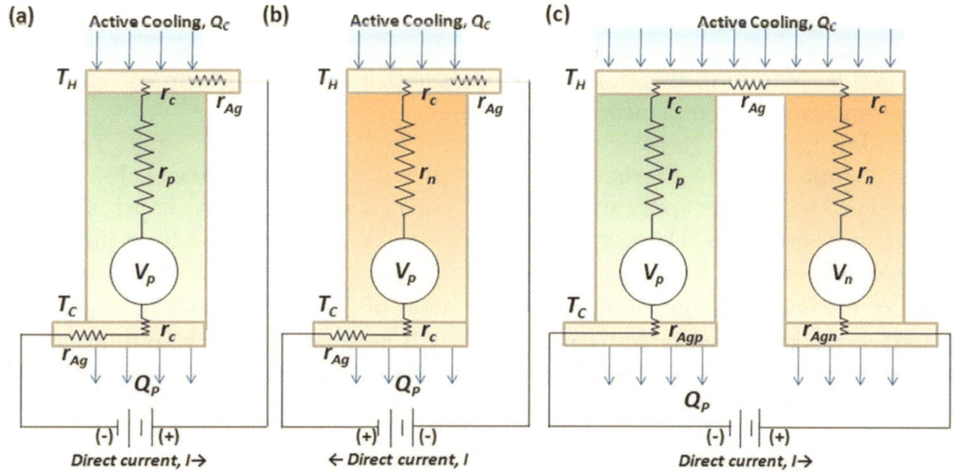

Figure 5.
Direct current, I is passing through the TE circuit, which results in active cooling on one side of the TE. Thermal circuit of TE in refrigeration mode (a) p-type TE leg, (b) p-type TE leg, (c) TE unicouple.

The current I flowing through the TE leg generates heat Qj due to the Joule heating [2].

$$\dot{Q}_J = \frac{l}{\sigma A} I^2 \tag{34}$$

This heat generated in the bulk distributes equally to both the hot and cold sides of the TE. Therefore, half of it will add to the heat absorbed at the cold side. The TE material within the TE leg will also contribute to the heat flow from the hot side to the cold side by conduction. As shown in **Figure 5(a)–(b)**, this heat flow is given by the thermal conduction Eqs. (13)–(14) for p-type and n-type legs, respectively.

The net rate of heat absorption or the cooling rate $\dot{Q}_{net}$ at the cold side of the p-type TE leg is given as;

$$\dot{Q}_{net} = \dot{Q}_{iP} - \frac{\dot{Q}_J}{2} - \dot{Q}_P \tag{35}$$

The use of TE coolers is worth it if maximum cooling is obtained on the cold side. To determine the maximum cooling rate or the maximum heat pumping capacity, Eq. (35) is differentiated with respect to I. By doing this, the maximum current and the maximum cooling rate $\dot{Q}_{P,max}$ are obtained as given in Eqs. (36) and (37).

$$I_{max} = \left(\frac{A}{l}\right) \Pi\sigma \tag{36}$$

$$\dot{Q}_{P,max} = \left(\frac{A}{l}\right)\left(\frac{1}{2}\Pi^2\sigma - k_P(T_H - T_C)\right) \tag{37}$$

Under steady-state conditions, ΔT is constant and the net cooling rate at the junction becomes zero. From Eq. (36), the maximum temperature difference achievable is calculated as;

$$\Delta T_{max} = \frac{\Pi^2\sigma}{2k_P} \tag{38}$$

From Eqs. (4) and (38) it can be summarized that Figure-of-Merit, ZT can be determined from the temperature difference.

By following Eqs. (33)–(38), similar relations can be derived for n-type TE leg and for the TE unicouple, the thermal circuit of which is shown in **Figure 5(c)**.

Based on the need or application, Peltier coolers come in various shapes, sizes, and performance categories. The shape and size of the TE majorly depends on the surface on which the TE cooler must be installed. Categorization based on the performance of the Peltier cooler is judged on the values of $\dot{Q}_{P,max}$, I_{max} and ΔT_{max}. From Eq. (37), it can be inferred that there is a trade-off between the heat pumping capacity and the temperature difference. With a small ΔT value, a large amount of heat will be transferred in the Peltier circuit, and the *COP* is the maximum at the lowest value of ΔT. An increase in current increases ΔT, but simultaneously the Joule heat increases, and after a threshold value, the Joule heat supersedes the absorbed heat. This is known as thermal runaway and hence now the sole purpose of using the Peltier cooler has vanished. To get rid of the thermal runaway, multistage or cascaded TE (discussed in Section 7.1) cooler can be used.

As shown in **Figure 6**, it can be interpreted that not all the TEs are as good as both Peltier or TE coolers and thermoelectric generators. Some are good as Peltier coolers, and some are good for TEG applications. But there is a class of TE materials that are good in both types of application areas, i.e., refrigeration and power generation. To make many of the TEs multidimensional, suitable measures need to be taken from two perspectives: (1) Joule heat to be kept as minimum as possible; (2) no increment in lattice thermal conductivity due to increasing temperature. This can be achieved by suitable use of doping, nano structuration, band engineering and use of proper set of materials (e.g. use of oxide dielectric layer in between the layers of TE material) [3, 7, 15].

One such example of TE leg synthesized and characterized is of Mn-doped p-type and Co-doped n-type thermoelectric (TE) β-iron disilicide ceramics [39]. These legs were fabricated for TEG using the powder metallurgy (PM) route. The reaction between iron and silicon to form TE β-iron disilicide is very sluggish [40], however, the PM route increases the kinetics of this reaction [30, 41]. The effects of variation in type and quantity of dopants (p-type dopant: Mn: stoichiometry $Fe_{1-x}Mn_xSi_2$ ($x = 0.04–0.12$) or n-type dopant: Co: stoichiometry $Fe_{1-y}Co_ySi_2$ ($y = 0.01–0.05$)) on thermoelectric properties were studied. The stoichiometric compositions studied in this research work were abbreviated as given in Table I to make the discussion more composed.

These prepared doped TE ceramic legs were studied for Seebeck coefficient, resistivity, thermal conductivity, and Figure-of-Merit with respect to temperature as shown in **Figure 7(I)–(IV)**. The p-type leg and n-type leg with the highest Seebeck coefficient were $Fe_{0.92}Mn_{0.12}Si_2$ and $Fe_{0.95}Co_{0.05}Si_2$, respectively. This was due to the intrinsic p-type nature of β-$FeSi_2$, which had a positive impact on the Seebeck coefficient of p-type compacts compared to n-type compacts. The maximum *ZT* values for $Fe_{0.92}Mn_{0.08}Si_2$ p-type and $Fe_{0.95}Co_{0.05}Si_2$ n-type compacts were 0.59 and 0.41, respectively, at 1150 K. Another important conclusion is that both p-type and n-type compacts had maximum *ZT* values at the same temperature, i.e., 1150 K. This

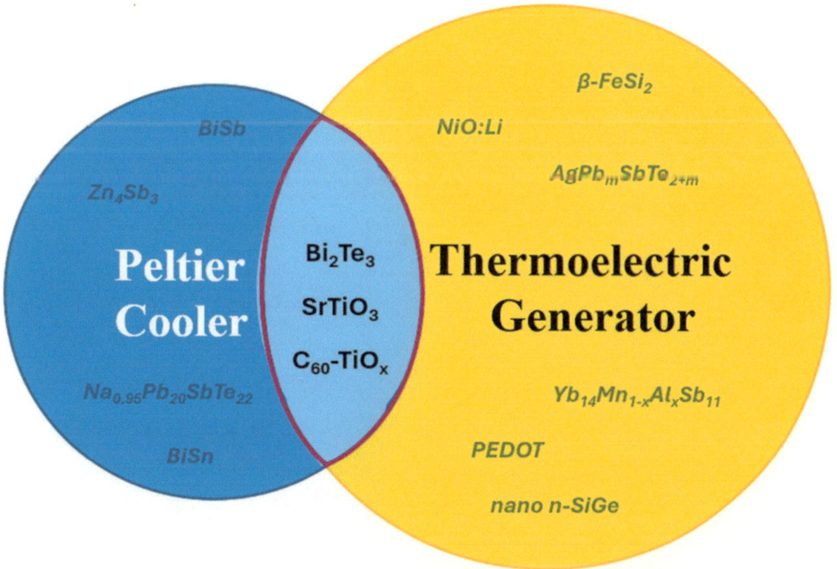

Figure 6.
Thermoelectric materials classification with respect to the power generation mode and refrigeration mode (Data was referred from Ref. [1, 11, 13, 21, 26, 38]).

p-type β-$FeSi_2$		*n-type* β-$FeSi_2$	
Stoichiometry composition	Abbreviation	Stoichiometry composition	Abbreviation
$Fe_{0.96}Mn_{0.04}Si_2$	Mn0.04	$Fe_{0.99}Co_{0.01}Si_2$	Co0.01
$Fe_{0.92}Mn_{0.08}Si_2$	Mn0.08	$Fe_{0.98}Co_{0.02}Si_2$	Co0.02
$Fe_{0.88}Mn_{0.12}Si_2$	Mn0.12	$Fe_{0.95}Co_{0.05}Si_2$	Co0.05

Table 1.
Abbreviations used for the various compositions of Mn-doped p-type and Co-doped n-type thermoelectric (TE) β-iron disilicide legs [24].

indicates that for a TEG developed from these legs, the TE properties would be optimal at 1150 K and that they both will perform the best at this temperature [24].

8.2 TE unicouple

A TE unicouple consists of one p-type and one n-type leg connected thermally parallel and electrically in series, as shown in **Figure 8**. The p-type and n-type legs are

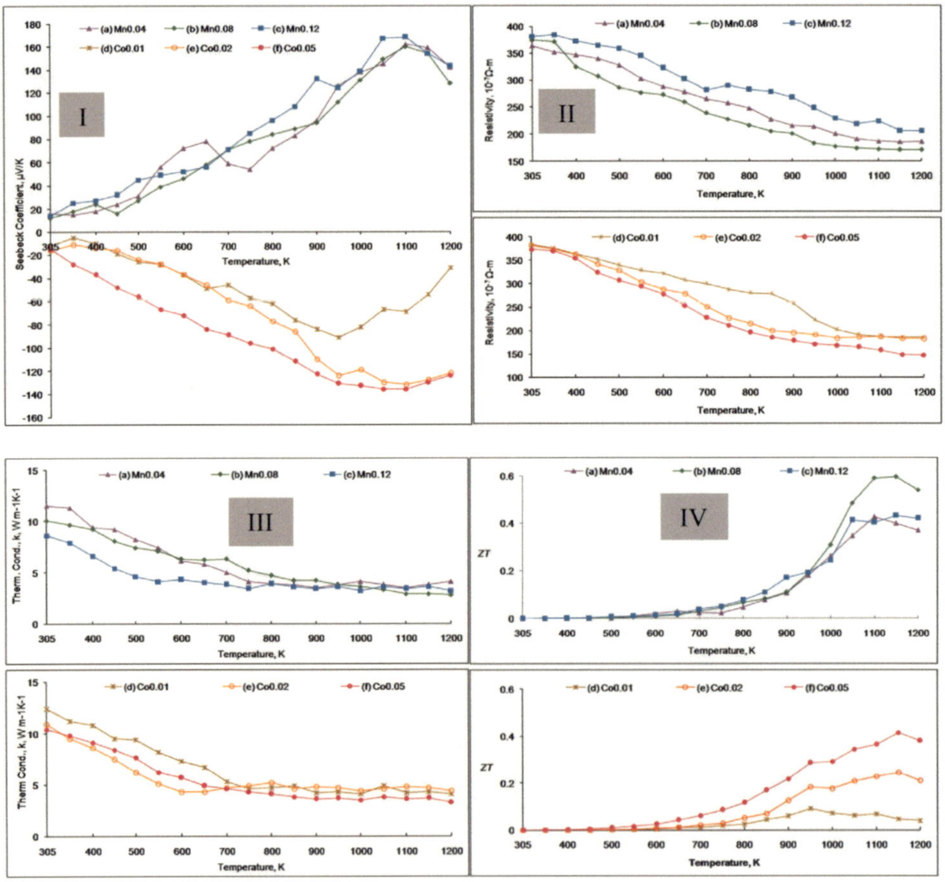

Figure 7.
Temperature dependence of I. Seebeck coefficient, II. resistivity, III. thermal conductivity, IV. Figure-of-Merit of doped compacts (a) Mn0.04, (b) Mn0.08, (c) Mn0.12, n-type: (d) Co0.01, (e) Co0.02, (f) Co0.05 [24].

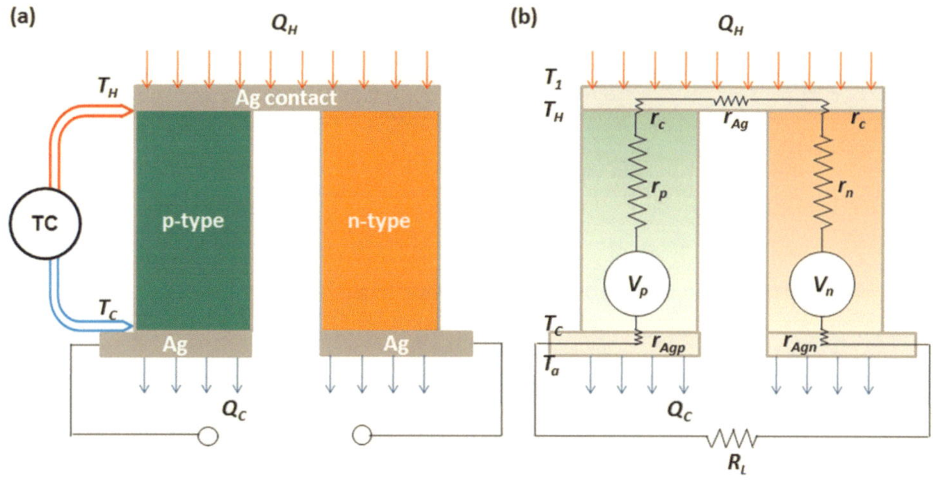

Figure 8.
Single thermoelectric (TE) unicouple made of p-type and n-type legs and under steady-state heat transfer conditions. The p-type TE leg is connected electrically in series to the n-type TE leg using a silver connector. The two joints are held at different temperatures to have thermal gradient. (a) Schematic showing physical setup of TE unicouple; (b) Thermal circuit representation of TE unicouple in power generation mode [34].

the two dissimilar TE materials that are connected using highly conducting film, silver here. The two joints are held at different temperatures. Assumptions here are:

1. being the same base material system, the temp distribution across p-type is same as that in n-type and

2. response from p-type and n-type legs to the change in temperature with respect to the physical properties is same.

The relations of Seebeck coefficient (S), thermal resistance (R), and electrical conductance (k) for a unicouple are given in Eqs. (39), (40), and (41) respectively.

$$S = S_p - S_n \tag{39}$$

$$R = \frac{\rho_p L_p}{A_p} + \frac{\rho_n L_n}{A_n} \tag{40}$$

$$k = \frac{k_p A_p}{L_p} + \frac{k_n A_n}{L_n} \tag{41}$$

The voltage generated by p-type and n-type legs is respectively given as:

$$\Delta V_p = S_p(T_H - T_C) \tag{42}$$

$$\Delta V_n = S_n(T_H - T_C) \tag{43}$$

where ΔV_p and ΔV_n are voltages generated, and S_p and S_n are Seebeck coefficients of p-type and n-type legs, respectively.

Thus, the voltage generated, V_T across TE unicouple, is determined from Eqs. (42) and (43) and is given as:

$$V_T = \Delta V_p - \Delta V_n = (S_p - S_n)(T_H - T_C) \tag{44}$$

The resulting ZT for the couple is given by Eq. (45).

$$ZT = \frac{(S_p - S_n)^2 T}{\left[(\rho_n k_n)^{1/2} + (\rho_p k_p)^{1/2}\right]^2} \tag{45}$$

The rate of heat flow through the TE unicouple is thus:

$$\dot{Q} = \frac{(k_p + k_n)(T_H - T_C)A}{l} \tag{46}$$

where k_p and k_n are thermal conductivities of p-type and n-type legs, respectively, and A and l are cross-sectional area and length of each leg, respectively. Consider that the thermal conductivities of p-type and n-type legs are nearly the same, represented as k_s ($k_s = k_p = k_n$), and so the rate of heat flow now becomes:

$$\dot{Q} = \frac{2k_s A(T_H - T_C)}{l} \tag{47}$$

Now let us consider that the same circuit is closed using a variable load resistance R_L and correspondingly an electric current I flows in the circuit. The current I flowing through the circuit is given as:

$$I = \frac{V_T}{r_{in} + R_L} \tag{48}$$

To evaluate the physical condition, let the circuit be divided into two halves, the upper hot side and the lower cold side. On the hot side and cold side, the rate of heat absorbed, $Q_{\dot{H}}$ and the rate of heat dissipated, $Q_{\dot{C}}$ are determined, respectively [36], as given in Eq. (48)–(49) as follows:

$\dot{Q}_H$ = Heat getting absorbed due to flow of current (Peltier effect)+
Rate of heat flow across the length (Seebeck effect)−
Heat transport boundary condition (Joules effect)

$$\dot{Q}_H = SIT_H + \frac{2k_s A(T_H - T_C)}{l} - \frac{1}{2}I^2 r_{in} \tag{49}$$

$\dot{Q}_C$ = Heat getting dissipated due to flow of current (Peltier effect)+
Rate of heat flow across the length (Seebeck effect)+
Heat transport boundary condition (Joules effect)

$$\dot{Q}_C = SIT_C + \frac{2k_s A(T_H - T_C)}{l} + \frac{1}{2}I^2 r_{in} \tag{50}$$

The power output, P_L, is given as:

$$P_L = \dot{Q}_H - \dot{Q}_C = SI(T_H - T_C) - I^2 r_{in} \tag{51}$$

Also, now let us determine the power output, P_L, across load resistance R_L.

$$P_L = I^2 R_L \tag{52}$$

From Eqs. (38), (47) and (51),

$$P_L = \frac{(S)^2 (T_H - T_C)^2}{(r_{in} + R_L)^2} R_L \tag{53}$$

The power of TE is proportional to the square of the temperature difference ($\Delta T = T_h - T_c$), as given by Eq. (53). To increase the power output, the temperature difference needs to be increased [38]. This can be done by using a suitable cooling system. Also, from Eq. (20) and Eq. (53), maximum power output from the TE unicouple is possible only when the applied load resistance, R_L, is equivalent to the internal resistance, r_{in}.

Eq. (53) now becomes:

$$P_L = \frac{S^2 (T_H - T_C)^2}{4 r_{in}} \tag{54}$$

Effective ZT determines the maximum energy conversion efficiency η, given in Eq. (55). The efficiency η of the TE unicouple is the ratio of power input to the load to the net heat flow rate. Here the heat flow is from the source to the heat sink, and so the net heat flow is positive.

$$\eta = \frac{P_L}{\dot{Q}_H} = \frac{T_h - T_c}{T_h} \frac{\sqrt{1 + (ZT)} - 1}{\sqrt{1 + (ZT)} + \frac{T_c}{T_h}} \tag{55}$$

The term $(1 + ZT)^{1/2}$ varies with the average temperature T. The important point that is concluded from efficiency η calculation is that η depends on two dimensionless quantities (T_c/T_h) and ZT [42]. Eq. (55) indicates that increasing efficiency requires both high ZT values and a large temperature gradient across the thermoelectric materials. A large temperature gradient can be established when the thermal conductivity is low. High thermal conductivity shorts the thermal circuit. The high Seebeck coefficient ensures a large potential/thermo voltage, the high electrical conductivity is needed to minimize the Joule heating effect, and the low thermal conductivity is needed to create a large temperature gradient [43].

8.3 TEG module

A TEG is made up of many legs of p-type and n-type TE materials that are connected electrically in series but thermally in parallel, as shown in **Figure 9(a)–(b)**. However, their Carnot overall efficiency is very poor, typically in the range of 5%, which is very low compared to other renewable energy sources like solar energy and wind energy [23]. To obtain optimum performance, it is important to take care of several design features when making the module to be used in an application [44]. The TEG module consists of unicouples connected in a circuit as shown in **Figure 10**. They can be fabricated such that a pair or an unicouple is manufactured with the same set of process parameters as one unit (as shown in **Figure 10(a)**) or individual p-type and n-type legs are manufactured and later assembled on the TEG ceramic board using metal contacts (as shown in **Figure 10(b)**) [6]. In both cases, p-type and n-type

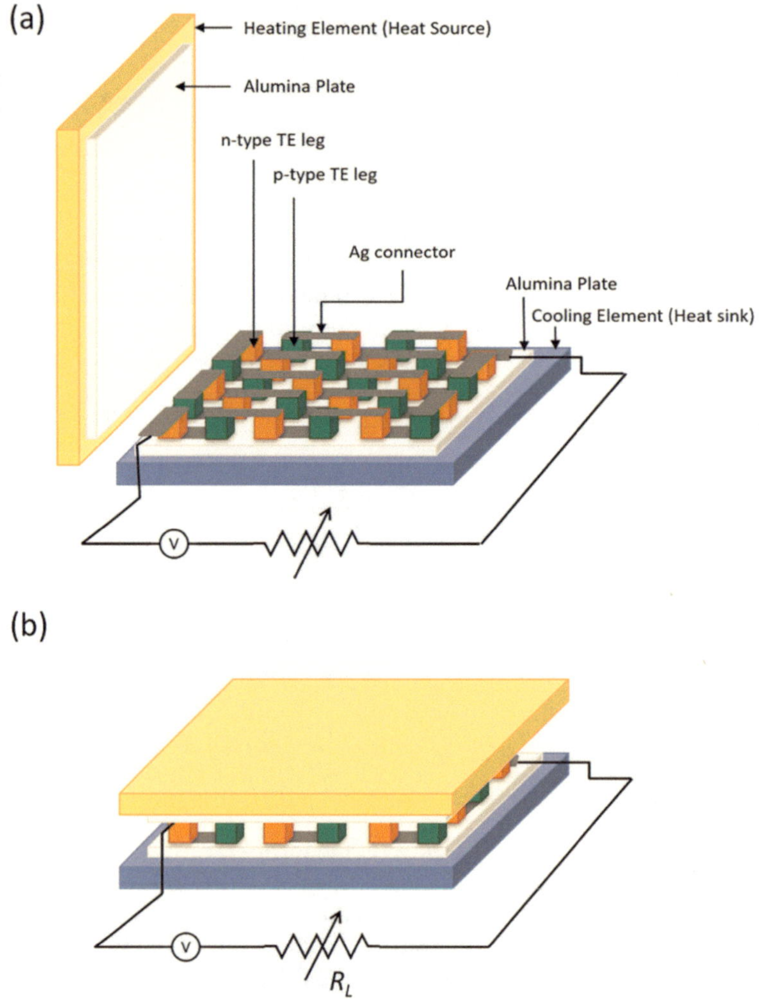

Figure 9.
Schematic of a thermoelectric generator module: (a) Various parts of the TE assembly; (b) Appearance of the complete TE assembly [34].

TE regions of the unicouple or legs are electrically in series and thermally in parallel [16].

The design process of a TEG module involves energy balance calculations across the heat source, across the thermoelectric elements, and importantly, systematic analysis of the circuit (e.g., contact resistances, radiation effects, etc.) to improve output power values. These calculations certify the usefulness and importance of a dissipated thermal energy converter in the consumer market [45].

The energy balance can be expressed as:

$$Q_{source} = Q_{legs} + Q_{loss} \tag{56}$$

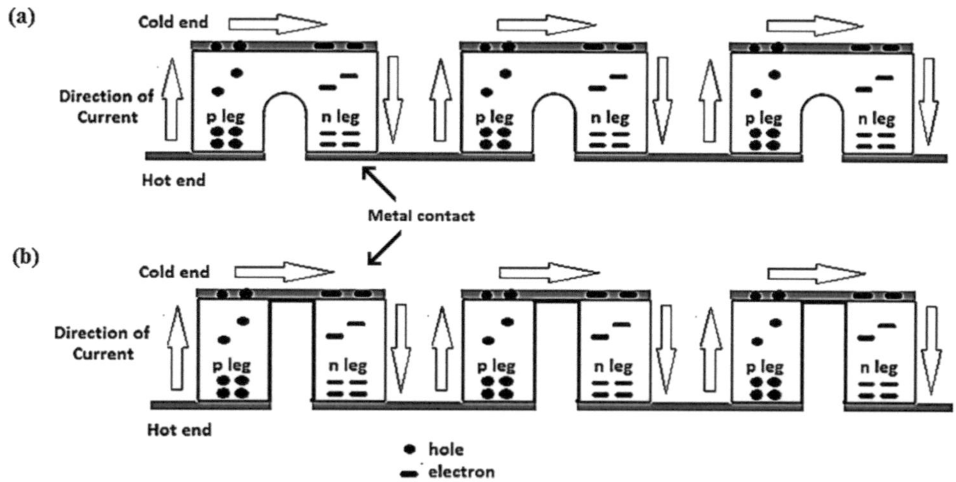

Figure 10.
Thermoelectric unicouple types: (a) p-type and n-type TE legs in physical contact; (b) p-type and n-type TE legs not in physical contact but are electrically in series via metal contact [6].

where, Q_{source}: heat absorbed from the heat source; Q_{legs}: heat transferred through TE legs; and Q_{loss}: heat losses due to radiations from the heat source and heat sink, due to convection.

When properly installed, the thermoelectric device can run for thousands of hours with consistency. For competent installation, however, attention must be given to the mechanical details along with thermoelectric performance. Some important design features that need special consideration are [46]:

1. Compressive loading: The sample must be assembled in the circuit using compressive loading so that proper electrical connections can be established and the contact resistance can be minimized. It has been observed that 1.4 MPa pressure is enough to have proper electrical contact. Also, the p-type and n-type TE samples need to have the same height to ease the circuit formation process. The decision of compressive load depends on the thermal expansion coefficient of the p-type and n-type TE samples and the connectors. If it is different for each TE device part, then a spring of suitable tension must be used to accommodate the dissimilarity.

2. Heat source overhang: Since the heat source will cool near its edges, the edge of the heat source should extend well beyond the edge of the module. In a laboratory test setup, an overhang of at least 0.5 inch is necessary to avoid most of the cooling effects caused by the proximity of the edge of the heat source.

3. Heat distribution: To avoid nonuniform temperature distribution or formation of hot spots or cold spots at the heat source, it is helpful to place a high thermal conductivity plate like Al for low-temperature heating and SiC for high-temperature applications. This plate is to be placed between the heat source and

TE device. Similarly, uniform temperature distribution is required in the heat sink, which can be achieved by using Al or Cu plate as a thermal spreader.

4. Area of contact: The power output is in DC form, and so area of contact is very important. The % area of contact of the conductor to the sample should be tending to 100% to have maximum power output.

5. The interface between the TE sample and the heat source and the interface between the sample and the heat sink are very critical components of any TE device. The surface of the device consists of the electrical conductors that join the n-type and p-type samples of the TE device. It is essential that these conductors are not electrically shorted together. Since most of the heat sources and heat sinks are electrically conducting, an insulating material must be placed at the interface.

Voltage detection is a very important aspect while determining TE module efficiency. This voltage detection is very sensitive to factors such as degree of homogeneity in the samples, position of the voltage probes, position of the thermocouples, method of contacting the sample, and precision in sample dimension determination [47].

The maximum power output, W_{max}, for TEG module is determined with the condition $r_{in} = R_L$ and is given as:

$$W_{max} = \Delta T^2 \, x \, \frac{nA}{l} \, x \, \sigma S^2 \tag{57}$$

where n is the number of legs connected electrically in series.

Eq. (57) expresses the dependency of power output on temperature difference (first term), TE leg dimensions (second term), and the type of TE material (third term or the power factor) [48].

The thermal resistance θ_h between the heat source and the TE module is given by Eq. (58). The thermal resistance θ_c between the TE module and heat sink is given by Eq. (59). The thermal resistance of the TE module can be expressed as in Eq. (60). T_1 and T_a are the heat source temperature and ambient temperature, respectively, as shown in **Figure 10**.

$$\theta_h = \frac{T_1 - T_h}{Q_{h,net}} \tag{58}$$

$$\theta_c = \frac{T_c - T_a}{Q_{c,net}} \tag{59}$$

$$\theta_{TEM} = \frac{1}{\frac{mA}{l}\left(k_p + k_n\right)} \tag{60}$$

$Q_{h,net}$ is the heat absorbed from the heat source by the TEG module, and $Q_{c,\,net}$ is the heat released from the cold side of the TEG module. Each unicouple is made of one p leg and one n leg, and m represents the number of couples present in the TE module. Using Newton's law of heat transfer, $Q_{h,\,net}$ and $Q_{c,\,net}$ are determined (Eqs. (61) and (62)).

$$Q_{h,net} = m\left(SIT_h + \frac{T_h - T_c}{m\theta_{TEM}} - 0.5I^2 r_{in}\right) \tag{61}$$

$$Q_{c,net} = m\left(SIT_c + \frac{T_h - T_c}{m\theta_{TEM}} + 0.5I^2 r_{in}\right) \tag{62}$$

The electrical resistivity, ρ_s, and voltage generated, V_T, by the TEG are given by Eq. (63) and Eq. (64) respectively.

$$r = \frac{2lm\rho_s}{A} \tag{63}$$

$$V_T = m\left(S_p - S_n\right)(T_h - T_c) \tag{64}$$

The power of TE is proportional to the square of the temperature difference ($\Delta T = T_h - T_c$), as given by Eq. (7). To increase the power output, this temperature difference can be increased by using thermal cascade or multistage TE. Depending on the design characteristics and TE material properties, two or more TEs are placed one above the other to form a thermal cascade and increase the ΔT. In such cases, the equivalent temperature difference, ΔT_e, is considered as given in Eq. (65).

$$\Delta T_e = \sqrt{\Delta T_1{}^2 + \Delta T_2{}^2 + \Delta T_3{}^2 + \ldots} \tag{65}$$

where ΔT_1, ΔT_2, ΔT_3, ... are the temperature differences across the cascade TE layers TE layer 1, TE layer 2, TE layer 3, ... respectively.

The TEG module was designed from p-type and n-type legs of β-$FeSi_2$, as shown in **Figure 11(a)**. The electrical analogy for the TEG module is shown in **Figure 11(b)**. On both the top and bottom surfaces of the TEG circuit, the SiC plate, alumina plate, and silver foil are represented as capacitors with their respective resistances. Across the TEG circuit, T_h and T_c temperatures are maintained, which results in charge flow. The electrical circuit with gray background summarizes the circuit to determine the Seebeck coefficient. The actual laboratory-scale working model of the TEG is shown in **Figure 11(c)**. Around 29 samples, 15 p-type and 14 n-type, were arranged alternately on a glass wool brick. Insight is the image of a red LED got lightened up by the power provided by the TEG module in the temperature range of around 650–870 K. To determine the voltage and hence power generation from the TEG module with respect to temperature, the red LED was replaced with a Keithley 8½-digit multimeter and the experiment was repeated [34].

The reported lifetime for TEG modules is 11–30 years, but the lifetime can be reduced if the modules are exposed to repeated hot side temperature changes. Repeated heating–cooling cycles may cause material deterioration of a TEG, so it is preferable to use TEGs in continuous heat flows [8].

9. Thermoelectric material applications in renewable

Photovoltaic-Thermoelectric Generator (PV-TE) is one of the emerging applications of TE in renewable energy resources. From last few decades, the usage of photovoltaic (PV) cells for electricity generation increased appreciably owing to its environmental sustainability, availability, and applicability. However, in the PV, most of the solar energy gets dissipated in the form of heat into the PV system. This waste heat causes an increase in temperature of the PV cell and degrades its efficiency and performance. In the current context, efforts are made to improve efficiency by

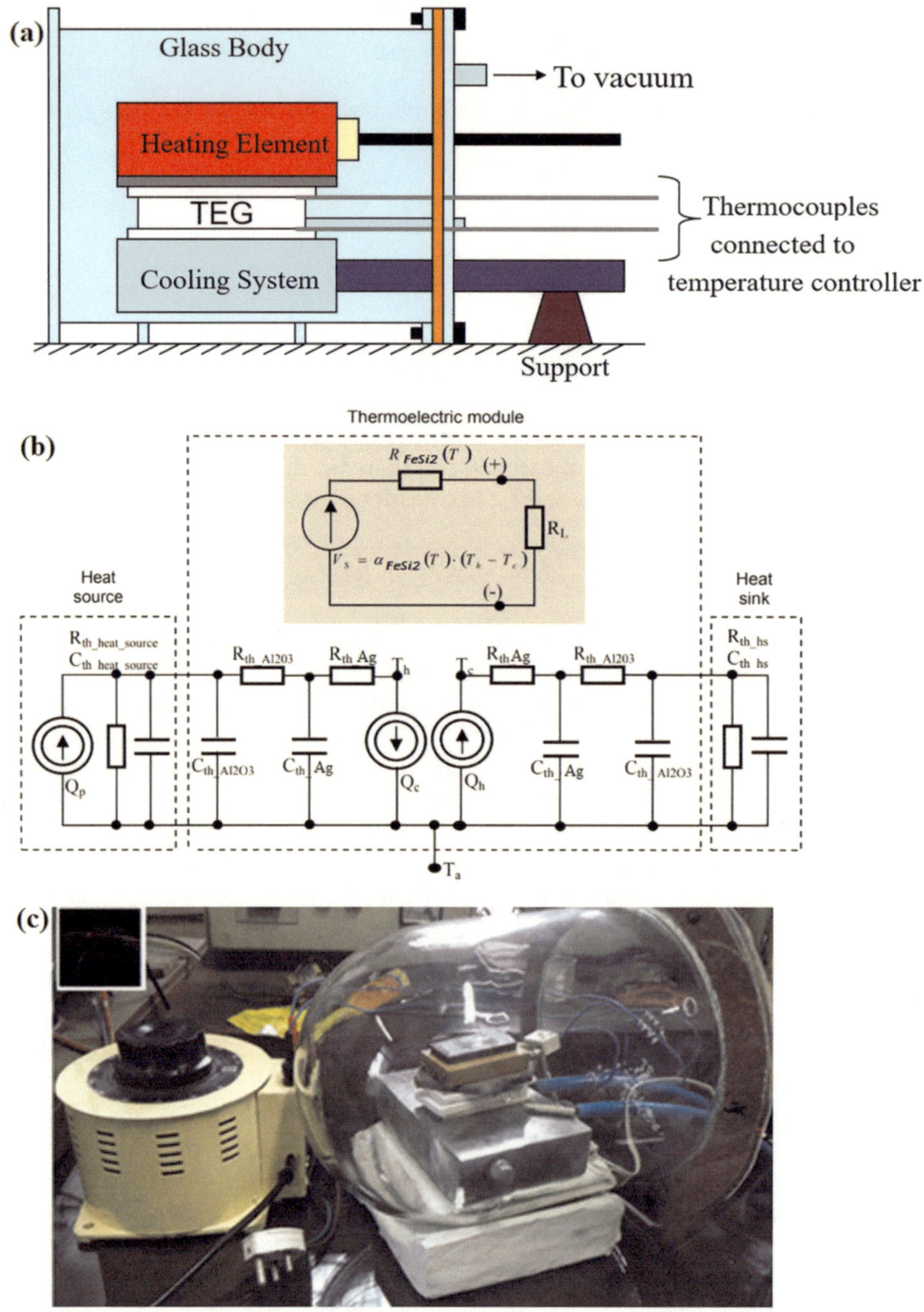

Figure 11.
TEG module (a) schematic representation of the TEG module, (b) electrical analogy of the TEG module, (c) working model of the TEG module [6].

utilizing this dissipated heat. One of the ways is to use low temperature thermoelectric (TE), based on the Seebeck effect, along with the PV cell. The overall efficiency with respect to temperature of hybrid Photovoltaic-Thermoelectric (PV-TE) is 0.6–1.91% more than PV alone [49].

Both PV and TE systems depend strongly on temperature, and thus, to reduce the temperature and improve the performance of the PV-TE hybrid system, an effective and affordable direct evaporative cooling (EC) system was used in this study. The mathematical model studied for performance evaluation was based on cascaded hybrid Photovoltaic-Thermoelectric-Direct Evaporative Cooling System (PV-TE-EC) combinations, i.e., PV-EC (without TE), PV-TE-EC, PV-2TE-EC (two cascaded layers of TE sandwiched between PV and EC), and PV-3TE-EC (three cascaded layers of TE sandwiched between PV and EC as shown in **Figure 12**) systems. The TE module placed adjacent to the lower surface of the PV is fabricated from p-type and n-type bismuth telluride ($(ZT) \sim 1$). Various hybrid systems made of PV and TE assisted or not assisted with EC are studied, e.g., PV, PV-TE, PV-EC, PV-TE-EC, PV-2TE-EC, and PV-3TE-EC. These hybrid systems used PV, TE, and EC in parallel arrangements [49].

The individual and overall power output contributions from PV and TE in the hybrid systems with respect to the concentration ratio are summarized in **Figure 13**. As the concentration ratio increased, the overall power output of all the hybrid systems increased. An increase in solar concentration ratio increases the overall temperature of the hybrid systems. In addition, the temperature rise augments the evaporative cooling effect from EC and results in a higher equivalent temperature difference across TEs. This reduces the power output from PVs but increases the power output from TEs. This leads to an increase in the overall efficiency of PV-TE-EC, PV-2TE-EC, and PV-3TE-EC in the ascending order.

Another emerging application of TE in renewable energy solutions is Solar Thermoelectric Generator (STEG). STEG can be used to generate electricity from tapped solar energy, and therefore they are an attractive renewable energy solution for standalone power conversion [50]. In addition, STEG has been used in power generation for health monitoring systems, wireless sensors, space applications, and several other low power applications [51]. In this, solar concentrators are placed over the heat source to conduct the thermal energy to TE and to perform energy conversion analysis

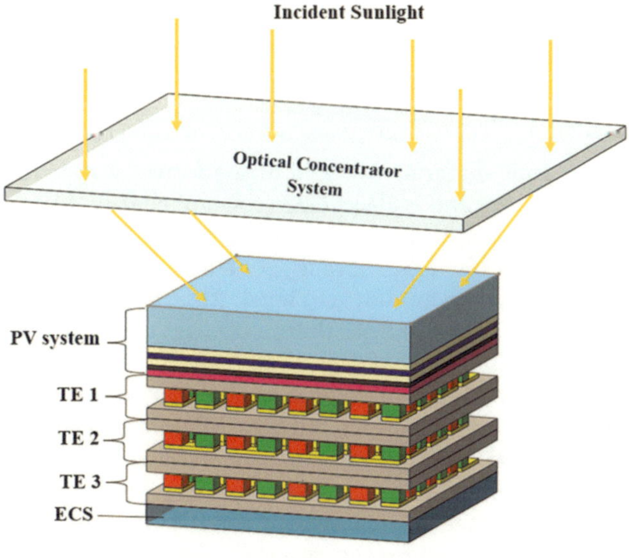

Figure 12.
Schematic representation of the structural diagram of the PV-3TE-EC system [49].

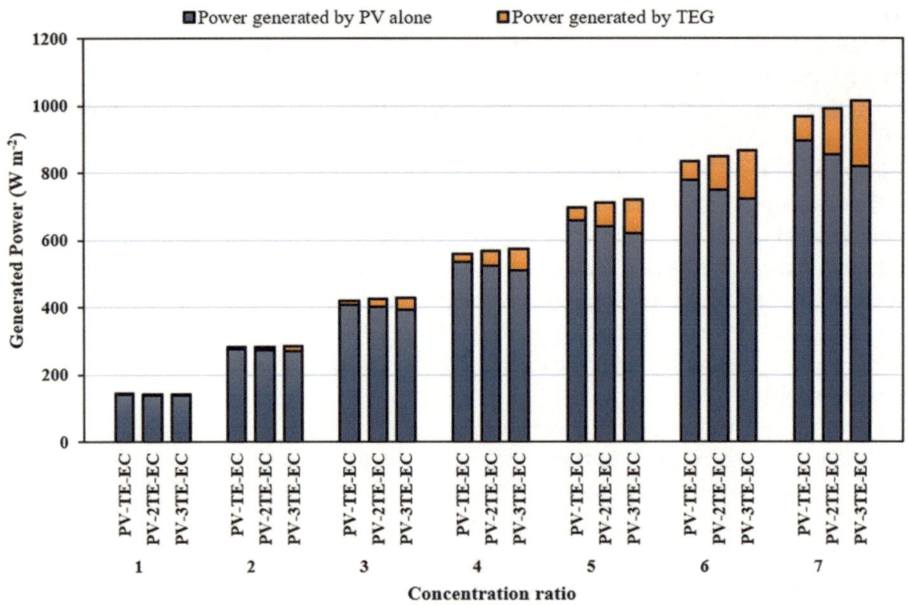

Figure 13.
Power output from the hybrid systems with respect to solar concentration ratios [49].

of the TE at various solar concentration ratios. The energy conversion efficiency was found to be around 15% and 23% for TE material with *ZT* 1 and 2, respectively. A modification to the setup of STEG was also tried in which the solar energy was concentrated on three-layered TE (first layer: Bismuth telluride, second layer: Lead telluride, third layer: silicon) using metamaterial. The Finned heat sink was used to increase the efficiency of conversion of solar energy to electricity to around 24%. However, the consideration over here was that all the energy falling on the metamaterial layer is reaching the TE top surface [2].

10. Conclusions

Thermoelectric ceramic material is a need of the future as it provides a means for harvesting waste heat energy and is also a clean form of energy. Many techniques have been employed for the synthesis of TE, a potential candidate for scavenging energy, but for many of the varieties of TEs, there is no such specific route that can lead us to the synthesis of this material on an industry level. Also, a unicouple or TEG or TEC needs to be formed of p-type and n-type thermoelectric material to achieve a higher value of Figure-of-Merit, *Z*.

Some TE materials contain toxic elements or less abundant elements, which narrow down their scope of application [52]. An ideal TE material should be a phonon glass and an electron crystal, which means that the materials should have a low lattice thermal conductivity just like glass and a high electrical conductivity just like in a crystal and at the same time should be environmentally friendly. Many types of TE have attracted attention in the last few years because of their technological potential in the fields of optoelectronics, light emitting diodes (LEDs), medical, aerospace, and defense applications due to the thermoelectric conversion properties, chemical stability, non-toxicity, and relatively low cost of production. There is a huge scope to do research in the choice of the element and its quantity that need to be added to get high ZT thermoelectric material.

Author details

Vaishali Poddar
Materials and Metallurgical Engineering Department, National Institute of Advanced Manufacturing Technology, Ranchi, India

*Address all correspondence to: vspoddar@niamt.ac.in

References

[1] Dey S, Sreenivasulu A, Veerendra GTN, Rao KV, Babu PSSA. Renewable energy present status and future potentials in India: An overview. Innovation and Green Development. 2022;**1**(1):100006. DOI: 10.1016/j.igd.2022.100006

[2] Riffat SB, Xiaoli M. Thermoelectrics: A review of present and potential applications. Applied Thermal Engineering. 2000;**23**:913-935. DOI: 10.1016/s1359-4311(03)00012-7

[3] Alam H, Ramakrishna S. A review on the enhancement of figure of merit from bulk to nano-thermoelectric materials. Nano Energy. 2013;**2**:190-212. DOI: 10.1016/j.nanoen.2012.10.005

[4] Chao Han, Zhen Li, S. X. Dou, Recent progress in thermoelectric materials, Australian Institute for Innovative Materials (Chinese Science Bulletin), 2014;59(18):2073-2091. DOI: 10.1007/s11434-014-0237-2

[5] Lange H. Electronic structure and interband optical properties of β-FeSi2. Thin Solid Films. 2001;**381**:171-175. DOI: 10.1016/S0040-6090(00)01739-9

[6] V.S. Poddar, Influence of additives on transformation characteristics and thermoelectric behaviour of β-iron Disilicide [thesis]. India: Dept. of Meta and Matls Sci.: College of Engineering Pune, SPPU University; 2019.

[7] Ohtaki M. Nanostructured oxide thermoelectric materials with enhanced phonon scattering. Oxide Thin Films, Multilayers, and Nanocomposites. 1st ed. Cham: Springer; 2015. pp. 109-122. DOI: 10.1007/978-3-319-14478-8_7

[8] Johansson MT, Söderström M. Electricity generation from low-temperature industrial excess heat—An opportunity for the steel industry. Energy Efficiency. 2013;**7**(2):203-215. DOI: 10.1007/s12053-013-9218-6

[9] Rowe DM. CRC Handbook of Thermoelectric. 1st ed. Boca Raton: CRC Press; 1994. pp. 1-20

[10] Abram Ioffe LS. Stil'bans, physical problems of thermoelectricity. Reports on Progress in Physics. 2002;**22**:167-203. DOI: 088/0034-4885/22/1/306

[11] Kishore RA, Priya S. A review on low-grade thermal energy harvesting: Materials, methods and devices. Materials (Basel). 2018;**11**(8):1433. DOI: 10.3390/ma11081433

[12] Jérôme D, Pierre A, Simon W, Ndjaka J. Influences of thermal cycles on thermoelectric Uni-leg modules made by Ca3Co4O9/Ca0.95Sm0.05MnO3 oxides. Journal of Taibah University for Science. 2014;**8**(4):1-10. DOI: 10.1016/j.jtusci.2014.05.003

[13] Poddar VS, Dhokey NB. Thermoelectric properties of iron Disilicide and manganese silicide: Synthesis and characterization. Transactions of the Indian Institute of Metals. 2019;**72**(10):2711-2719. DOI: 10.1007/s12666-019-01743-8

[14] Kasap S. Principles of Electronic Materials and Devices. 3rd ed. New York: McGraw-Hill; 2005. pp. 373-443. DOI: 10.5555/1594045

[15] Gibbs Z, Hyun-Sik K, Heng W, Jeff S. Band gap estimation from temperature dependent Seebeck measurement—Deviations from the 2e|S|maxTmax relation. Applied Physics Letters. 2015;**106**:022112. DOI: 10.1063/1.4905922

[16] Tritt T. Thermoelectric materials, phenomena, and applications: A Bird's eye view. Annual Review of Materials Research. 2011;**41**:433-448. DOI: 10.1557/mrs2006.44

[17] Kim HJ, Skuz J, Park Y, King G, Choi S, Nagavalli A, System to Measure Thermal Conductivity and Seebeck Coefficient for Thermoelectrics, NASA STI program - Technical Memorandum, NASA/TM–2012-217791

[18] Resistivity by Four Probe Method [Internet]. 2013. Available from: http://vlab.amrita.edu/?sub=1&brch=282&sim=1512&cnt=1 [Accessed: March 12, 2022]

[19] Zhao L-D, Dravid VP, Kanatzidis MG. The panoscopic approach to high performance thermoelectrics. Energy and Environmetal Science. 2014;**7**:251-268. DOI: 10.1039/c3ee43099e

[20] Sootsman J, Chung DY, Kanatzidis M. New and old concepts in thermoelectric materials. Angewandte Chemie International Edition. 2009;**48**: 8616-8639. DOI: 10.1002/anie.200900598

[21] Klochko NP, Klepikova KS, Zhadan DO, Petrushenko SI, Kopach VR, Khrypunov GS, et al. Structure, optical, electrical and thermoelectric properties of solution-processed Li-doped NiO films grown by SILAR. Materials Science in Semiconductor Processing. 2018;**83**: 42-49. DOI: 10.1016/j.mssp.2018.04.010

[22] Fedorov M. Thermoelectric silicides: Past present and future. Journal of Thermoelectricity. 2009;**2**:51-60

[23] Zabek D, Morini F. Solid state generators and energy harvesters for waste heat recovery and thermal energy harvesting. Thermal Science and Engineering Progress. 2019;**9**:235-247. DOI: 10.1016/j.tsep.2018.11.011

[24] Poddar VS, Dhokey NB. Evaluation of thermoelectric properties of doped β-iron Disilicide prepared by the powder metallurgy technique. Transactions of the Indian Institute of Metals. 2021;**74**: 399-410. DOI: 10.1007/s12666-020-02167-5

[25] He J, Sootsman JR, Girard SN, Zheng J-C, Wen J, Zhu Y, et al. On the origin of increased phonon scattering in nanostructured PbTe based thermoelectric materials. Journal of the American Chemical Society. 2012; **132**(25):8669-8675. DOI: 10.1021/ja1010948

[26] Balout H, Boulet P, Record MC. Thermoelectric properties of Mg2Si thin films by computational approaches. The Journal of Physical Chemistry C. 2001; **118**(34):19635-19645. DOI: 10.1021/jp506323r

[27] Salam L, Matthews R, Robertson H. Fabrication of iron disilicide (FeSi2) thermoelectric generator by the tape casting method. Materials and Design. 1999;**20**:223-228. DOI: 10.1016/S0261-3069(99)00031-X

[28] Chen HY, Zhao XB, Lu YF, Mueller E, Mrotzek A. Microstructures and thermoelectric properties of Fe0.92Mn0.08Six alloys prepared by rapid solidification and hot pressing. Journal of Applied Physics. 2003;**94**(10): 6621-6626. DOI: 10.1063/1.1622773

[29] Suryanarayana C. Mechanical alloying and milling. Progress in Materials Science. 2001;**46**:119-184. DOI: 0079-6425/01/S

[30] Poddar VS, Dhokey NB, Garbade RR, Butee SP, Deep Prakash RD, Purohit. Rapid production

of iron disilicide thermoelectric material by hot press sintering route. Materials Science in Semiconductor Processing Journal. 2017;**71**:477-481. DOI: 10.1016/j.mssp.2017.06.027

[31] Zhou Z, Uher C. Apparatus for Seebeck coefficient and electrical resistivity measurements of bulk thermoelectric materials at high temperature. Review of Scientific Instruments. 2005;**76**:023901-023901-5. DOI: 10.1063/1.1835631

[32] Bergman TL, Lavine AS, Incropera FP, DeWitt DP. Introduction to Heat Transfer. 6th ed. Jefferson City: Wiley; 2011. p. 917. Available from: books.google.co.in/books?id=YBaNa LurTD4C

[33] Cangel Y, Ghajar A. Heat and Mass Transfer- Fundamentals and Applications. 5th ed. New York: Mc-Graw Hill; 2015. p. 939

[34] Poddar VS. Eco-friendly transition metal silicide for high temperature thermoelectricity. Journal of Materials Science: Materials in Electronics. 2023; **34**:1-11. DOI: 10.1007/s1085F164-023-11472-w

[35] Skomedal G, Holmgren L, Hugh Middleton IS, Eremin GNI, Jaegle M, Tarantik K, et al. Design, assembly and characterization of silicide-based thermoelectric modules. Energy Conversion and Management. 2016;**110**: 13-21. DOI: 10.1016/j.enconman.2015.11.068

[36] Bonin R, Boero D, Chiaberge M, Tonoli A. Design and characterization of small thermoelectric generators for environmental monitoring devices. Energy Conversion and Management. 2013;**73**:340-349. DOI: 10.1016/j.enconman.2013.05.016

[37] Zhou Z-G, Zhu D-S, Hong-Xia W, Zhang H-S. Modeling, experimental study on the heat transfer characteristics of thermoelectric generator. Journal of Thermal Science. 2013;**22**(1):48-54. DOI: 10.1007/s11630-013-0591-4

[38] Liu Y, Xu L, Zhao C, Shao M, Hu B. Tuning the Seebeck effect in C60-based hybrid thermoelectric devices through temperature-dependent surface polarization and thermally-modulated interface dipoles. Physical Chemistry Chemical Physics. 2017;**19**(22):14793-14800. DOI: 10.1039/c7cp01736g

[39] Katsuyama S, It M, Nagai H. Preparation of the composite thermoelectric materials with small particles dispersion by MA. Kona Powder and Particle Journal. 2004;**22**: 186-194. DOI: 10.14356/kona.2004022

[40] Vining C. Thermoelectric Properties of Silicides. 1st ed. California: CRC Press; 1995. pp. 1-18. DOI: 10.1201/9781420049718-23

[41] Vaishali Poddar NB, Dhokey SP, Butee NB, Revade MM, Thombre RD, Purohit DP. Evolution of thermoelectric β – FeSi2 phase by Cryo milling and sintering. Transaction of Indian Institute of Metals. 2017;**70**(1):167-174. DOI: 10.1007/s12666-016-0873-0

[42] Ware RM, McNeill DJ. Iron disilicide as a thermoelectric generator material. Journal IEE. 1964;**111**(1):518. DOI: 10.1049/piee.1964.0029

[43] Gayner C, Kar KK. Recent advances in thermoelectric materials. Progress in Materials Science. 2016;**83**:330-382. DOI: 10.1016/j.pmatsci.2016.07.002

[44] Sahin AZ, Yilbas BS. The thermoelement as thermoelectric power generator: Effect of leg geometry on the efficiency and power generation. Energy

Conversion and Management. 2013;**65**: 26-32. DOI: 10.1016/j.enconman.2012.07.020

[45] Rowe DM, Min G. Design theory of thermoelectric modules for electrical power generation. IEEE Proceedings - Science, Measurement and Technology. 1996;**143**(06):351-356. DOI: 10.1049/ip-smt_19960714

[46] Leavitt F, Elsner N, Bass J. Use, application and testing of hi-Z thermoelectric modules. In: Proceedings of the 15th International Conference on Thermoelectrics. Pasadena, CA, USA: ICT '96; 1996. pp. 378-382. DOI: 10.1109/ICT.1996.553508

[47] Benedict TS, Angello SJ, Berman R, Adam G, Apker L, et al. Methods in experimental physics. In: Lark-Horovitz K, Johnson VA, editors. 1st ed. UK: Academic Press; 1959. pp. 1-143

[48] Namir D, Beck J. On the significance of the thermoelectric figure of merit Z. Journal of Electronic Materials. 2012;**39** (9):1897-1901. DOI: 10.1007/s11664-009-1060-4

[49] Poddar VS, Ranawade VA, Dhokey NB. Study of synergy between photovoltaic, thermoelectric and direct evaporative cooling system for improved performance. Renewable Energy. 2021; **182**:817-826. DOI: 10.1016/j.renene.2021.10.040

[50] Li G, Shittu S, Ma X, Zhao X. Comparative analysis of thermoelectric legs optimum geometry between photovoltaic-thermoelectric and solar thermoelectric. Energy. 2019;**171**:599-610. DOI: 10.1016/j.energy.2019.01.057

[51] Chen X, Zhou Z, Lin Y-H, Nan C. Thermoelectric thin films: Promising strategies and related mechanism on boosting energy conversion performance. Journal of Materiomics. 2020;**6**(3):494-512. DOI: 10.1016/j.jmat.2020.02.008

[52] Wills BA, Napier-Munn T. Mineral Processing Technology. 7th ed. UK: Elsevier; 2006. p. 3. DOI: 10.1016/B978-0-7506-4450-1.X5000-0

Chapter 14

Clay-Based Ceramic Membranes

Khedidja Makhloufi, Issa Samb and Fayeda Srarfi

Abstract

Faced with water shortages and the need to recycle wastewater, ceramic membranes stand out as an efficient, environmentally friendly solution for filtration. Although their manufacturing cost is relatively high, the use of clay, an abundant and inexpensive material, reduces these costs. This chapter examines the importance of flat clay-based ceramic membranes, highlighting their durability and cost-effectiveness. The chapter covers membrane-manufacturing steps, including raw material conditioning, powder preparation, shaping, and heat treatment or sintering. It highlights the incorporation of pore-forming agents to improve membrane porosity and permeability. All in all, the use of clay makes it possible to manufacture cost-effective, high-performance ceramic membranes tailored to today's water treatment needs.

Keywords: ceramic membrane, clay, filtration, pore-forming agent, sintering

1. Introduction

Over the last century, water consumption has increased at more than twice the rate of human population growth, making water scarcity one of the most pressing challenges facing humanity [1]. The problems associated with water scarcity and climate change make wastewater treatment essential to guarantee sufficient, high-quality water resources. Among treatment processes, membrane filtration has proven to be one of the most effective methods of water purification. This process, based on the passage of a liquid through a porous structure under the action of various driving forces (thermal, osmotic, electrical, or pressure), is widely used in the treatment of industrial effluents and wastewater [2].

Membranes can be classified into two broad categories: organic and inorganic. Organic membranes, mainly made of polymers, dominate the market due to their relatively low production costs [3]. Although they cover a wide range of membrane processes, they have significant limitations, such as low resistance to temperature and corrosion and limited service life [4]. These drawbacks restrict their use under severe conditions.

On the other hand, inorganic membranes, notably ceramic membranes made from metal oxides such as alumina, titanium, and zirconia, offer better thermal and mechanical resistance, as well as long service life [5]. These membranes, particularly those made from alumina [6–8], are among the most commonly used ones. However, their production costs are high, mainly due to the high sintering temperatures required (in excess of 1500°C) to achieve good mechanical strength and porosity [9].

 IntechOpen

In recent years, the use of low-cost raw materials as precursors for ceramic membranes has attracted increasing attention. One promising solution for reducing the production costs of ceramic membranes lies in the use of abundant, low-cost materials such as clay [10]. Clay is a natural sedimentary rock composed mainly of phyllosilicate minerals rich in silica and alumina. It is characterized by its plasticity when mixed with water and its ability to harden when heated [11]. Clay, available in various forms such as kaolin, fire clay, bending clay, and bentonite, offers significant economic advantages. In addition, clays can be molded into various shapes and are durable [12].

Membrane technology has long been considered one of the most efficient methods for industrial effluent filtration and wastewater treatment [13]. Ceramic membranes have been widely used in a variety of fields, including food and beverage, biotechnology, chemistry, and waste recovery [9, 14–16]. They can be regenerated for multiple use cycles and are particularly effective for separations of aqueous and non-aqueous solutions, including in the petrochemical sector where organic membranes cannot be used [17]. Numerous ceramic membranes are already available on the market for industrial separation applications [18].

This chapter will focus on flat clay-based ceramic membranes, highlighting their growing importance in today's context of sustainable water resource management. We will examine the advantages and challenges associated with their manufacture and use and detail the processes involved in manufacturing and characterizing these membranes.

2. Ceramic membrane manufacture

The manufacture of ceramic membranes is based on proven technologies from the ceramics industry, which are organized into several essential steps as recommended by Fantozzi et al. [19].

2.1 Raw material conditioning

The process begins with the conditioning of raw materials, where the necessary components are meticulously prepared and measured to ensure optimum quality and homogeneity of the materials used in membrane preparation. This initial stage is of paramount importance, as it includes the rigorous selection of clays, renowned for their abundance and suitability for the economical manufacture of ceramic membranes. These choices are guided by their physicochemical properties, such as plasticity and drying behavior, which are essential for obtaining robust, efficient membranes.

2.2 Material preparation

Membrane performance is closely linked to the materials of which they are made. Membrane composition directly influences characteristics such as rejection rate (selectivity), clogging tendency, mechanical strength, and reactivity [20]. To meet the growing demand for cost-effective ceramic membranes, several researchers have turned their attention to the use of abundant natural clays such as kaolinitic clay, kaolinitic clay, smectic clay, and Moroccan pozzolan clay [16, 21–24]. These materials offer a promising alternative due to their abundance in nature and properties suitable for the manufacture of efficient, cost-effective ceramic membranes.

2.3 Clay

Clays occupy a prominent place among the minerals mostly present on the earth's surface, reaching around 16%, as reported by Al-Ani and Sarapää, [25]. Due to their physicochemical properties, clays possess a wide variety of uses [26–28]. In particular, the plasticity of clays, their behavior during drying, and the mechanical strength of products after drying are also of great importance [29].

All these clays have different compositions due to different geological conditions [2], for example, a Chinese kaolinite clay will not have the same composition as a Tunisian kaolinite clay. Clay is considered one of the most widely used industrial minerals, with increasing demand in a wide range of industrial applications, such as ceramics, paints, rubbers, plastics, and refractory industries [30, 31]. Clay is the preferred raw material for the manufacture of porous ceramics [10].

2.4 Agent porogène

A porogène can be added to the ceramic powder to increase the porosity of the final material, thus improving membrane permeability [32]. Porosity, defined as the volume of voids within a porous substance [33], is essential for filtration applications. Obada et al. [34] reported that the porosity of ceramic membranes increased in proportion to the percentage of the pore-forming agent added.

Many materials can be used as pore formers, such as potato starch [35], sago starch [36], rice bran [37], sawdust [38], phosphate washing sludge [24, 39], cattle bone ash [40], rice husk waste [41], coconut shells and eggshells [42], sugarcane bagasse [43], and banana peel powder [44]. Pure substances such as urea can also be used as pore formers [45].

However, increasing the percentage of the porogen agent leads to a decrease in the mechanical strength of the ceramic membrane and an increase in shrinkage behavior [46]. Higher porosity, resulting from an increased percentage of porogène, makes the membrane less mechanically robust, and the sample size decreases considerably.

The porogène also has a significant impact on the sintering process. A notable phenomenon can occur due to the combustion of this porous material and the loss of moisture during sintering. As the percentage of the pore-forming agent used increases, so does membrane shrinkage. For zero percent pore-forming agent, shrinkage is mainly due to moisture loss [46]. At the same time, the density of the membrane decreases as the pore content increases. This decrease in density is caused by the removal of the pore-forming agent, leading to the formation of pores in the material's structure, which affects its physical properties.

2.5 Manufacturing

There are a number of different processes for manufacturing ceramic membranes, among which dry pressing stands out as an efficient and environmentally friendly method. Unlike wet methods, dry pressing does not require the use of water, making it environmentally friendly. In addition, uniaxial pressing specifically densifies the membrane structure by compressing the ceramic powder in a mold under uniaxial pressure. This increased compression promotes close contact between particles, reducing porosity and facilitating the formation of intermolecular bonds and solid bridges between grains [47, 48].

The uniaxial pressing process involves several key steps: first, the ceramic powder is carefully distributed and compacted in the mold, determining the final thickness of the membrane. Next, uniaxial pressure is applied for a set time, consolidating the particles and improving the cohesion of the structure. After this compression phase, decompression is performed by slowly withdrawing the plunger from the mold, allowing the elastic energy of the powder to recover. However, this process can sometimes lead to defects such as pellet cleavage, a problem that can be mitigated by the judicious addition of plasticizers to improve flexibility and resistance to deformation.

In comparison, isostatic pressing uses uniform hydrostatic pressure applied equally to all surfaces of the mold filled with ceramic powder. This process is preferred for achieving uniform, controlled densification of the membranes, ensuring homogeneous distribution of physical properties throughout the structure.

In short, the choice between uniaxial and isostatic pressing depends on the specific density, porosity, and mechanical strength requirements of ceramic membranes. These techniques play a crucial role in the economical and efficient manufacture of clay-based membranes, meeting the growing needs of water treatment and other industrial applications.

2.6 Thermal treatment or sintering

Before a ceramic membrane achieves its final properties, it must undergo the drying and sintering stages essential for its consolidation and functionality.

2.6.1 Drying

Drying of ceramic membranes is generally characterized by measuring mass loss as a function of time but can also be expressed as a function of humidity [49]. In general, drying corresponds to the removal of solvent from the membranes [19, 49, 50]. This step is crucial to avoid crack formation during sintering, as proper drying ensures that the membrane is prepared for the high temperatures of the sintering process.

2.6.2 Sintering

Sintering is the final stage in the production of ceramic membranes. This process transforms a solvent-bonded powder mass into a dense, coherent part, under the effect of heat treatment while remaining below the melting temperature of the main constituent [49, 50]. The smaller the particle size of the powders, the faster the sintering process, as the contact points between individual particles fuse more easily when the temperature reaches a sufficiently high level. This leads to a transformation from solid-vapor to solid-solid interfaces [50].

The two main mechanisms during sintering are densification and granular growth/coalescence [19]. A typical sintering thermal program consists of several temperature steps. The first stage corresponds to the desorption of physisorbed water. This step is essential to prevent the formation of fissures during sintering, requiring the ceramic membrane to be dried at a temperature of between 100 and 110°C before the temperature is increased [51]. The second stage is designed to remove organic additives and other volatiles from the membrane. The third stage is generally for final consolidation of the material.

The consolidation temperature is essential and must be adapted to the materials used to avoid defects in the final product. Too high a sintering temperature can lead to excessive densification of the ceramic material, causing complete pore closure [52].

On the other hand, too low a temperature can produce an unconsolidated substrate with low mechanical strength. Precise control of the temperature and duration of each stage is therefore essential to obtain ceramic membranes with optimum properties.

3. Characterization of ceramic materials (clay and pore-forming agent)

Prior to use, raw materials need to be characterized to gain in-depth knowledge of the samples. This characterization is divided into several essential aspects.

3.1 Mineralogical characterization by X-ray diffraction (XRD)

The mineralogy of clays is studied to identify the different minerals present. This analysis is crucial to understanding the physical and mechanical properties of ceramic materials, as well as their behavior during sintering.

3.2 Chemical analysis using X-ray fluorescence (XRF)

It determines the elemental composition of clay in the form of oxides, which is important for understanding its physicochemical properties. For example, iron oxide (Fe_2O_3) gives a reddish coloration to clay membranes after sintering. The color of ceramics can also be influenced by the CaO content, which can give yellowish and pinkish hues [53, 54], and by titanium oxide (TiO_2), which contributes to a brownish color. The oxides K_2O, Na_2O, $Fe2O_3$, CaO, and MgO are fluxing constituents that promote vitrification and influence the densification of ceramic materials during firing, thus reducing the temperature required [55]. A geochemical analysis of clays is decisive in predicting chemical reactions during firing. Silica-rich clay behaves better than silica-poor clay [56]. Silica plays a fundamental role in the densification of the ceramic product during high-temperature firing.

3.3 Thermal characterization

Clays and pore-formers need to be thermally characterized to study their crystalline structure and transformations. Differential thermal analysis (DTA) and thermogravimetric analysis (TGA) are necessary, because during the sintering process, several temperature steps must be respected to avoid manufacturing defects and ensure product homogeneity. During heating, materials undergo physical or chemical transformations, such as phase changes, structural modifications, decomposition, and volume variations [57, 58].

3.4 Determination of organic matter and carbonates

The determination of organic matter and carbonates in clays is crucial for controlling membrane porosity. The presence of these materials can generate gases during firing, leading to the formation of undesirable pores and cracks in the membrane structure. The decomposition of organic matter and carbonates at specific temperatures must be taken into account to respect the temperature steps necessary for the transformation of clay into ceramic. This ensures the optimum chemical composition of the clay, avoiding the introduction of undesirable elements that could alter membrane properties.

3.5 Determination of humidity

Determining the humidity content of samples enables precise control of the ceramic membrane manufacturing process. Adequate drying is essential to ensure optimum conditions throughout production.

3.6 Determination of plasticity

In the field of ceramics, Atterberg limits are important for the manufacture of membranes, especially for shaping. These limits depend on the mineralogy and particle size distribution of the materials. The more fine particles the raw material contains, the more water it absorbs. Plasticity gives an idea of the amount of water required to manufacture a ceramic paste, especially in the case of wet manufacturing.

In short, complete characterization of raw materials is essential to guarantee the quality and efficiency of the ceramic membranes produced.

4. Some examples of recent industrial and other applications

Ceramic membranes have found diversified applications in many industrial sectors due to their unique properties, such as resistance to high temperatures, chemical and mechanical resistance, as well as their ability to filter fine particles. Ceramic membranes are widely used in drinking water and industrial wastewater treatment. Their robustness enables them to filter out particles, bacteria, and organic substances without deteriorating rapidly. Ceramic microfiltration and ultrafiltration systems are commonly used in water treatment plants to ensure effective purification. Currently, several studies have demonstrated the effectiveness of these membranes for water desalination [59], turbidity reduction [60], dye retention [60], and bacterial retention [42].

5. Conclusion

The objective of this chapter is to demonstrate the importance and benefits of flat clay-based ceramic membranes in water treatment. Although experimental results are lacking, we have provided a detailed description of the manufacturing and characterization processes, as well as the factors influencing membrane properties. This content is essential to convince publishers of the importance and relevance of this research in the current context of sustainable water resource management.

Acknowledgements

I would like to express my sincere gratitude to all the people and institutions who have contributed to this work. Special thanks go to the Centre de Recherche et des Technologies des Eaux (CRTE) in Borj Cédria, Tunisia.

I would also like to thank my colleagues and collaborators for their invaluable help, sound advice, and continued support throughout this project.

Conflict of interest

The authors declare no conflict of interest.

Author details

Khedidja Makhloufi[1*], Issa Samb[2] and Fayeda Srarfi[3]

1 Faculty of Sciences of Tunis, Department of Geology, Laboratory of Mineral Resources and Environment, University of Tunis El Manar, Campus Universitaire, Tunis, Tunisia

2 Organic Chemistry and Therapeutics Team (ECOT), Université Alioune Diop, Bambey, Senegal

3 Laboratory 3G LR18ES37, University of Tunis El Manar, Tunis, Tunisia

*Address all correspondence to: makhloufikhadidja12@gmail.com

References

[1] Food and Agriculture Organization of the United Nations (FAO). Coping with water scarcity: An action framework for agriculture and food security. In: FAO Water Reports No. 38. 2012. Available from: https://www.fao.org/4/i3015e/i3015e.pdf

[2] Samhari O. Ceramic and Polymer Membranes Modified by Graphene Oxide for the Removal of Organic Molecules and the Desalination of Brackish and Seawater Thesis. France: University of Rennes 1; 2021

[3] Huang SC, Huang CT, Lu SY, Chou KS. Ceramic/polyaniline composite porous membranes. Journal of Porous Materials. 1999;**6**:153-159. DOI: 10.1023/A:1009687523387

[4] Azaman F, Nor MAAM, Abdullah WRW, Razali MH, Zulkifli RC, Zaini MAA, et al. Review on natural clay ceramic membrane: Fabrication and application in water and wastewater treatment. Malaysian Journal of Fundamental and Applied Sciences. 2021;**17**:62-78

[5] Jana S, Saikia A, Purkait MK, Mohanty K. Chitosan based ceramic ultrafiltration membrane: Preparation, characterization and application to remove Hg (II) and As (III) using polymer enhanced ultrafiltration. Chemical Engineering Journal. 2011;**170**(1):209-219

[6] Li K. Ceramic Membranes for Separation and Reaction. John Wiley & Sons; 2007

[7] Li W, Ling GQ, Huang P, Li K, Lu HQ, Hang FX, et al. Performance of ceramic microfiltration membranes for treating carbonated and filtered remelt syrup in sugar refinery. Journal of Food Engineering. 2016;**170**:41-49. DOI: 10.1016/j.jfoodeng.2015.09.012

[8] Wei Z, Hou J, Zhu Z. High-aluminum fly ash recycling for fabrication of cost-effective ceramic membrane supports. Journal of Alloys and Compounds. 2016;**683**:474-480. DOI: 10.1016/j.jallcom.2016.05.088

[9] Li L, Chen M, Dong Y, Dong X, Cerneaux S, Hampshire S, et al. A low-cost alumina-mullite composite hollow fiber ceramic membrane fabricated via phase-inversion and sintering method. Journal of the European Ceramic Society. 2016;**36**(8):2057-2066. DOI: 10.1016/j.jeurceramsoc.2016.02.020

[10] Ganesh I, Ferreira JM. Influence of raw material type and of the overall chemical composition on phase formation and sintered microstructure of mullite aggregates. Ceramics International. 2009;**35**(5):2007-2015. DOI: 10.1016/j.jeurceramsoc.2016.02.020

[11] Caillère S. S, Henin et M. Rautureau, Les argiles, éd. Paris: Septima; 1989. p. 126

[12] Elma M, Yacou C, Diniz da Costa JC, Wang DK. Performance and long term stability of mesoporous silica membranes for desalination. Membranes. 2013;**3**(3):136-150. DOI: 10.3390/membranes3030136

[13] Hube S, Eskafi M, Hrafnkelsdóttir KF, Bjarnadóttir B, Bjarnadóttir MÁ, Axelsdóttir S, et al. Direct membrane filtration for wastewater treatment and resource recovery: A review. Science of the Total Environment. 2020;**710**:136375. DOI: 10.1016/j.scitotenv.2019.136375

[14] Jedidi I, Khemakhem S, Larbot A, Amar RB. Elaboration and characterisation of fly ash based mineral supports for microfiltration and ultrafiltration membranes. Ceramics International. 2009;**35**(7):2747-2753. DOI: 10.1016/j.ceramint.2009.03.021

[15] Khemakhem S, Larbot A, Amar RB. New ceramic microfiltration membranes from Tunisian natural materials: Application for the cuttlefish effluents treatment. Ceramics International. 2009;**35**(1):55-61. DOI: 10.1016/j.ceramint.2007.09.117

[16] Baraka NE, Saffaj N, Mamouni R, Laknifli A, Younssi SA, Albizane A, et al. Elaboration of a new flat membrane support from Moroccan clay. Desalination and Water Treatment. 2014;**52**(7-9):1357-1361. DOI: 10.1080/19443994.2013.797542

[17] Guizard C, Ayral A, Julbe A. Potentiality of organic solvents filtration with ceramic membranes. A comparison with polymer membranes. Desalination. 2002;**147**(1-3):275-280. DOI: 10.1016/S0011-9164(02)00552-0

[18] Jeong Y, Lee S, Hong S, Park C. Preparation, characterization and application of low-cost pyrophyllite-alumina composite ceramic membranes for treating low-strength domestic wastewater. Journal of Membrane Science. 2017;**536**:108-115. DOI: 10.1016/j.memsci.2017.04.068

[19] Fantozzi G, Niepce JC, Bonnefont G. Les céramiques industrielles: Propriétés, mise en forme et applications. Dunod; 2013

[20] Rose J, Bottero JY, Levard C, Masion A, Cortalezzi MM, Barron AR, et al. Les membranes céramiques formées à partir de nanoparticules. Actualite Chimique. 2009;**331**:36

[21] Ali MB, Hamdi N, Rodriguez MA, Mahmoudi K, Srasra E. Preparation and characterization of new ceramic membranes for ultrafiltration. Ceramics International. 2018;**44**(2):2328-2335. DOI: 10.1016/j.ceramint.2017.10.199

[22] Misrar W, Loutou M, Saadi L, Mansori M, Waqif M, Favotto C. Cordierite containing ceramic membranes from smectetic clay using natural organic wastes as pore-forming agents. Journal of Asian Ceramic Societies. 2017;**5**(2):199-208. DOI: 10.1016/j.jascer.2017.04.007

[23] Achiou B, Elomari H, Ouammou M, Albizane A, Bennazha J, Younssi SA, et al. Elaboration and characterization of flat ceramic microfiltration membrane made from natural Moroccan pozzolan (central middle atlas). Journal of Materials and Environmental Science. 2016;**7**(1):196-204

[24] Makhloufi K, Moussi B, Srarfi F, Tagorti MA. Development of ceramic filtering membranes based on Tunisian clay and phosphate sludge in wastewater treatment. Solid State Sciences. 2024;**148**:107430. DOI: 10.1016/j.solidstatesciences.2023.107430

[25] Al-Ani, Sarapää O. Clay and Clay Mineralogy. Geologian Tutkuskeskus. M19/3232/41 30.6; 2008

[26] Diawara SO. Caractéristiques géotechniques, chimiques et mécaniques des matériaux argileux utilisés dans la construction de l'actuelle cité de Djenné: essais de renforcement. These de l'université de Bamako (Mali). 2009

[27] Sorgho B. Caractérisation et valorisation de quelques argiles du Burkina Faso: application au traitement des eaux et aux géomatériaux de construction (Doctoral dissertation, Thèse de Doctorat). Universite de Ouagadougou; 2013

[28] Keita I. Géomatériaux argileux du Mali pour la construction. Propriétés mécaniques, durabilité et rôle des tanins (Doctoral dissertation, Thèse de Doctorat). Universite des Sciences Techniques et Technologiques de Bamako; 2014

[29] Magniont C. Contribution à la formulation et à la caractérisation d'un écomatériau de construction à base d'agroressources (Doctoral dissertation). Toulouse 3; 2010

[30] White CE, Provis JL, Riley DP, Kearley GJ, Van Deventer JS. What is the structure of kaolinite? Reconciling theory and experiment. The Journal of Physical Chemistry B. 2009;**113**(19):6756-6765

[31] Al-Shameri A, Lei XR. Characterization and evaluation of algaof kaolin deposits of yemen for industrial application. American Journal of Engineering and Applied Sciences. 2009;**2**(2):292-296. DIO: 10.3844/ajeassp.2009.292.296

[32] Elomari H, Achiou B, Karim A, Ouammou M, Albizane A, Bennazha J, et al. Influence of starch content on the properties of low cost microfiltration membranes. Journal of Asian Ceramic Societies. 2017;**5**(3):313-319

[33] Youmoue M, Fongang RT, Sofack JC, Kamseu E, Melo UC, Tonle IK, et al. Design of ceramic filters using clay/sawdust composites: Effect of pore network on the hydraulic permeability. Ceramics International. 2017;**43**(5):4496-4507

[34] Obada DO, Dodoo-Arhin D, Dauda M, Anafi FO, Ahmed AS, Ajayi OA. Potentials of fabricating porous ceramic bodies from kaolin for catalytic substrate applications. Applied Clay Science. 2016;**132**:194-204

[35] Jamaludin AR, Kasim SR, Abdullah MZ, Ahmad ZA. Sago starch as binder and pore-forming agent for the fabrication of porcelain foam. Ceramics International. 2014;**40**(3):4777-4784

[36] Lorente-Ayza MM, Orts MJ, Pérez-Herranz V, Mestre S. Role of starch characteristics in the properties of low-cost ceramic membranes. Journal of the European Ceramic Society. 2015;**35**(8):2333-2341

[37] Hasan MM, Shafiquzzaman M, Azam MS, Nakajima J. Application of a simple ceramic filter to membrane bioreactor. Desalination. 2011;**276**(1-3):272-277

[38] Bose S, Das C. Role of binder and preparation pressure in tubular ceramic membrane processing: Design and optimization study using response surface methodology (RSM). Industrial & Engineering Chemistry Research. 2014;**53**(31):12319-12329

[39] Loutou M, Misrar W, Koudad M, Mansori M, Grase L, Favotto C, et al. Phosphate mine tailing recycling in membrane filter manufacturing: Microstructure and filtration suitability. Minerals. 2019;**9**(5):318. DOI: 10.3390/min9050318

[40] Mouafon M, Dayirou N, Mohamed H, André N, Gisèle Laure LN, Daniel N. Effect of porogenic agent type and firing temperatures on properties of low-cost microfiltration membranes from kaolin. Transactions of the Indian Ceramic Society. 2020;**79**(1):1-12

[41] Hubadillah SK, Othman MHD, Ismail AF, Rahman MA, Jaafar J, Iwamoto Y, et al. Fabrication of low cost, green silica based ceramic hollow fibre membrane prepared from waste rice husk for water filtration application. Ceramics International. 2018;**44**(9):10498-10509. DOI: 10.1016/j.ceramint.2018.03.067

[42] Kamgang SP. Membranes céramiques à base d'argiles kaolinitiques, de coques de noix de coco et de coquilles d'œufs: Elaboration, caractérisation et mise en œuvre pour la désinfection des eaux destinées à la consommation. Thèse de Doctorat. Cameroun: Faculté des sciences; 2022

[43] Jamalludin MR, Harun Z, Othman MHD, Hubadillah SK, Yunos MZ, Ismail AF. Morphology and property study of green ceramic hollow fiber membrane derived from waste sugar cane bagasse ash (WSBA). Journal of Ceramics International. 2018;**44**:18450-18461. DOI: 10.1016/j.ceramint.2018.07.063

[44] Mouiya M, Abourriche A, Bouazizi A, Benhammou A, El Hafiane Y, Abouliatim Y, et al. Flat ceramic microfiltration membrane based on natural clay and Moroccan phosphate for desalination and industrial wastewater treatment. Desalination. 2018;**427**:42-50. DOI: 10.1016/j.desal.2017.11.005

[45] Vijayan S, Narasimman R, Prabhakaran K. A urea crystal templating method for the preparation of porous alumina ceramics with the aligned pores. Journal of the European Ceramic Society. 2013;**33**(10):1929-1934

[46] Bazin MM, Ahmat MA, Zaidan N, Ismail AF, Ahmad N. Effect of starch addition on microstructure and strength of ball clay membrane. Jurnal Teknologi. 2014;**69**(9):117-120. DOI: 10.11113/jt.v69.3408

[47] Saffaj N, Younssi SA, Albizane A, Messouadi A, Bouhria M, Persin M, et al. Elaboration and properties of TiO2–ZnAl2O4 ultrafiltration membranes deposited on cordierite support. Separation and Purification Technology. 2004;**36**(2):107-114. DOI: 10.1016/S1383-5866(03)00203-X

[48] Jana S, Purkait MK, Mohanty K. Preparation and characterization of low-cost ceramic microfiltration membranes for the removal of chromate from aqueous solutions. Applied Clay Science. 2010;**47**(3-4):317-324

[49] Dejou J. Les céramiques, Société Francophone de Biomatériaux Dentaires. 2010. 27 p

[50] Julian A. Elaboration par coulage en bande et cofrittage de réacteurs catalytiques membranaires multicouches-performances. Thèse de Doctorat. Université de Limoges-France; 2008. 144 p

[51] Saffaj N, Persin M, Younsi SA, Albizane A, Cretin M, Larbot A. Elaboration and characterization of microfiltration and ultrafiltration membranes deposited on raw support prepared from natural Moroccan clay: Application to filtration of solution containing dyes and salts. Applied Clay Science. 2006;**31**(1-2):110-119

[52] Malik N, Bulasara VK, Basu S. Preparation of novel porous ceramic microfiltration membranes from fly ash, kaolin and dolomite mixtures. Ceramics International. 2020;**46**(5):6889-6898. DOI: 10.1016/j.ceramint.2019.11.184

[53] Dondi M, Raimondo M, Zanelli C. Clays and bodies for ceramic tiles: Reappraisal and technological classification. Applied Clay Science. 2014;**96**:91-109. DOI: 10.1016/j.clay.2014.01.013

[54] Eliche-Quesada D, Felipe-Sesé MA, López-Pérez JA, Infantes-Molina A. Characterization and evaluation of rice husk ash and wood ash in sustainable clay matrix bricks. Journal of Ceramics International. 2017;**43**:463-475. DOI: 10.1016/j.ceramint.2016.09.181

[55] Moussi B. Mode de genèse et valorisation de quelques argiles de

la région de Nefza-Sejnane (Tunisie septentrionale) [Doctoral dissertation, PhD thesis]. Carthage University; 2012

[56] Jouenne CA. Traité de céramiques et matériaux minéraux, [Nouv. Éd. Modifiée, Complétée et Actualisée]. Septima; 1990. p. 62

[57] Caillère S, Hénin S, Rautureau M. Les argiles. 2e édition ed. Paris: Ed. Septima; 2004

[58] Kujawa J, Cerneaux S, Kujawski W, Knozowska K. Hydrophobic ceramic membranes for water desalination. Applied Sciences. 2017;**7**:402. DOI: 10.3390/app7040402

[59] Mouiya M, Bouazizi A, Abourriche A, Benhammou A, El Hafiane Y, Ouammo M, et al. Fabrication and characterization of a ceramic membrane from clay and banana peel powder: Application to industrial wastewater treatment. Journal Materials Chemistry and Physics. 2019;**227**:291-301. DOI: 10.1016/j.matchemphys.2019.02.011

[60] Kumar CM, Roshni M, Vasanth D. Treatment of aqueous bacterial solution using ceramic membrane prepared from cheaper clays: A detailed investigation of fouling and cleaning. Journal Water Process Engineering. 2019;**29**:100797. DOI: 10.1016/j.jwpe.2019.100797

Chapter 15

Advanced Ceramics for Photonic Applications: A Rich Landscape

Hamid-Reza Bahari, Ali Karatutlu, Bülend Ortaç and Faisal Rafiq Adikan

Abstract

This comprehensive book chapter delves into cutting-edge advancements in the field of ceramics for photonic applications, a field poised to revolutionize light manipulation and control. The chapter explores the unique properties and synthesis methods of these advanced ceramic materials, which make them ideal for developing innovative photonic devices. The chapter highlights critical areas like photonic crystals, nonlinear optics, integrated photonics, and biophotonic ceramics, showcasing their applications in high-performance optics, sensing, energy harvesting, and biomedicine. We also delve into the potential of transparent ceramics, ceramic upconversion nanoparticles (UCNP), transparent glass/ceramics, rare-earth doped ceramics, and ceramic metamaterials, highlighting the diverse applications of these advanced ceramic materials. By examining the latest research and developments in this rapidly evolving field, the chapter aims to provide a detailed overview of how ceramics can impact photonic technologies and shape the future direction of light-based technologies. Through a thorough review of these materials' properties, synthesis techniques, and applications, this chapter serves as a valuable resource for researchers, scientists, and professionals interested in the intersection of ceramics and photonics.

Keywords: transparent ceramics, nonlinear optical materials, biophotonic ceramics, rare-earth doped ceramics, integrated photonics, ceramic periodic structures, upconversion nanoparticles

1. Introduction

The world of photonics, harnessing the power of light for communication, sensing, energy, and healthcare, is undergoing a transformative shift [1–5]. From revolutionizing telecommunications with high-speed optical fibers to enabling precise medical diagnostics with biophotonic devices, photonics is rapidly expanding its reach across diverse fields. This growth is driven by the relentless pursuit of miniaturization, improved performance, and the discovery of new functionalities [6–9].

While semiconductors and polymers have long dominated the photonic landscape, a new era is dawning with the emergence of advanced ceramics as promising contenders [10]. These materials, traditionally associated with ruggedness and strength, and renowned for their exceptional properties such as high melting points, mechanical strength, and chemical durability, are now being engineered to interact with light in unprecedented ways [10, 11]. Their unique properties such as high refractive index, exceptional hardness, thermal stability, and chemical inertness offer a compelling combination for developing advanced photonic devices and applications [12, 13].

This chapter delves into the fascinating world of advanced ceramic photonics, revealing a treasure trove of possibilities that arise from the unlikely pairing of the rugged, often opaque ceramic and the delicate world of light [1, 10]. We will embark on a journey through the key aspects of this rapidly evolving field, starting with a deep dive into the fundamental properties such as refractive index, transparency, and nonlinearity, which are key to their interaction with light. Next, we delve into the strategies and techniques used to modify and optimize ceramics, including controlled processing methods and precise material composition. We explore the various fabrication techniques employed to create ceramic photonic structures [3, 9]. From traditional powder processing and sintering to advanced sol-gel methods and thin-film deposition, each technique offers unique advantages and challenges [11, 14]. Understanding these fabrication techniques is essential for realizing complex ceramic photonic structures and devices with precise optical properties.

This rich landscape encompasses diverse approaches, from tailoring the transparency and functionality of ceramics through various strategies to controlling light with carefully crafted periodic structures like photonic crystals and metamaterials [3, 9, 15–17]. We will explore the potential of nonlinear optics in ceramics, where the interaction of light with matter becomes more complex and leads to fascinating phenomena like second harmonic generation and saturable absorbing properties.

Furthermore, this chapter will delve into the emerging field of integrated photonics where ceramics are integrated with miniaturized structures, paving the way for high-performance optical devices [18, 19]. Finally, we will explore the exciting prospect of harnessing light for biomedicine and beyond, using biocompatible ceramics to develop innovative biophotonic devices and open new avenues for diagnostics and therapeutics [4, 20, 21].

This chapter will provide a comprehensive overview of the latest advancements in advanced ceramic photonics, highlighting the key challenges and future directions for this rapidly growing field. This chapter serves as a roadmap, guiding us through the exciting landscape of advanced ceramic photonics and its potential to unlock a new wave of innovation in photonics and beyond. Join us as we explore the exciting possibilities of harnessing the power of light through the versatility and resilience of advanced ceramics.

2. Key areas of advanced ceramic photonics

2.1 Transparent ceramics: The transparency of strength

Transparent ceramics (TCs) present a fascinating convergence of material science and optics [6]. While the term "ceramic" often evokes images of robust, opaque materials, TCs defy this perception by exhibiting remarkable optical clarity alongside exceptional strength and durability. These attributes make TCs ideal candidates for a wide range of applications in photonics, pushing the boundaries of what can be achieved with light [6, 7]. TCs are crystalline materials capable of transmitting a significant portion of visible light. Unlike traditional ceramics, they lack scattering centers such as pores, grain boundaries, and inclusions, which typically deflect and absorb light, leading to opacity (**Figure 1**). The transparency of a ceramic is intricately linked to several key factors, including its crystal structure, grain size and distribution, porosity, and the presence of inclusions and impurities [22, 23]. A highly ordered, defect-free crystal structure is paramount for transparency. It allows light to pass through with minimal scattering. Similarly, smaller, uniformly sized grains contribute to reduced scattering. Conversely, large grain boundaries or uneven grain size increase light scattering, diminishing transparency. Pores act as scattering centers, further reducing transparency. Consequently, densification techniques are crucial for producing transparent ceramics with minimal porosity. Furthermore, foreign particles and impurities within the ceramic matrix can also scatter light and compromise transparency. Therefore, strict purity and meticulous control during processing are essential [23, 24]. The interaction of light with a material is governed by its refractive index, which determines the speed of light in the medium. In transparent materials, the refractive index is relatively uniform throughout, enabling light to pass through with minimal deviation [22, 25]. When the refractive index of the material closely matches that of the surrounding medium, such as air, minimal reflection occurs, maximizing transparency. However, variations in refractive index encountered at grain boundaries or pores cause light to scatter, thereby reducing transparency. Despite their potential, achieving high transparency in ceramics poses significant challenges due to the inherent nature of these materials. High-temperature sintering, a crucial step for densification and achieving good transparency, can introduce defects and impurities [26]. Additionally, precise control over grain growth is essential to minimize scattering. Achieving a uniform, fine-grained microstructure requires careful processing. Moreover, trace impurities can significantly impact transparency, necessitating rigorous purification methods to ensure the presence of

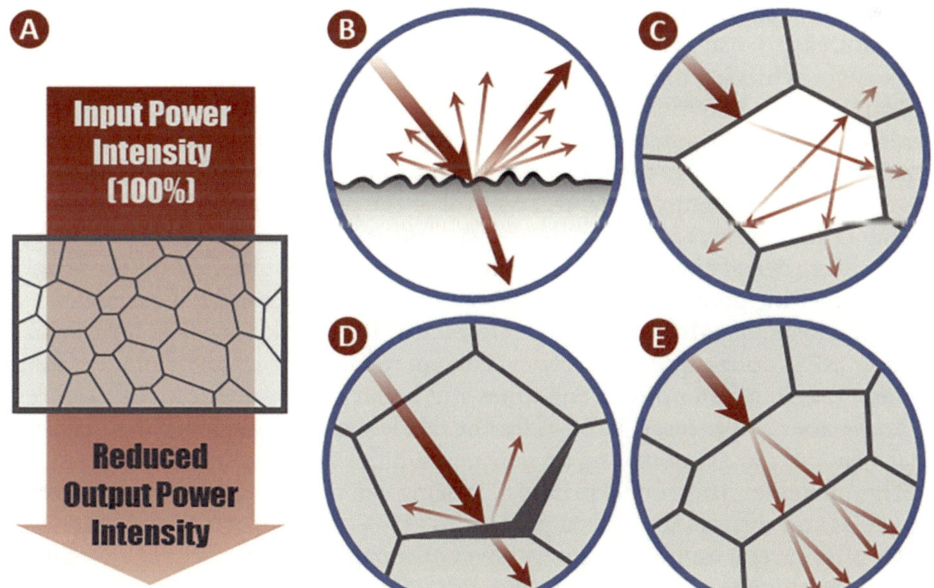

Figure 1.
(A) Light power intensity reduces when passing through ceramics by possible sources of scattering such as (B) surface roughness, (C) pores, (D) secondary phase (impurity) at the grain boundaries, and (E) double refraction.

only the desired elements [27, 28]. Additionally, the random arrangement of grains in ceramics is detrimental as it diminishes transparency. Crystallographic orientation plays a crucial role in improving transparency, particularly in birefringent ceramics, by minimizing birefringence at grain boundaries [29].

TCs can be broadly categorized based on their primary material composition. Following the successful demonstration of laser oscillation in Nd^{3+}-doped $Y_3Al_5O_{12}$ (Nd:YAG) transparent ceramics by Ikesue et al. in 1995 [30], YAG transparent ceramics containing a range of rare-earth ions such as Nd^{3+}, Yb^{3+}, and Er^{3+} have established themselves as the optimal laser host material, attributed to their straightforward manufacturing process and exceptional physical and chemical attributes [28, 31]. Yttria (Y_2O_3), a highly refractive material, finds applications in optical windows, lenses, and high-power lasers. Its exceptional mechanical strength, high melting point, and chemical stability make it a highly desirable material [32–34]. Alumina (Al_2O_3), known for its high hardness and strength, also exhibits transparency [12, 35]. It is frequently used for windows, lenses, and protective coatings due to its high optical performance as well as resistance to physical damages [23]. Other noteworthy materials include magnesium aluminate spinel ($MgAl_2O_4$), known for its high refractive index and thermal stability, used in optical windows and high-power laser applications [36]. Spinel ($MgAl_2O_4$) offers excellent optical properties and thermal shock resistance and a wide transparency window from UV to MIR, making it suitable for optical windows and protective coatings [37]. Zinc sulfide (ZnS) holds promise for infrared optics due to its high transmittance in the infrared region [38]. Finally, calcium fluoride (CaF_2), known for its wide transmission range and its performance as a host for lanthanides, finds applications in ultraviolet and infrared optics especially in laser technology [39, 40].

Various methods are employed for synthesizing TCs, each offering distinct advantages and drawbacks. Traditional powder processing involves synthesizing fine powders of the desired material, followed by pressing and sintering at high temperatures to densify the material [41]. While cost-effective, this method presents challenges in achieving precise control over grain size and porosity [31]. Sol-gel processing, on the other hand, involves the controlled hydrolysis and condensation of precursor materials to form a gel [42]. The gel is then dried and sintered at lower temperatures, resulting in high purity and homogeneity. However, this process can be time-consuming and requires careful control of the precursor chemistry. Hot isostatic pressing (HIP) involves applying high pressure at elevated temperatures, allowing the ceramic to densify and achieve high transparency [15]. It is effective for materials with high melting points but necessitates specialized equipment. Recent advances in fabrication techniques have further propelled the development of TCs. Spark plasma sintering (SPS) utilizes pulsed electrical current to rapidly heat and sinter ceramic powders, enabling faster processing times and better control over microstructure [15, 43]. Laser-assisted processing allows for the fabrication of complex, three-dimensional structures in TCs, opening up new possibilities for photonic devices [44]. Additive manufacturing, or 3D printing, is being explored to create TCs with intricate geometries and customized optical properties [45–47].

TCs offer several unique advantages over other optical materials, making them essential for photonics [6, 22]. Their high refractive indices enable the creation of compact optical devices. Their high hardness, thermal stability, and chemical resistance make them suitable for demanding environments. Some TCs can transmit light across a wide range of wavelengths, from the ultraviolet to the infrared [39]. Additionally, their properties can be fine-tuned by varying the material composition and processing techniques [15]. The applications of TCs in photonics are diverse and rapidly expanding. Their high transparency, strength, and resistance to harsh environments make them ideal for optical windows in spacecraft, lasers, and high-intensity lighting applications [22]. TCs can be fabricated into high-performance lenses for various optical instruments, including microscopes, telescopes, and laser systems. They are also used as host materials for solid-state lasers, providing high-power output and excellent thermal stability [5]. TC fibers offer unique advantages for high-power laser transmission and sensing applications due to their exceptional strength and durability [32]. Transparent ceramics represent a rapidly evolving field with immense potential to revolutionize photonics. Their unique combination of optical clarity, strength, and durability unlocks new possibilities for developing advanced optical devices for various applications. As research and development continue, we can expect even more exciting breakthroughs in TC fabrication and applications, pushing the boundaries of what is possible with light.

2.2 Transparent glass/ceramics: Combining the best of both worlds

Transparent glass/ceramics (TGC) are a unique class of materials that combine the amorphous nature of glass with the crystalline properties of ceramics [16]. This hybrid nature allows TGCs to exhibit superior mechanical, thermal, and optical properties, making them highly suitable for various photonic applications. The formation of TGC involves controlled crystallization within a glass matrix, resulting in a material that maintains transparency while gaining the advantageous properties of ceramics [16, 48].

The formation of TGCs begins with the preparation of a glass precursor, which is then subjected to a controlled heat treatment process known as ceramization [49]. During this process, nucleation and growth of crystalline phases occur within the glass matrix [50]. The key to maintaining transparency lies in the size and distribution of these crystalline phases. If the crystallites are smaller than the wavelength of visible light (typically less than 100 nm), they do not scatter light significantly, thus preserving the material's transparency [48, 49]. The transparency of TGCs is also influenced by the refractive index matching between the glass matrix and the crystalline phases [50]. A close match minimizes light scattering at the interfaces, further enhancing transparency. The absence of impurities and defects is also crucial, as these can introduce scattering centers and reduce optical clarity. The controlled crystallization process ensures that the crystalline phases are uniformly distributed with a control of their size below the critical threshold for light scattering. The mechanism of controlled crystallization in transparent glass/ceramics involves a series of carefully controlled steps that influence the formation of crystalline phases within a glassy matrix [50]. Several parameters, including heat treatment conditions and material composition, play crucial roles in this process [50]. The controlled heating and cooling cycle allow for the nucleation and growth of crystalline phases within the glassy matrix. The temperature and duration of the heat treatment are critical in determining the size, distribution, and composition of the crystalline phases in the TGC. The composition of the glassy matrix in TGC plays a significant role in controlling crystallization [48]. The composition of the glass material can be tailored to promote the formation of specific crystalline phases during heat treatment. Nucleation is the initial formation of small crystalline nuclei within the glassy matrix, which serve as the starting points for crystal growth [48]. Therefore, nucleating agents, such as titanium dioxide or zirconium dioxide, can be added to the glass composition to facilitate the nucleation of crystals. Adding nucleating agents such as zirconia or titania promotes the nucleation process and helps achieve a higher density of crystalline phases. The growth of crystals occurs as the material is further heated and cooled, leading to the development of larger crystalline structures within the glassy matrix [51]. So, crystal growth inhibitors, such as aluminum oxide or magnesium oxide, or bore oxide, can be used to control the growth of crystals and prevent their excessive growth, ensuring a fine and uniform distribution of crystalline phases [49, 52, 53]. By carefully adjusting control parameters, researchers can optimize the formation of crystalline phases and tailor the properties of the material to meet specific requirements, such as transparency, strength, and thermal stability.

Combining the characteristics of glass and ceramics, TGCs offer several unique properties. One of the most significant advantages is their mechanical strength [44, 54]. TGCs exhibit higher mechanical strength and toughness than conventional glasses, making them more resistant to mechanical damage. This is particularly important in applications where durability and longevity are critical. Another advantage is the enhanced thermal stability provided by the crystalline phases within TGCs. This allows TGCs to withstand higher temperatures without deformation, making them suitable for high-temperature applications [54]. Additionally, TGCs maintain high optical clarity, which is essential for photonic applications [51]. The controlled crystallization process ensures that the material remains transparent while gaining the beneficial properties of ceramics. Chemical durability is another important property of TGCs [48, 55]. They are often more resistant to chemical attack than pure glasses, extending their lifespan in harsh environments. This makes TGCs suitable for applications in chemically aggressive environments where conventional glasses would degrade.

TGCs can be classified based on their material composition. Some crucial types include tellurites [49], phosphates [54], and silicates [51, 56]. Tellurite-based TGCs are known for their high refractive indices and low phonon energies, making them excellent for nonlinear optical applications. Phosphate-based TGCs offer good chemical durability and are used in laser and optical amplifier applications. Silicate-based TGCs are widely used due to their excellent mechanical properties and thermal stability. Several methods are used to synthesize TGCs, including melt quenching [56], sol-gel process [11], and spark plasma sintering (SPS) [51]. Melt quenching involves melting the raw materials and rapidly cooling them to form a glass, followed by controlled heat treatment to induce crystallization. This method is widely used due to its simplicity and scalability. The sol-gel process is a chemical method that involves the transition of a solution into a gel, which is then dried and heat-treated to form TGCs [57]. The sol-gel method offers better control over composition and microstructure, allowing for the synthesis of TGCs with tailored properties. However, it is more complex and time-consuming compared to melt quenching. Spark plasma sintering (SPS) is a technique that uses pulsed electric current to sinter the material at lower temperatures and shorter times, preserving transparency [51]. The SPS technique also provides rapid processing and allows for synthesizing TGCs with high density and uniform microstructure. However, it requires specialized equipment and is less scalable compared to melt quenching. Recent advances in fabrication techniques, such as additive manufacturing and advanced sintering methods, have expanded the possibilities for TGCs [58, 59]. Additive manufacturing allows for the precise control of material composition and microstructure, enabling the synthesis of TGCs with tailored properties. Advanced sintering methods, such as SPS, provide rapid processing and high-density materials with uniform microstructure.

TGCs are crucial for photonic applications due to their unique combination of properties. Significant applications include high-performance optics, displays, and optical fibers [16, 48]. In high-performance optics, TGCs are used in lenses, windows, and other optical components that require high transparency and durability. Their optical clarity and mechanical strength make TGCs ideal for protective covers and substrates in display technologies. In optical fibers, TGCs with tailored refractive indices are used in the core and cladding, enhancing signal transmission and durability [14, 18, 60]. The ability to precisely control the refractive index and maintain high optical clarity makes TGCs suitable for advanced optical fiber applications. State-of-the-art applications of TGCs include laser systems, photonic integrated circuits, and biomedical optics [61, 62]. In laser systems, TGCs are used in high-power laser components due to their excellent thermal and optical properties. The integration of TGCs in photonic circuits enhances performance and miniaturization, enabling the development of compact and efficient photonic devices. In biomedical optics, TGCs are being explored for use in medical imaging and diagnostic devices due to their biocompatibility and optical clarity [62].

In conclusion, transparent glass/ceramics represent a vital class of materials in photonics, offering a unique blend of properties that make them indispensable for advanced optical applications. Continued research and development in synthesis methods.

2.3 Rare-earth doped ceramics: Illuminating photonics applications

Rare-earth (RE) elements, renowned for their unique electronic configurations and sharp, narrow emission lines, have profoundly impacted photonics. Their integration into ceramic matrices, resulting in rare-earth doped ceramics (RE-Cs), has unlocked a plethora of opportunities for enhancing and customizing optical properties, leading to innovative photonic devices [8, 15].

The core principle behind RE doping lies in introducing specific RE ions into the host ceramic lattice. These ions, typically trivalent (RE^{3+}), possess partially filled 4f orbitals shielded by outer 5s and 5p orbitals. This unique electronic structure is responsible for the characteristic sharp emission bands and long fluorescence lifetimes exhibited by RE ions, rendering them ideal for a variety of photonic applications [8, 63]. RE doping significantly influences the optical properties of ceramics, manifesting in several ways. Notably, RE^{3+} ions act as luminescent centers, absorbing energy and emitting light at specific wavelengths. The emitted color is determined by the specific RE^{3+} ion and the host material. Certain RE^{3+} such as Er^{3+}, Tm^{3+}, Ho^{3+} and Nd^{3+} ions exhibit upconversion photoluminescence, a process where they absorb multiple low-energy photons and emit a higher-energy photons, effectively converting longer wavelengths to shorter ones [8, 15, 62, 63]. This capability has immense potential for applications like bioimaging and solar energy conversion. Conversely, some RE ions such as Eu^{3+}, Tb^{3+}, Ce^{3+} and Dy^{3+} display downconversion luminescence, where they absorb high-energy photons and emit lower-energy photons, which is advantageous for improving the efficiency of light-emitting diodes (LEDs) by shifting blue light to longer wavelengths within the visible spectrum [15, 25, 57, 61, 62]. Additionally, RE doping can influence the refractive index of the host ceramic, enabling the development of optical components with specific refractive index profiles [62]. A few common mechanisms of down and upconversion emissions are demonstrated schematically in **Figure 2**.

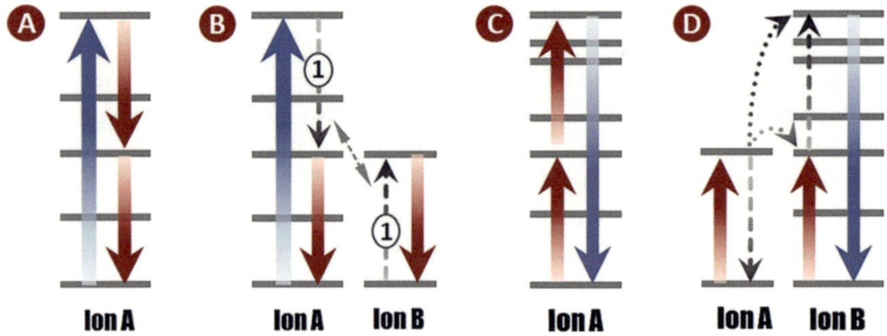

Figure 2.
Schematic view of a few common downconversion mechanisms occurs based on (A) one RE^{3+} ion, and (B) cooperative two RE^{3+} ions, as well as upconversion mechanism observed in (C) one activator ion, and (D) assisted by a sensitizer RE^{3+} ion.

RE-Cs play a pivotal role in various photonic applications. They are widely used as gain media in solid-state lasers due to their high optical gain, narrow emission linewidth, and tunability [62]. These characteristics make them suitable for a wide array of applications, including telecommunications, medical diagnostics, and scientific research [4, 8, 18, 62, 64, 65]. RE-Cs contribute to energy-efficient solid-state lighting (SSL) by providing white light emission. They act as phosphors in white LEDs, converting blue light emitted from the semiconductor chip into longer wavelengths to produce white light [61, 62]. RE-Cs are also essential components in optical amplifiers, which boost signal strength in optical communication systems [18, 62]. They amplify optical signals without significant signal distortion, enabling long-distance communication.

The unique properties of RE-Cs have led to exciting state-of-the-art applications. They are being explored for their potential in quantum computing and communication due to their long coherence times and narrow emission lines, making them suitable for storing and processing quantum information [18, 65]. In biomedical applications, RE-Cs, especially in nanoparticle form, serve as biocompatible luminescent probes for bioimaging and diagnostics, their narrow emission bands and high quantum yields enable high sensitivity and specificity in biological applications [8, 21, 66–68]. RE-Cs are also being incorporated into advanced displays, offering enhanced color purity and efficiency, leading to broader color gamut and improved display quality [25, 50, 69].

In conclusion, rare-earth doped ceramics have revolutionized photonics, offering a wide array of optical functionalities and applications. Their unique luminescent properties and advancements in fabrication techniques continue to drive innovation in lasers, solid-state lighting, optical amplifiers, and emerging fields such as quantum computing and biomedicine. As research progresses, RE-Cs are poised to play an increasingly crucial role in shaping the future of photonics.

2.4 Advanced ceramics for nonlinear optics: Gateway beyond linearity

The realm of photonics is being reshaped by the burgeoning field of nonlinear optics (NLO), which explores the fascinating interactions of light with materials that exhibit nonlinear optical responses. Unlike linear optical phenomena, where the light's behavior is directly proportional to the strength of the applied field, nonlinear optics involves complex interactions where the response of the material is dependent on the intensity of the light. This opens up a vast array of possibilities for manipulating and controlling light in ways previously inconceivable, leading to the development of advanced photonic devices with unique functionalities.

At the heart of nonlinear optics lies the concept of optical nonlinearity, a phenomenon where the polarization of a material changes non-linearly with the electric field strength of the incident light. This nonlinear response gives rise to various intriguing effects, including second or third harmonic generation (SHG, THD), sum and difference of frequency, two or three-photon absorption (2PA, 3PA) and saturable absorption (SA) [2, 70]. In SHG, light interacting with a non-centrosymmetric material is converted to a new frequency, double the original frequency. This process is crucial for generating new light sources and finds applications in frequency conversion, laser design, and microscopy. In 2PA (3PA), two (three) photons at a similar frequency are simultaneously absorbed by the NLO material. Saturable absorbers, on the other hand, have the unique property of absorbing light at low intensities but becoming transparent at high intensities. This property arises from the energy level structure of the dopant ions. This phenomenon is critical for laser Q-switching and mode-locking, enabling the generation of ultrashort laser pulses.

The search for materials possessing specific nonlinear optical properties has ignited intense research efforts, particularly focusing on ceramics. These materials offer several advantages over conventional materials for nonlinear optics. Ceramics exhibit a high resistance to laser-induced damage, making them ideal for high-power laser applications. Their robust chemical stability ensures their performance over time, even under harsh environmental conditions. Moreover, their composition and microstructure can be precisely controlled, allowing for fine-tuning of their optical properties, including nonlinear refractive index and SHG efficiency.

Significant progress has been made in developing new ceramic materials with enhanced nonlinear optical properties. Minerals, particularly perovskite oxides, like strontium barium niobate (SBN), lithium niobate ($LiNbO_3$), barium titanate ($BaTiO_3$), and $KTiOPO_4$ (KTP), have demonstrated remarkable SHG efficiencies and are under investigation for frequency conversion applications in telecommunications and laser technologies [1, 71, 72]. Rare-earth compounds with noncentrosymmetric crystal structures, like $La_3Ga_5SnO_{14}$, $LiCs_2Y_2(PO_4)_3$, $Cd_4LuO(BO_3)_3$, $Na_8Lu_2(CO_3)_6F_2$, $Na_4La_2(CO_3)_5$, $Y(OH)_2NO_3$ and much more, continue to garner widespread attention as an important class of nonlinear optical (NLO) materials [1]. This is because the highly distorted structural arrangements centered around the rare-earth ions can significantly enhance the materials' second harmonic generation (SHG) properties. Chromium (Cr^{3+}) and Cobalt (Co^{2+}) are popular dopants for SA due to their strong absorption in the visible and near-infrared (NIR) region for a variety of laser applications [13, 73]. Cr:YAG, Cr:Sapphire, Cr:ZnS as well as Co:$MgAl_2O_4$, Co:ZnSe, Co:$ZnAl_2O_4$ are well-established materials for passive Q-switching and high-power applications with good thermal properties, enabling the generation of nanosecond, picosecond and femtosecond pulses and also used in integrated optics for pulse shaping and switching applications [13, 73].

The potential applications of nonlinear ceramic materials span various fields within photonics. Frequency conversion using SHG enables efficient signal generation in telecommunications, enhancing bandwidth and data transmission rates. SHG microscopy allows for label-free imaging of biological samples, providing insights into cellular structure and dynamics. Nonlinear ceramics are explored for advanced imaging modalities, like nonlinear optical microscopy, for non-invasive diagnostics and drug delivery [1, 70]. SA ceramics play a vital role in laser Q-switching, enabling the production of high-energy, short laser pulses used in diverse applications like laser cutting, marking, and micromachining [2, 13, 70, 73]. Nonlinear ceramics with sensitivity to specific parameters like temperature, pressure, or chemical species are being explored for developing robust optical sensors.

2.5 Ceramic in periodic structures: Controlling light with periodicity

Ceramic materials have emerged as pivotal components in the design and fabrication of periodic structures such as photonic crystals and metamaterials. These structures are engineered to manipulate electromagnetic waves in ways impossible with conventional materials. Photonic crystals, characterized by their periodic dielectric structures, influence the propagation of photons similarly to how semiconductor crystals affect electron movement [3]. These artificial crystals have a photonic bandgap, which is a range of frequencies where light cannot propagate through the material. This property allows for the control and manipulation of light at the nanoscale level, enabling the creation of devices with unique optical properties. Metamaterials, conversely, are artificially engineered to exhibit properties not found in nature, enabling unprecedented control over electromagnetic waves. Metamaterials are typically composed of artificially engineered structures made from a combination of metals, insulators, or both [74, 75]. The structure of metamaterials consists of subwavelength unit cells that are arranged in a periodic pattern. These unit cells can be made from various materials, such as metals like gold, silver, or copper, or insulators like dielectric materials. The choice of materials and the design of the unit cells determine the electromagnetic properties of the metamaterial, allowing it to exhibit unique optical characteristics. By designing the structure of these materials at the subwavelength scale, it is possible to manipulate light in ways that are not achievable with natural materials. Metamaterials can exhibit negative refractive index, cloaking abilities, and other extraordinary optical properties that go beyond what is possible with traditional materials [74]. Integrating ceramics into these periodic structures has opened new avenues in photonics, offering enhanced performance and novel functionalities [75].

The use of ceramics in fabricating periodic structures is important due to their exceptional properties, such as high-temperature resistance, chemical stability, mechanical strength, and diverse dielectric properties [3, 74, 75]. These properties make ceramics ideal for applications where traditional materials may not be suitable, especially for environments that demand high durability and performance under extreme conditions, such as in high-temperature environments or corrosive conditions. Furthermore, ceramics can be engineered to possess specific refractive indices, such as high refractive index contrast, which is crucial for the precise manipulation of light. The ability to tailor the optical properties of ceramics enhances their utility in creating photonic structures that require exact control over light propagation and interaction [76].

Recent advancements in fabrication techniques have significantly improved the ability to create complex ceramic photonic structures [77]. Techniques such as lithography, etching, and deposition processes have been optimized to create intricate structures with subwavelength features [76]. Additive manufacturing methods, such as 3D printing, allow for the layer-by-layer construction of ceramic materials with nanoscale precision and have also been explored for rapid prototyping of complex photonic structures [78]. This technique enables the production of intricate geometries that are essential for the functionality of photonic crystals and metamaterials. Sol-gel processing is another versatile approach, offering the ability to create ceramics with tailored porosity and refractive indices [17]. Additionally, laser-assisted fabrication techniques provide a means to directly write photonic structures with sub-micron resolution [79]. These methods collectively enhance the precision and functionality of ceramic photonic devices.

Ceramic photonic crystals and metamaterials have found extensive applications in various photonic devices, including optical filters, lasers, and sensors [3, 75, 80]. In optical filters, the periodic structure of photonic crystals creates photonic band gaps that can be engineered to selectively block or transmit specific wavelengths of light. This capability is crucial in applications such as wavelength division multiplexing in optical communications, where precise wavelength control is essential for efficient data transmission. In the realm of lasers, ceramic photonic crystals and metamaterials are employed to manipulate light within the laser cavity [77]. By incorporating these materials, it is possible to achieve enhanced control over the emission properties of the laser, such as wavelength, beam shape, and coherence [3]. The high sensitivity of photonic crystals to environmental changes makes them ideal for sensor applications [17]. This sensitivity is particularly beneficial in fields such as environmental monitoring, medical diagnostics, and industrial process control, where precise and reliable sensing is paramount.

2.6 Integrated photonics: Combining ceramics with miniaturization

Integrated photonics, the integration of optical functions onto a single chip, is revolutionizing various fields, including telecommunications, sensing, and computing [9]. The inherent properties of ceramics make them particularly well-suited for integrated photonic applications. Additionally, ceramics excel in heat dissipation, effectively mitigating thermal effects that can degrade device performance. This is crucial for high-power applications like lasers and optical amplifiers, where heat management is critical [9, 19]. Furthermore, ceramics are robust, resistant to harsh chemicals and environmental conditions, ensuring the long-term reliability and durability of photonic devices. Finally, many ceramics exhibit transparency across a wide range of wavelengths, enabling the development of devices operating in various spectral regions, from visible to infrared [6].

These advantages are already being harnessed in the development of various photonic devices. Ceramic-based waveguides with minimal propagation losses have been demonstrated for applications like high-speed data transmission, optical sensing, and optical interconnects [9]. Integrating multiple optical components like waveguides, splitters, and couplers on a single ceramic chip enables compact, efficient photonic integrated circuits (PICs) for various applications [19]. The high thermal conductivity of ceramics has also enabled the fabrication of high-power lasers with enhanced stability and efficiency.

However, challenges remain in the fabrication and integration of ceramic-based photonic devices. Achieving high-quality, low-loss waveguides with precise dimensions requires advanced microfabrication techniques, which can be complex and costly [10]. Furthermore, finding ceramics with the ideal combination of optical, thermal, and mechanical properties for specific applications can be challenging. Additionally, combining ceramic components with other materials like polymers or silicon requires careful consideration of compatibility and fabrication processes.

Despite these challenges, the potential of ceramics in integrated photonics is significant [9, 10]. Ongoing research focuses on developing new fabrication techniques, exploring novel ceramic materials, and enhancing integration with other photonic components. As this field progresses, we can expect to see miniaturized, high-performance photonic devices based on advanced ceramics, revolutionizing various sectors and opening up exciting possibilities for the future.

2.7 Ceramic-based biophotonics: A new frontier in harnessing light for biomedicine

The burgeoning field of biophotonics seeks to harness the power of light for biomedical applications, enabling advancements in diagnostics, therapeutics, and personalized medicine [8, 64]. While traditional silicon-based photonics has seen considerable success, ceramic materials are increasingly recognized for their unique advantages in biophotonic applications. Their biocompatibility, chemical inertness, and tailored optical properties make them promising candidates for developing advanced biophotonic devices.

Biocompatible ceramics, like hydroxyapatite (HA), bioactive glass, and alumina (Al_2O_3), are gaining traction in biophotonics due to their excellent biocompatibility and favorable optical properties [36, 81]. These materials can be engineered to possess specific optical properties like bioluminescence, fluorescence, or even upconversion, making them suitable for various biomedical applications. For example, HA, with its biocompatibility and excellent bioactivity, has shown potential for optical sensing platforms in bone regeneration and drug delivery applications [81].

Ceramic-based biophotonic devices are designed to leverage the unique optical properties of these materials for biomedical applications. These devices can be designed for optical diagnostics, drug delivery, and therapeutic applications [66, 68]. Upconversion nanoparticles (UCNPs), especially, composed of lanthanide-doped ceramics, exhibit remarkable properties. They can convert low-energy photons (e.g., near-infrared light) into higher-energy photons (e.g., visible light). This upconversion phenomenon has opened new possibilities for biophotonics, particularly in bioimaging and sensing (**Figure 3**) [8, 68]. The use of UCNPs allows

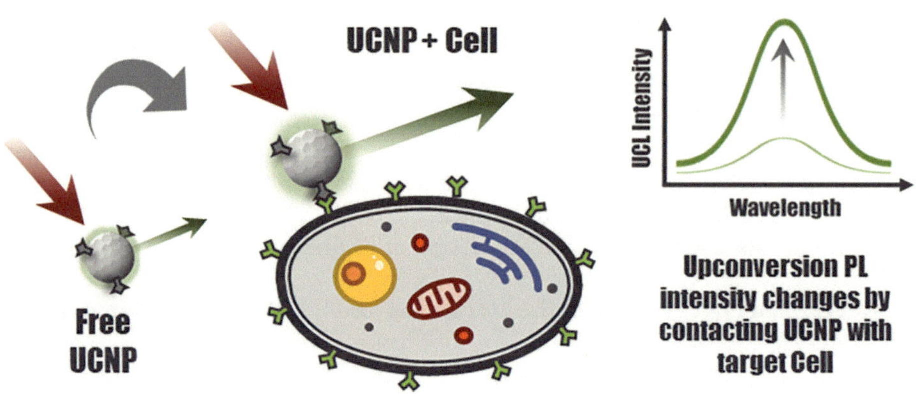

Figure 3.
Schematic demonstration of the application of UCNPs for cellular bioimaging. By excitation of UCNPs by NIR light (red arrows), a visible light (green arrows) can be generated and the corresponding intensity is different between two scenarios of free UCNP and UCNP attached to a target cell.

for deep tissue imaging with minimal autofluorescence and improved signal-to-noise ratio. Moreover, UCNPs can be used to deliver light to specific targets for photodynamic therapy, where they generate reactive oxygen species for cancer cell destruction [66].

Biophotonic ceramics are poised to play a crucial role in the future of personalized medicine and advanced healthcare [68]. The ability to tailor their optical properties and biocompatibility opens up exciting avenues for developing novel biophotonic devices. Integrating ceramics with other materials like polymers and metals can create hybrid devices with enhanced functionality, enabling advancements in personalized medicine, targeted drug delivery, and minimally invasive therapies [8, 82]. The development of biocompatible ceramic micro- and nanostructures, coupled with advances in upconversion technology, could revolutionize optical diagnostics and therapeutics.

In conclusion, biophotonic ceramics represent a promising frontier in biomedical applications, offering enhanced biocompatibility, tailored optical properties, and a wide range of functionalities. Their integration into biophotonic devices holds immense potential for revolutionizing optical diagnostics, therapies, and personalized medicine. Further research and development in this field can pave the way for transformative innovations in healthcare.

3. Conclusion: Challenges and future directions

Advanced ceramic photonics promises to revolutionize light-based technologies, but faces challenges in fabrication, material development, integration, scalability, and standardization. Creating high-quality, transparent, and defect-free ceramic materials at scale remains a hurdle. The search for new ceramics with tailored optical properties like high refractive indices and low losses continues. Integration with other materials and cost-effective scaling are also crucial. To overcome these challenges, researchers are exploring novel fabrication techniques like 3D printing and high-throughput screening for new materials. Advanced characterization methods and hybrid integration strategies are also key. Focusing research on specific applications like high-speed communications and bioimaging will accelerate the transition of these promising materials into real-world solutions.

Acknowledgements

Sincere gratitude is expressed by the authors to the management, editorial, and processing team of IntechOpen for their editing services, which made this publication possible. The use of a complimentary (free version) AI toolbox, including ChatGPT and Grammarly AI assistants, is acknowledged for the language polishing of the manuscript, which was carefully conducted by human drivers to enhance the clarity and effectiveness of the manuscript for the readers.

Author details

Hamid-Reza Bahari[1]*, Ali Karatutlu[2], Bülend Ortaç[2] and Faisal Rafiq Adikan[3,4]

1 CIGHTech, Center of Innovation for Green and High Technologies, Tehran, Iran

2 UNAM, National Nanotechnology Research Center, Bilkent University, Ankara, Turkey

3 Office of the Vice President of Research and Development, Monash University Malaysia, Bandar Sunway, Malaysia

4 School of Electrical and Electronic Engineering, Universiti Sains Malaysia (USM), Engineering Campus, Penang, Malaysia

*Address all correspondence to: hr.bahari.59@gmail.com; hr.bahari@cightech.com

References

[1] Zhao J, Mei D, Wang W, Wu Y, Xue D. Recent advances in nonlinear optical rare earth structures. Journal of Rare Earths. 2021;**39**(12):1455-1466. DOI: 10.1016/j.jre.2021.07.005

[2] Zhang M, Chen H, Yin J, Wang J, Wang J, Yan P. Recent development of saturable absorbers for ultrafast lasers [invited]. Chinese Optics Letters. 2021;**19**(8):081405. DOI: 10.3788/COL202119.081405

[3] Gangwar RK, Pathak AK, Kumar S. Recent progress in photonic crystal devices and their applications: A review. Photonics. 2023;**10**(11):1199. DOI: 10.3390/photonics10111199

[4] Bhardwaj H, Rajesh, and G. Sumana. Recent advances in nanomaterials integrated immunosensors for food toxin detection. Journal of Food Science and Technology. 2022;**59**(1):12-33. DOI: 10.1007/s13197-021-04999-5

[5] Tian F, Ikesue A, Li J. Progress and perspectives on composite laser ceramics: A review. Journal of the European Ceramic Society. 2022;**42**(5):1833-1851. DOI: 10.1016/j.jeurceramsoc.2021.12.061

[6] Akinribide OJ et al. A review on optical properties and application of transparent ceramics. Journal of Materials Research and Technology. 2022;**21**:712-738. DOI: 10.1016/j.jmrt.2022.09.027

[7] Orera VM, Merino RI. Ceramics with photonic and optical applications. Boletin de la Sociedad Espanola de Ceramica y Vidrio. 2015;**54**(1):1-10. DOI: 10.1016/j.bsecv.2015.02.002

[8] Bahari H, Mousavi Khaneghah A, Eş I. Upconversion nanoparticles-modified aptasensors for highly sensitive mycotoxin detection for food quality and safety. Comprehensive Reviews in Food Science and Food Safety. 2024;**23**(3):1-24, e13369. DOI: 10.1111/1541-4337.13369

[9] Teng Y, Sharafudeen K, Zhou S, Qiu J. Glass-ceramics for photonic devices. Journal of the Ceramic Society of Japan. 2012;**120**(1407):458-466. DOI: 10.2109/jcersj2.120.458

[10] Al-Amri AM. Recent progress in printed photonic devices: A brief review of materials, devices, and applications. Polymers (Basel). 2023;**15**(15):3234. DOI: 10.3390/polym15153234

[11] Lam Tran TN et al. Sol-gel-derived transparent glass-ceramics for photonics. Optical Materials. 2022;**130**:112577. DOI: 10.1016/j.optmat.2022.112577

[12] Jin X, Gao L, Sun J. Highly transparent alumina spark plasma sintered from common-grade commercial powder: The effect of powder treatment. Journal of the American Ceramic Society. 2010;**93**(5):1232-1236. DOI: 10.1111/j.1551-2916.2009.03544.x

[13] Chaika M. Advancements and challenges in sintering of Cr4+:YAG: A review. Journal of the European Ceramic Society. 2024;**44**(13):7432-7450. DOI: 10.1016/j.jeurceramsoc.2024.05.050

[14] Quandt A, Ferrari M, Righini GC. Advancement of glass-ceramic materials for photonic applications. In: Sol-Gel Based Nanoceramic Materials: Preparation, Properties and Applications. Cham: Springer International Publishing; 2017. pp. 133-155. DOI: 10.1007/978-3-319-49512-5_5

[15] Qiao J, Xia Z. Luminescent transparent ceramic. In: Advanced Ceramics for Energy Storage, Thermoelectrics and Photonics. Amsterdam, Netherlands: Elsevier; 2023. pp. 463-490. DOI: 10.1016/B978-0-323-90761-3.00001-2

[16] Dymshits O, Shepilov M, Zhilin A. Transparent glass-ceramics for optical applications. MRS Bulletin. 2017;**42**(03):200-205. DOI: 10.1557/mrs.2017.29

[17] Li B et al. Ordered ceramic microstructures from butterfly bio-template. Journal of the American Ceramic Society. 2006;**89**(7):2298-2300. DOI: 10.1111/j.1551-2916.2006.00978.x

[18] Tick PA. Are low-loss glass–ceramic optical waveguides possible? Optics Letters. 1998;**23**(24):1904. DOI: 10.1364/OL.23.001904

[19] Tran TNL et al. Assessment of SnO_2-nanocrystal-based luminescent glass-ceramic waveguides for integrated photonics. Ceramics International. 2021;**47**(4):5534-5541. DOI: 10.1016/j.ceramint.2020.10.137

[20] DaCosta MV, Doughan S, Han Y, Krull UJ. Lanthanide upconversion nanoparticles and applications in bioassays and bioimaging: A review. Analytica Chimica Acta. 17 June 2014;**832**:1-33. DOI: 10.1016/j.aca.2014.04.030

[21] Wang C, Tao H, Cheng L, Liu Z. Near-infrared light induced in vivo photodynamic therapy of cancer based on upconversion nanoparticles. Biomaterials. 2011;**32**(26):6145-6154. DOI: 10.1016/j.biomaterials.2011.05.007

[22] Zhou X, Xia Z. Fundamentals of ceramics for photonics applications. In: Advanced Ceramics for Energy Storage, Thermoelectrics and Photonics. Amsterdam, Netherlands: Elsevier; 2023. pp. 365-394. DOI: 10.1016/B978-0-323-90761-3.00003-6

[23] Tyagi J, Mishra SK, Ahmad S. Transparent ceramics: The material of next generation. In: Metal Oxides for Next-Generation Optoelectronic, Photonic, and Photovoltaic Applications. Amsterdam, Netherlands: Elsevier; 2024. pp. 45-75. DOI: 10.1016/B978-0-323-99143-8.00003-1

[24] Shahbazi H, Tataei M, Enayati MH, Shafiite A, Azizi M, Malekabadi. Structure-transmittance relationship in transparent ceramics. Journal of Alloys and Compounds. 2019;**785**:260-285. DOI: 10.1016/j.jallcom.2019.01.124

[25] Birkel A et al. Rapid microwave preparation of highly efficient Ce^{3+}-substituted garnet phosphors for solid state white lighting. Chemistry of Materials. 2012;**24**(6):1198-1204. DOI: 10.1021/cm3000238

[26] Coble RL. Sintering crystalline solids. II. Experimental test of diffusion models in powder compacts. Journal of Applied Physics. 1961;**32**(5):793-799. DOI: 10.1063/1.1736108

[27] Ikesue A, Aung YL. Ceramic laser materials. Nature Photonics. 2008;**2**(12):721-727. DOI: 10.1038/nphoton.2008.243

[28] Zhang L et al. Defects and solarization in YAG transparent ceramics. Photonics Research. 2019;**7**(5):549. DOI: 10.1364/PRJ.7.000549

[29] Ashikaga T, Kim B-N, Kiyono H, Suzuki TS. Effect of crystallographic orientation on transparency of alumina prepared using magnetic alignment and SPS. Journal of the European Ceramic

Society. 2018;**38**(7):2735-2741. DOI: 10.1016/j.jeurceramsoc.2018.02.006

[30] Ikesue A, Kinoshita T, Kamata K, Yoshida K. Fabrication and optical properties of high-performance polycrystalline Nd:YAG ceramics for solid-state lasers. Journal of the American Ceramic Society. 1995;**78**(4):1033-1040. DOI: 10.1111/j.1151-2916.1995.tb08433.x

[31] Cavalli E et al. YAG:Pr^{3+} transparent ceramics for applications in photonics: Synthesis and characterization. Materials Research Express. 2014;**1**(4):045903. DOI: 10.1088/2053-1591/1/4/045903

[32] Kim W et al. Overview of transparent optical ceramics for high-energy lasers at NRL. Applied Optics. 2015;**54**(31):F210. DOI: 10.1364/AO.54.00F210

[33] Gan L, Park Y-J, Kim H, Kim J-M, Ko J-W, Lee J-W. The effects of the temperature and pressure on ZrO_2-doped transparent yttria ceramics fabricated by a hot-pressing method. Optical Materials. 2017;**71**:109-116. DOI: 10.1016/j.optmat.2016.05.014

[34] Wang J et al. Yb:Y2O3 transparent ceramics processed with hot isostatic pressing. Optical Materials. 2017;**71**:117-120. DOI: 10.1016/j.optmat.2016.04.029

[35] Ghanizadeh S et al. Improved transparency and hardness in α-alumina ceramics fabricated by high-pressure SPS of nanopowders. Ceramics International. 2017;**43**(1):275-281. DOI: 10.1016/j.ceramint.2016.09.150

[36] Talimian A, Pouchly V, El-Maghraby HF, Maca K, Galusek D. Transparent magnesium aluminate spinel: Effect of critical temperature in two-stage spark plasma sintering. Journal of the European Ceramic Society. 2020;**40**(6):2417-2425. DOI: 10.1016/j.jeurceramsoc.2020.02.012

[37] Goldstein A. Correlation between $MgAl_2O_4$-spinel structure, processing factors and functional properties of transparent parts (progress review). Journal of the European Ceramic Society. 2012;**32**(11):2869-2886. DOI: 10.1016/j.jeurceramsoc.2012.02.051

[38] Chen Y et al. Fabrication of transparent ZnS ceramic by optimizing the heating rate in spark plasma sintering process. Optical Materials. 2015;**50**:36-39. DOI: 10.1016/j.optmat.2015.03.058

[39] Li W, Huang H, Mei B, Song J, Xu X. Effect of Y3+ ion doping on the microstructure, transmittance and thermal properties of CaF2 transparent ceramics. Journal of Alloys and Compounds. 2018;**747**:359-365. DOI: 10.1016/j.jallcom.2018.03.059

[40] Liu Z, Mei B, Song J, Yuan D, Wang Z. Microstructure and optical properties of hot-pressed Er:CaF2 transparent ceramics. Journal of Alloys and Compounds. 2015;**646**:760-765. DOI: 10.1016/j.jallcom.2015.05.272

[41] Hostaša J, Picelli F, Hříbalová S, Nečina V. Sintering aids, their role and behaviour in the production of transparent ceramics. Open Ceramics. 2021;7:100137. DOI: 10.1016/j.oceram.2021.100137

[42] Guo K, Chen H-H, Guo X, Yang X-X, Xu F-F, Zhao J-T. Morphology investigation of yttrium aluminum garnet nano-powders prepared by a sol–gel combustion method. Journal of Alloys and Compounds. 2010;**500**(1):34-38. DOI: 10.1016/j.jallcom.2010.03.037

[43] Chaim R, Marder R, Estournès C. Optically transparent ceramics by spark plasma sintering of oxide nanoparticles. Scripta Materialia. 2010;**63**(2):211-214. DOI: 10.1016/j.scriptamat.2010.03.056

[44] Li P, Yan Z, Wang L, Xu Y, Lu P, Zhang J. Advanced transparent glass-ceramics via laser anisotropic nanocrystallization. Laser & Photonics Reviews. Sep 2024;**18**(9):1-10, 2301403. DOI: 10.1002/lpor.202301403

[45] Chen Q, Li H, Han W, Yang J, Xu W, Zhou Y. Three-dimensional printing of yttrium oxide transparent ceramics via direct ink writing. Materials. 2024;**17**(13):3366. DOI: 10.3390/ma17133366

[46] Wang H et al. 3D printing of transparent spinel ceramics with transmittance approaching the theoretical limit. Advanced Materials. 2021;**33**(15):1-9, 2007072. DOI: 10.1002/adma.202007072

[47] Zhang G, Wu Y. Three-dimensional printing of transparent ceramics by lithography-based digital projection. Additive Manufacturing. 2021;**47**:102271. DOI: 10.1016/j.addma.2021.102271

[48] Tarafder A, Molla AR, Karmakar B. Advanced glass-ceramic nanocomposites for structural, photonic, and optoelectronic applications. In: Glass Nanocomposites. Amsterdam, Netherlands: Elsevier; 2016. pp. 299-338. DOI: 10.1016/B978-0-323-39309-6.00013-4

[49] Patra P, Annapurna K. Transparent tellurite glass-ceramics for photonics applications: A comprehensive review on crystalline phases and crystallization mechanisms. Progress in Materials Science. 2022;**125**:100890. DOI: 10.1016/j.pmatsci.2021.100890

[50] Chen D et al. Advances in transparent glass–ceramic phosphors for white light-emitting diodes—A review. Journal of the European Ceramic Society. 2015;**35**(3):859-869. DOI: 10.1016/j.jeurceramsoc.2014.10.002

[51] Singarapu B, Galusek D, Durán A, Pascual MJ. Glass-ceramics processed by spark plasma sintering (SPS) for optical applications. Applied Sciences. 2020;**10**(8):2791. DOI: 10.3390/app10082791

[52] Benitez T, Gómez SY, de Oliveira APN, Travitzky N, Hotza D. Transparent ceramic and glass-ceramic materials for armor applications. Ceramics International. 2017;**43**(16):13031-13046. DOI: 10.1016/j.ceramint.2017.07.205

[53] Liu X, Zhou J, Zhou S, Yue Y, Qiu J. Transparent glass-ceramics functionalized by dispersed crystals. Progress in Materials Science. 2018;**97**:38-96. DOI: 10.1016/j.pmatsci.2018.02.006

[54] Zhang Q et al. Mechanical properties of transparent sodium phosphosilicate glass-ceramics. Physical Review Materials. 2023;7(6):063606. DOI: 10.1103/PhysRevMaterials.7.063606

[55] Yu L, Zhang K, Liu K, Luo B. Chemical durability of Gd2Zr2O7 transparent ceramics under different pH conditions. Ceramics International. 2023;**49**(18):30755-30762. DOI: 10.1016/j.ceramint.2023.07.031

[56] Wang SM, Kuang FH, Yan QZ, Zhang QC, Ge CC. Crystallization behavior of a new transparent glass-ceramics. Advanced Materials Research. 2010;**105-106**:597-599. DOI: 10.4028/www.scientific.net/AMR.105-106.597

[57] Pawlik N, Szpikowska-Sroka B, Goryczka T, Pisarski WA. Sol-gel glass-ceramic materials containing CaF2:Eu3+ fluoride nanocrystals for reddish-orange photoluminescence applications. Applied Sciences. 2019;**9**(24):5490. DOI: 10.3390/app9245490

[58] Moore DG, Barbera L, Masania K, Studart AR. Three-dimensional printing of multicomponent glasses using phase-separating resins. Nature Materials. 2020;**19**(2):212-217. DOI: 10.1038/s41563-019-0525-y

[59] Singh SP, Sontakke AD. Transparent glass ceramics. Crystals (Basel). 2021;**11**(2):156. DOI: 10.3390/cryst11020156

[60] Ferrari M, Righini GC. Glass-ceramic materials for guided-wave optics. International Journal of Applied Glass Science. 2015;**6**(3):240-248. DOI: 10.1111/ijag.12129

[61] Lin H, Hu T, Cheng Y, Chen M, Wang Y. Glass ceramic phosphors: Towards long-lifetime high-power white light-emitting-diode applications—A review. Laser & Photonics Reviews. 2018. pp 1-3, 1700344. DOI: 10.1002/lpor.201700344

[62] de Araújo CB, Kassab LRP, da Silva DM. Optical properties of glasses and glass-ceramics for optical amplifiers, photovoltaic devices, color displays, optical limiters, and random lasers. Optical Materials. 2022;**131**:112648. DOI: 10.1016/j.optmat.2022.112648

[63] Bahari Poor H-R, Sidek HAA, Zamiri R. Ultrasonic and optical properties and emission of Er^{3+}/Yb^{3+} doped lead bismuth-germanate glass affected by Bi^{+}/Bi^{2+} ions. Journal of Luminescence. 2013;**143**:526-533. DOI: 10.1016/j.jlumin.2013.05.053

[64] Markose KK, Anjana R, Jayaraj MK. Upconversion Nanophosphors: An Overview. In: Jayaraj M, editor. Nanostructured Metal Oxides and Devices. Materials Horizons: From Nature to Nanomaterials. Singapore: Springer; 2020. pp. 47-102. DOI: 10.1007/978-981-15-3314-3_2

[65] Kunkel N et al. Rare-earth doped transparent ceramics for spectral filtering and quantum information processing. APL Materials. 2015;**3**(9):1-6, 096103. DOI: 10.1063/1.4930221

[66] Chen G, Qiu H, Prasad PN, Chen X. Upconversion nanoparticles: Design, nanochemistry, and applications in Theranostics. Chemical Reviews. May 2014;**114**(10):5161-5214. DOI: 10.1021/cr400425h

[67] Niu X, Chen H, Yunqing W, Wang XS, Chen L. Upconversion fluorescence-SERS dual-mode tags for cellular and in vivo imaging. ACS Applied Materials & Interfaces. 2014;**6**(7):5152-5160. DOI: 10.1021/am500411m

[68] Chen B, Wang F. Emerging frontiers of upconversion nanoparticles. Trends in Chemistry. 2020;**2**(5):427-439. DOI: 10.1016/j.trechm.2020.01.008

[69] Azlan MN et al. Red emission, upconversion and intensity parameters of erbium oxide doped tellurite glass for laser glass. Journal of Materials Science: Materials in Electronics. 2021;**32**(19):24415-24428. DOI: 10.1007/s10854-021-06917-z

[70] Zaręba JK, Nyk M, Samoć M. Nonlinear optical properties of emerging nano- and microcrystalline materials. Advanced Optical Materials. 2021;**9**(23):1-34, 2100216. DOI: 10.1002/adom.202100216

[71] Cardinal T et al. Chapter 19: Glass-ceramics for engineering optical properties and nonlinear optics for engineering glass ceramics. In: From Glass to Crystal. Les Ulis, France: EDP Sciences; 2020. pp. 441-458. DOI: 10.1051/978-2-7598-1997-3.c026

[72] Gopalakrishnan J, Ramesha K, Kasthuri Rangan K, Pandey S. In search

of inorganic nonlinear optical materials for second harmonic generation. Journal of Solid State Chemistry. 1999;**148**(1):75-80. DOI: 10.1006/jssc.1999.8353

[73] Malyarevich AM, Denisov IA, Yumashev KV, Dymshits OS, Zhilin AA, Kang U. Cobalt-doped transparent glass ceramic as a saturable absorber Q switch for erbium:glass lasers. Applied Optics. 2001;**40**(24):4322. DOI: 10.1364/AO.40.004322

[74] Khurgin JB. Replacing noble metals with alternative materials in plasmonics and metamaterials: How good an idea? Philosophical Transactions of the Royal Society A: Mathematical, Physical and Engineering Sciences. 2017;**375**(2090):20160068. DOI: 10.1098/rsta.2016.0068

[75] Jahani S, Jacob Z. All-dielectric metamaterials. Nature Nanotechnology. 2016;**11**(1):23-36. DOI: 10.1038/nnano.2015.304

[76] Mori H, Kirihara S, Miyamoto Y. Fabrication of three-dimensional ceramic photonic crystals and their electromagnetic properties. Journal of the European Ceramic Society. 2006;**26**(10-11):2195-2198. DOI: 10.1016/j.jeurceramsoc.2005.09.079

[77] Lv X et al. Research progress in preparation and application of photonic crystals. Chinese Journal of Mechanical Engineering. 2023;**36**(1):39. DOI: 10.1186/s10033-023-00836-2

[78] Minas C, Carnelli D, Tervoort E, Studart AR. 3D printing of emulsions and foams into hierarchical porous ceramics. Advanced Materials. 2016;**28**(45):9993-9999. DOI: 10.1002/adma.201603390

[79] Henriques B, Fabris D, Voisiat B, Boccaccini AR, Lasagni AF. Direct laser interference patterning of zirconia using infra-red picosecond pulsed laser: Effect of laser processing parameters on the surface topography and microstructure. Advanced Functional Materials. 2024;**34**(2):1-9, 2307894. DOI: 10.1002/adfm.202307894

[80] Xiong Y et al. Photonic crystal enhanced fluorescence: A review on design strategies and applications. Micromachines (Basel). 2023;**14**(3):668. DOI: 10.3390/mi14030668

[81] Kabilan N, Babu KD, Karthikeyan N, Chinnakali K. Optical nonlinear properties of hydroxyapatite based materials. Optik (Stuttgart). 2022;**265**:169562. DOI: 10.1016/j.ijleo.2022.169562

[82] Saboktakin M et al. Metal-enhanced upconversion luminescence tunable through metal nanoparticle–nanophosphor separation. ACS Nano. 2012;**6**(10):8758-8766. DOI: 10.1021/nn302466r

Chapter 16

Processing and Applications of Composite Ceramic Materials for Emerging Technologies

Olusegun Adigun Afolabi and Oludolapo Akanni Olanrewaju

Abstract

Ceramics are a material that has been used for hundreds of years. The uses range from pottery to complicated aerospace components. They are valued for their toughness, strength, resilience to wear, electrical and thermal insulation, and chemical durability. Furthermore, these materials have a variety of unique optical, chemical, electrical, magnetic, mechanical, and thermal properties, making them suitable for contemporary investigations and developing technologies in medicine, aerospace, communications, electronics, energy, transportation, and chemical manufacturing. There are numerous processing methods for creating ceramic matrix composites. These approaches include the powder metallurgy, sol-gel method, freeze-casting, additive manufacturing, hot pressing and hot isostatic pressing (HIP), and slip-casting processes. This chapter focused on discussing how the different processing methods of ceramic composites have been employed in their manufacturing for various industrial applications.

Keywords: composite material, processing techniques, ceramic materials, emerging technologies, applications

1. Introduction

The term "ceramics" is derived from the Greek word "keramicos," which means burnt materials. They have been around for over a thousand decades and are made of naturally existing basic ingredients. Ceramics are typically inorganic and non-metallic solids (made from powdered components), with relatively low melting points that require high temperatures for production and application. They are typically a compound of various substances or a mixture of compounds containing metallic and non-metallic elements and might be oxides, carbides, borides, or silicides [1].

Composite ceramic materials (CCM) are classes of advanced materials that integrate the exceptional properties of ceramics such as high hardness, electrical insulation, thermal stability, wear resistance, high-temperature stability, and chemical inertness and have emerged as vital components in the development of advanced technologies. Since the early 1990s, Astrium has been developing a leading-edge essential technology known as ceramic matrix composites (CMCs).

Carbon-reinforced silicon carbide (C/SiC) is an intriguing representative of the category of CMCs, which is manufactured utilizing the liquid polymer infiltration (LPI) process [2].

There are two types of ceramics: traditional and advanced. The first group contains inorganic non-metallic substances (either non-metallic or metallic compounds), such as clay (plastic materials), silica (filler), and feldspar (fluxes), whereas the other group includes oxides, nitrate compounds, carbides, and nonsilicate glass. Ceramics are historically manufactured using oxide compositions, which can be amorphous or crystalline. In many circumstances, they fail to be entirely dense and incorporate permeability in the smaller to larger size categories [1].

Traditional ceramics are made using oxide compositions that might be amorphous or crystalline. As a result, they are not entirely dense in their formations and have permeability in both small and large size categories. Advanced ceramics differentiate themselves from typical ceramics in that they have more strength, adaptability, hardness, and higher temperature ratings. **Table 1** compares the benefits and drawbacks of standard and innovative composite ceramic materials. **Table 2** also includes some examples of advanced ceramic materials, showing their structural phases and characteristics.

These materials, which combine the benefits of ceramics with those of other materials, are increasingly being utilized in sectors ranging from aerospace and electronics

Items	Traditional ceramics	Advanced ceramics
Raw materials	Their formations are derived from natural materials and not processed, they include: clay, feldspar, and quartz.	They are made of synthetic, excellent-quality powders. These chosen raw materials are accountable for offering a variety of unique functions and features.
Structures	The clay's composition determines its structure. For example, pottery from distinct regions has varying textures. Traditional ceramics have more complex chemical structures and compositions since they are made from different raw materials.	Advanced ceramics possess simple chemical structures that are highly pure and clear. Also, their formation was done using manually calculated ingredients, which shows that the raw materials are under control. Thus, their microstructure is often fine and consistent.
Manufacturing processes	Minerals for conventional ceramics can be utilized directly for wet moldings, such as slurry grouting or mud plastic molding. Their temperatures change from 900 to 1400°C after sintering, which implies that the green body is ready for use.	The use of dry and wet molding for advanced ceramics is applicable once organic additives have been introduced to enhance the raw materials of high-purity powders. Additional processing is required after fire at higher sintering temperatures (ranging from 1200 to 2200°C) depending on the material used. In terms of preparation techniques, sophisticated ceramics surpass the limitations of traditional ceramics.
Functions and applications	Traditional ceramic materials often get produced where they are daily used as materials for building purposes.	Advanced ceramics outperform traditional ceramics in terms of both performance and innovative uses. They exhibit a variety of Physical and mechanical qualities include large toughness, durability, thermal shock resistance, durability against wear, durability against corrosion, and extreme-temperature resistance.

Table 1.
Advantages and disadvantages of traditional and advanced composite ceramic materials [1].

Ceramic materials	Phases/Bonds/Forms	Properties
Aluminum nitride (AlN)	Has a hexagonal crystal shape and is covalently linked.	Chemical stability, spectrum gap (62 eV, direct), hardness (1100 kg/mm2), threshold electrical field (12MV/cm), resistivity to oxidation, thermal conductivity (320 W/mK), etc.
Aluminum oxynitride (AlON)	Transparent polycrystalline ceramic material with a spinel-type structure.	Excellent thermomechanical and optical transparency properties.
Titanium diboride (TiB_2)	Their structure is hexagonal lattice in shape (C_{32}) with a space group of P6/mmm.	High hardness, excellent corrosion, thermal oxidation, high chemical stability, and wear resistance.
Zirconia toughened alumina	It contains around 7–25 wt.% unstabilized ZrO_2 or Y-TZP incorporated into an Al_2O_3 matrix.	They have compressive strength of 2500 MPa, elastic modulus of 330 GPa, igh hardness, fracture toughness of 7.3 MPa $m^{1/2}$, coefficient of thermal expansion of 8 x $10^{-6o}C^{-1}$, low cost, and simple processing procedures.
Titanium silicocarbide (Ti_3SiC_2)	A hexagonal structure having a P63/mmc space group. It is composed of three interaction methods, all of which are identical to the van der Waals force (Ti and Si).	Vickers hardness (4 GPa), density (4520 kg/m^3), Young's modulus (322 GPa), high-temperature oxidation and corrosion resistance, easy machinability, better stability.
Silicon dioxide or silica (SiO_2)	They can be classified into two essential compositional forms (crystalline and amorphous) and exist in three major phases (quartz, tridymite, and cristobalite).	They are defined by excellent stability in chemical form, a reasonable cost, a non-toxic dielectric coefficient (50–100), and a favorable effect on the surroundings.
Hafnium nitride (HN)	They show a rock-salt structure with a lattice constant of 4.525 A.	They possess a good melting point of 3653.15 K, high thermal stability and chemical inertness, and a hardness of 16.3 GPa.
Calcium copper titanate (CCTO)	They possess body-centered cubic (bcc) with a Lm3 space group.	They have a bandgap of (3.4 eV), high dielectric losses, and a low breakdown voltage.
Magnesium dititanate ($MgTi_2O_5$)	Their structure is Rthorhombic pseudobrookite-type.	They are thermally stable (< 1873.15 K), low CTE, low thermal conductivity, and high thermal shock resistance.
Cadmium sulfide (CdS)	They are in the form of n-type semiconductors.	They have a direct bandgap of (2.4 eV), superior light absorption, suitable flat-band potential, and easy agglomeration.

Table 2.
Advanced composite ceramic materials, structural phases, and properties [1, 3–6].

to biomedical engineering and energy storage. This chapter explores the processing techniques and diverse applications of composite ceramic materials, highlighting their significance in driving innovation in emerging technologies.

The worldwide market for ceramic materials with diverse uses in the fields of ecology, specialized tools, biomedical and electronics, and environmental domains is increasing. Numerous ceramic substances and production techniques are currently developed with task-specific properties. The material, production processes, and manufacturing conditions all have an impact on attributes such as durability against corrosion, superior optical and electrical performance, hardness, and wrinkle prevention [1].

Different studies have been done regarding ceramic composites, which include ceramic coating of both heat-resistant, biomedical, corrosion-resistant, improving microstructural and electromagnetic wave absorption performances using the atmospheric plasma spraying method [7–11], the use of ceramic composites as filters in the reduction of non-metallic inclusions in molten steel to improve the quality of steel products [12], as biomaterials with bioglasses for bone replacement applications [13], and it was also used as a coating on rubber by aerosol deposition with cryogenic substrate cooling to show the optical transparency and mechanical properties of the ceramic film formed [3, 4, 14–18]. Moreover, ceramic and fluoropolymer composites were used as coating materials for controlling friction and wear for tribological purposes [19]. Furthermore, Sawunyama et al. [20] reported that a coal fly-ash based ceramic membrane was used in wastewater treatment. This chapter discussed the processes and applications of ceramic composites for various emerging technological uses.

2. Processing techniques for composite ceramic materials

Composite ceramic materials can be made in a variety of ways. Composite ceramic materials can be arranged in a number of shapes, including hollow fibers, tubes, and flat discs. Furthermore, composite ceramic materials can be manufactured in four steps: (i) material preparation, (ii) processing, (iii) sintering, and (iv) finishing. Some of the processing techniques involved in composite ceramic materials include powder metallurgy, sol-gel techniques, freeze-casting, additive manufacturing, hot pressing, and slip casting.

2.1 Powder metallurgy

Powder metallurgy is a widely used technique for producing composite ceramics. It involves the mixing of ceramic powders with other materials, followed by compaction and sintering at high temperatures. This method allows for precise control over the composition and microstructure of the material, making it possible to engineer composites with desired properties such as enhanced mechanical strength and thermal conductivity. The process provides a better way of manufacturing high-quality sintered and structural components for a large array of applications in various industries. This is due to its low cost, unusual versatility, and the ability to blend with high melting points. The flexibility of powder metallurgy in composite ceramics helps to control porosity, reduce noise or damping vibrations, enhance good surface finishing, and provide unique magnetic properties. There are four methods involved, which include powder production, compacting, sintering, and mixing and blending [21, 22].

2.2 Sol-gel processing techniques

The sol-gel process is a versatile method for synthesizing composite ceramics at relatively low temperatures. This technique involves the transition of a solution containing metal alkoxides or other precursors into a gel, which is then dried and calcined to form a ceramic material. The sol-gel process is particularly advantageous for producing nanocomposites with uniform particle distribution and fine microstructures, which are essential for applications requiring high precision and performance, high quality with defined threshold strength, near-net-shape, and complex ceramic parts [1].

In sol-gel manufacturing, ceramic precursors are important to the gel-forming mechanism (through polycondensation, hydrolysis, etc.). The gel network's backbone is made up of hydroxides and metal alkoxides, which are eventually processed into ceramics. When employing sol-gel casting to create ceramic components, there are two key factors to consider. First, calculate the ideal gelation speed, which must be quick enough to keep the foam from collapsing. Second, the gel rheology must enable all aspects of the mold to be filled; hence, gels with high fluidity are necessary, especially for manufacturing very complicated structures [1].

2.3 Freeze casting

Freeze-casting was invented as a "near net-shape" forming method for robust ceramics, but the porous character caused by frozen crystalline melting during the casting process was considered problematic at the time. The goal was carried out to avoid the formation of voids caused by evaporating crystallized ice, which was developed as a "near net-shape" forming method for strong ceramics; however, the porous nature induced by frozen crystal sublimation during the casting process was deemed problematic at the time. An attempt was made to prevent the voids created by evaporating crystallized ice. Subsequently, through an alternate viewpoint, the regulated formation, growth, and evaporation of crystalline ice was recognized as an important benefit of the freeze-casting technique for producing coordinated and functional porosity.

It demonstrated the utility of freeze-casting technology in the production of macroporous ceramics. Freeze-cast frameworks with significant permeability (~90%) and vertically aligned, hierarchically structured pores were originally created for biomedical applications [23]. **Figure 1** depicts the procedures of freeze-casting utilizing aqueous slurry that is layered over a three-phase structure made up of water, and **Figure 2** depicts the schematic depiction of freeze-casting for composite ceramic components.

Freeze-casting involves creating an enduring aqueous or non-aqueous colloidal solution, putting it into a mold, freezing the molded suspension, sublimating the solidified phase at reduced pressure, and sintering to form a porous structure. The sintering process happens by connecting the surfaces of particulates before heating

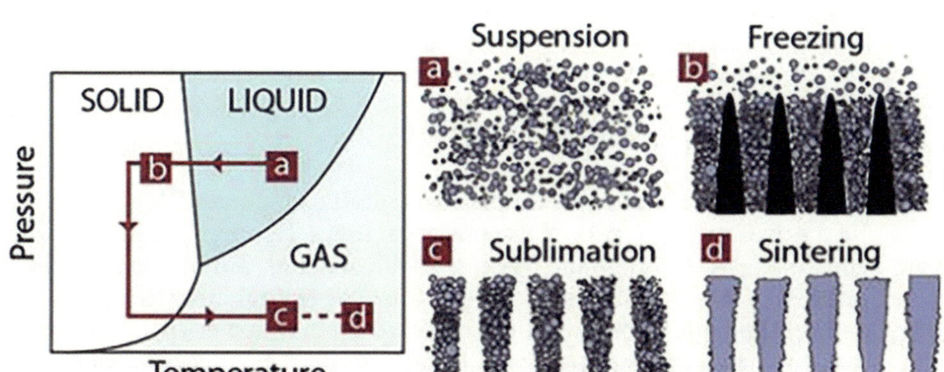

Figure 1.
Preparation methods for freeze-cast sintered sections: (a) initial slurry, (b) solidified body, (c) green body following sublimation, and (d) sintered body [23].

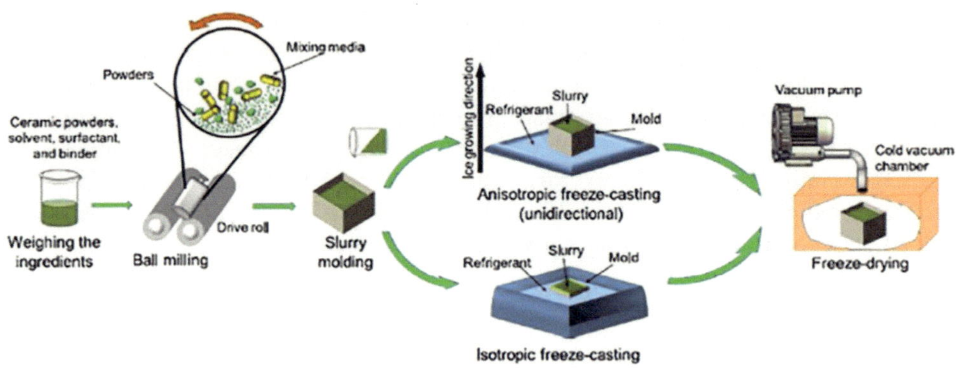

Figure 2.
Freeze-casting process for composite ceramic components: schematic representation [1, 23].

through substantial mass transmission and surface-transport systems, with the primary objective becoming to generate a completely dense solid constituent (some remnant porous persists in most circumstances) [1]. Sintering technologies were divided into three mainstreams: (i) modified spark plasma sintering (SPS) route, (ii) flash sintering-like processes, and (iii) hydro-consolidation (sintering in the presence of water) [24, 25].

2.4 Additive manufacturing (AM)

Additive manufacturing, or 3D printing, has emerged as a transformative technology for the fabrication of composite ceramics. AM enables the layer-by-layer construction of complex geometries with precise control over the material's composition and microstructure. Techniques such as stereolithography, selective laser sintering, and binder jetting are being adapted for ceramic composites, allowing for the production of customized components with intricate designs. The flexibility of AM is particularly beneficial for emerging technologies in electronics, where miniaturization and integration are key. Traditional ceramics are notoriously difficult to produce, making it difficult to manufacture objects with intricate shapes. The new method of three-dimensional (3D) printing provides substantial opportunities for growth in the area of molding and creating outstanding durability ceramics. It is predicted to break past the technological limitations of traditional ceramic processing and manufacture, opening up new paths for the use of essential ceramic components [25].

2.5 Hot pressing and hot isostatic pressing (HIP)

Hot pressing and hot isostatic pressing are advanced techniques that enhance the density and mechanical properties of composite ceramics. Hot pressing involves the simultaneous application of heat and pressure to the material, promoting densification and reducing porosity. HIP uses a high-pressure gas environment to uniformly apply pressure, eliminating internal voids and resulting in near-net-shape components. These techniques are crucial for producing high-performance composites required in demanding applications such as aerospace and defense. The process involves the use of high-pressure, low strain rate, and high temperature for synthesis in the design process to produce a dense, compact material using powder metallurgy. It is often used to synthesize carbides, borides, nitride, and oxide materials [26].

2.6 Slip casting

Slip casting is one of the most widely utilized industrial procedures for ceramic processing because it allows for precise microstructure control in cast porous materials as well as a high degree of ceramic phase dispersion. Slip-casting is the procedure of immersing a permeable shape (often a mold) containing ceramic slurry. Slip casting is developed primarily for hollow components/forms (though it can additionally be utilized to create concave and flattened tiles), is affordable, and integrates form elegance with sophistication as contrasted with alternative constituent fabrication methods, as illustrated in **Figure 3** [1, 27].

A slip-casting slurry requires four primary elements: ceramic particles (ball-milled to sub-micron small category), desalinated water (in order to assist in meticulous control of dispersion settings), chemical deflocculant (allows to optimize the proportions of solids and preserve colloidal consistency in the slurry), and organic adhesive (to provide green toughness) [1, 28].

3. Applications of composite ceramic materials in emerging technologies

3.1 Aerospace and defense

Composite ceramics are used in aerospace and military to make components with high thermal stability, strength, and wear resistance. Ceramic matrix composites (CMCs) are used in turbine blades, heat shields, and missile nose cones to endure severe temperatures and mechanical strains. These materials are lightweight, which adds to fuel efficiency and increased performance in aeronautical applications. Composite ceramic materials possess better durability, make the structure lighter weight, and ensure the highest safety level when used in aerospace industries [29]. Composite ceramic materials have good thermal conductivity, high-temperature

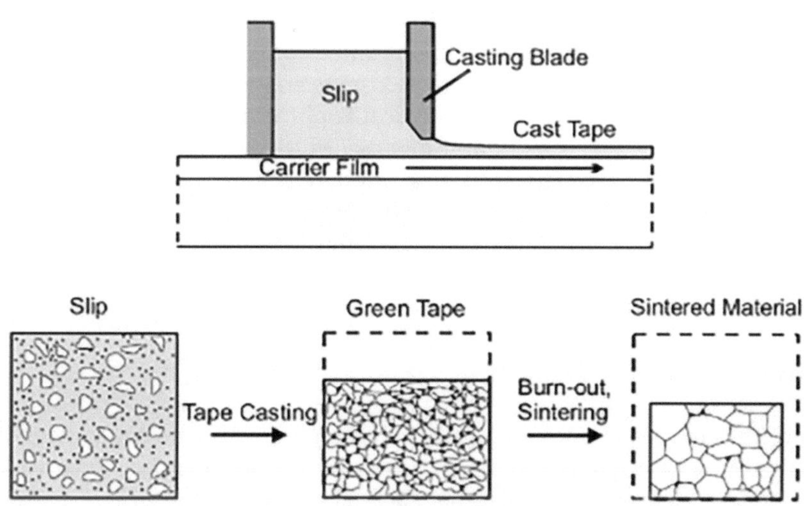

Figure 3.
The different stages of slip-casting containing H_2O, ceramic particles, and adhesive, the molded item, dried green sheet, and ultimately, the internal makeup of the sintered components [1, 27].

stability, high hardness, high corrosion resistance, and good versatility and are mainly used in the high-temperature section of the aircraft exhaust nozzle [30].

3.2 Electronics and telecommunications

Composite ceramics' electrical insulating characteristics aid the electronics sector, which uses them in substrates, capacitors, and insulating layers for integrated circuits. The high thermal conductivity of certain ceramic composites also helps with heat dissipation, which is critical for electronic device reliability. Ceramic composites are used in telecommunications components like filters, resonators, and antennas because of their stability and low dielectric loss, which improve signal clarity and transmission efficiency. Ceramic components are widely used in electronic applications, including automotive, communications, electrical components, semiconductor devices, Filters, circulators, oscillators, thermally resilient resonators, isolators, phasing changers, and antennas made from dielectric resonators are used in microwave and millimeter wave interactions, as well as radar and sensor systems. They can operate at extreme temperatures, frequencies, and power ratings in hostile environments [1].

Power electronics packaging involves adequate distribution of heat. Copper linked to ceramic components serves as a heat sink. The housing additionally delivers electrical linkage to the outside world, protects the integrated circuit from dangerous chemical species, and allows for chip manipulation. Glass ceramic is the most common type of ceramic component used in microelectronic packages, Al_2O_3, beryllia, zirconia, or magnesia. Copper's strong thermal conductivity allows for fast heat sinking, while Al_2O3 has a high dielectric capacity and so provides good electrical insulation [1, 4–6, 31].

3.3 Biomedical engineering

Composite ceramics are used extensively in biomedical engineering, particularly in the production of implants and prostheses. Bio-ceramics, such as hydroxyapatite composites, are utilized in bone transplants and dental implants because they are biocompatible and encourage bone formation [9]. Ceramic composites' wear resistance and inertness make them ideal for joint replacements and other load-bearing implants. Additionally, the development of bioactive ceramic composites is opening the door to enhanced delivery mechanisms for drugs and tissue engineering applications. Ceramic components are used in a variety of biomedical applications, including dental and hip prostheses, sensors, and implantable devices [9]. Because ceramic nanomaterials are not subjected to high temperatures, low-temperature adhesive technologies such as glass ionomer cements, zinc phosphate mortars, and composite resin mortars are widely utilized to bond to dental ceramics. Ceramic components have gained appeal in prostheses such as dental implants and orthopedics because of their durability against wear, strength, biocompatibility, and the ideal balance of stability [1].

3.4 Energy storage and conversion

Composite ceramics play an important role in energy storage and conversion technologies. Energy storage technologies are used to store solar and wind power, such as electricity, thermal energy, or mechanical energy, in the form of batteries, pumped hydro storage, and compressed air energy storage [32]. Solid oxide fuel cells (SOFCs), which use ceramic electrolytes, provide a highly efficient and ecologically benign means of generating energy from a wide range of fuels. Ceramic composites are also utilized in batteries, capacitors, and supercapacitors, where their high ionic conductivity and stability improve energy density and charge/discharge cycles. These materials are critical for advancing renewable energy technology and developing sustainable energy systems.

3.5 Environmental applications

Composite ceramics are increasingly used in environmental technologies, including filtration and catalysis. Ceramic membranes, for example, are utilized in water and gas filtration systems because they are chemically resistant and can tolerate extreme circumstances. Composite ceramics are used in catalytic converters to support catalysts that minimize hazardous emissions from vehicles and industrial activities. These materials' longevity and effectiveness make them indispensable for resolving environmental issues and fostering sustainability.

Membrane-based components are preferable for environmental applications because they are faster and less expensive, more adaptable to other processes, and more selective. Ceramic membranes typically consist of an upper level, a middle level, and a base layer, resulting in high flexibility and penetration. In membrane applications, the ceramic support offers mechanical strength to the upper level, a minimal barrier to filtrate flow, and a middle layer that inhibits penetration [1].

4. Conclusion

The production and application of composite ceramic materials are critical to the growth of new technologies. These materials provide outstanding performance and reliability in a variety of applications, including aerospace and biomedical engineering. As processing techniques progress and new applications emerge, composite ceramics will become increasingly crucial in influencing the future of innovative materials and technology. Their adaptability and outstanding qualities make them indispensable in the pursuit of improved performance, sustainability, and innovation across multiple industries.

Composite ceramics are an appealing material, resulting in their multiple applications in biomedical, environmental, and electrical fields. Composite ceramics in power electronics are a newly developed field with promising applications in high-temperature cycling due to their reduced lack of switching and increased voltage failure. They supply a low-resistance transmitting route, a thermal pathway to the energy sink, and insulation in the wiring and the heat sink.

Acknowledgements

The authors are grateful for the enabling support from the National Research Foundation (NRF) postdoctoral grant (PSTD 23041090962) and the Durban University of Technology, South Africa.

Conflict of interest

The authors declare no conflict of interest regarding this submission.

Author details

Olusegun Adigun Afolabi* and Oludolapo Akanni Olanrewaju
Industrial Engineering Department, Durban University of Technology, South Africa

*Address all correspondence to: oluseguna@dut.ac.za

References

[1] Ayode Otitoju T et al. Advanced ceramic components: Materials, fabrication, and applications. Journal of Industrial and Engineering Chemistry. 2020;**85**:34-65. DOI: 10.1016/j.jiec.2020.02.002

[2] Schmidt S et al. Advanced ceramic matrix composite materials for current and future propulsion technology applications. Acta Astronautica. 2004;**55**(3-9):409-420. DOI: 10.1016/j.actaastro.2004.05.052

[3] Bulbul F, Efeoglu I. Pulsed-dc bias magnetron sputtered TiB_2 ceramic coating. International Journal of Refractory Metals and Hard Materials. 2024;**120**:106624. DOI: 10.1016/j.ijrmhm.2024.106624

[4] Soylu A, Karaahmet O, Cicek B. Exploring luminescence in transparent glass–ceramic coating: Study on Pr^{3+}/ Dy^{3+} Co-doped $SrO–Al_2O_3–SiO_2$ glass–ceramic particles. Materials Chemistry and Physics. 2024;**318**:129250. DOI: 10.1016/j.matchemphys.2024.129250

[5] Afolabi OA, Ndou N. Synergy of hybrid fillers for emerging composite and nanocomposite materials-a review. Polymers (Basel). 2024;**16**(13):1907. DOI: 10.3390/polym16131907

[6] Vijaya G et al. Nano indentation studies on ceramic thinfilms coatings deposited using sputtering process for energy applications. Materials Science for Energy Technologies. 2024;7:115-123. DOI: 10.1016/j.mset.2023.08.001

[7] Niu X et al. Preparation and characterization of a composite ceramic coating containing absorber LaFe12O19 by plasma spraying. Ceramics International. 2023;**49**(16):27079-27085. DOI: 10.1016/j.ceramint.2023.05.252

[8] Wang Y et al. Characterization of the interfacial structure and fracture behavior of in situ synthesized ceramics to reinforce Ni-based composite coatings. Journal of Materials Science & Technology. 2024;**208**:11-25. DOI: 10.1016/j.jmst.2024.03.079

[9] Nilawar S, Uddin M, Chatterjee K. Surface engineering of biodegradable implants: Emerging trends in bioactive ceramic coatings and mechanical treatments. Materials Advances. 2021;**2**(24):7820-7841. DOI: 10.1039/d1ma00733e

[10] Kaya E. Ceramic reinforced composite coatings on cold work tool steel fabricated by wired direct energy deposited plasma. Surface and Coatings Technology. 2023;**475**:130127. DOI: 10.1016/j.surfcoat.2023.130127

[11] Shishir R et al. Development of corrosion-resistant and bioactive ceramic-polymer hybrid coating over Zn-1Mg biodegradable implant material. Surface and Coatings Technology. 2024;**487**:131031. DOI: 10.1016/j.surfcoat.2024.131031

[12] Peng W et al. Impact of functional coatings on microstructure and properties of periclase-magnesium aluminate spinel ceramic filter and its purification capacity on molten steel. Ceramics International. 2024;**50**(14):25293-25303. DOI: 10.1016/j.ceramint.2024.04.259

[13] Punj S, Singh J, Singh K. Ceramic biomaterials: Properties, state of the art and future prospectives. Ceramics International. 2021;**47**(20):28059-28074. DOI: 10.1016/j.ceramint.2021.06.238

[14] Goto T, Matsubayashi Y, Akedo J. Ceramic coating on rubber by aerosol deposition with cryogenic substrate cooling. Ceramics International. 2024;**50**(1, Part A):892-896. DOI: 10.1016/j.ceramint.2023.10.173

[15] Cheng Z et al. In situ formation of micro arc oxidation ceramic coating on refractory high entropy alloy. International Journal of Refractory Metals and Hard Materials. 2024;**120**:106563. DOI: 10.1016/j.ijrmhm.2024.106563

[16] Zhao Z et al. Preparation and properties of micro-arc oxidation ceramic coatings using machine vision technology. Materials Chemistry and Physics. 2024;**320**:129427. DOI: 10.1016/j.matchemphys.2024.129427

[17] Ling X et al. Organic-ceramic hybrid lubricating coating from thermal-sprayed ceramic templates. Surface and Coatings Technology. 2024;**481**:130625. DOI: 10.1016/j.surfcoat.2024.130625

[18] Kumar S, Singh AP. Characterization of surface properties of TiC ceramic coating developed on AISI 1020 steel. Surfaces and Interfaces. 2024;**46**:103960. DOI: 10.1016/j.surfin.2024.103960

[19] He C et al. Thermal-sprayed ceramic/fluoropolymer coatings with tight bond and self-lubrication: Microstructure, tribological properties, and lubrication mechanism. Applied Surface Science. 2024;**660**:159954

[20] Sawunyama L et al. Application of coal fly ash based ceramic membranes in wastewater treatment: A sustainable alternative to commercial materials. Heliyon. 2024;**10**(2):e24344. DOI: 10.1016/j.heliyon.2024.e24344

[21] Schloder M. Powder Metallurgy Process: From Start to Finish, in Alpha Precision Group: A Division of Nicholas Portland Inc. St. Marys, PA: Alpha Precision Group; 2023

[22] Narayanan RG. Powdermetallurgy. Guwahati: Indian Institute of Technology; 2024

[23] Du Y et al. Freeze-casting for the fabrication of solid oxide fuel cells: A review. Materialia. 2018;**1**:198-210. DOI: 10.1016/j.mtla.2018.07.005

[24] Biesuz M, Grasso S, Sglavo VM. What's new in ceramics sintering? A short report on the latest trends and future prospects. Current Opinion in Solid State & Materials Science. 2020;**24**:100868

[25] Wang Y, Wu T, Huang G. State-of-the-art research progress and challenge of the printing techniques, potential applications for advanced ceramic materials 3D printing. Materials Today Communications. 2024;**40**:110001. DOI: 10.1016/j.mtcomm.2024.110001

[26] Hu C et al. Developments in hot pressing (HP) and hot isostatic pressing (HIP) of ceramic matrix composites. In: Advances in Ceramic Matrix Composites. Sawston, Cambridge: Woodhead Publishing, Elsevier; 2014. pp. 177-202. DOI: 10.1016/b978-0-08-102166-8.00008-6

[27] Madou MJ. Fundamentals of Microfabrication: The Science of Miniaturization. 2nd ed. New York: CRC Press; 2002

[28] Studart AR, Amstad E, Gauckler LJ. Colloidal stabilization of nanoparticles in concentrated suspensions. Langmuir. 2007;**23**:1081-1090

[29] Skoczylas J, Samborski S, Kłonica M. The application of composite materials in the aerospace industry. Journal

of Technology and Exploitation in Mechanical Engineering. 2019;**5**(1):1-6. DOI: 10.35784/jteme.73

[30] Shrinivasa D et al. Application of ceramic and polymer composites in aerospace industries. International Journal of Mechanical and Production Engineering Research and Development. 2020;**10**(3):11011-11016

[31] Li Q et al. Enhanced surface wettability modification of Al_2O_3 for laser cladding ceramic-metal composite coatings. Materials Today Communications. 2024;**40**:109746. DOI: 10.1016/j.mtcomm.2024.109746

[32] Randhawa KS. Advanced ceramics in energy storage applications: Batteries to hydrogen energy. Journal of Energy Storage. 2024;**98**:113122. DOI: 10.1016/j.est.2024.113122